Introduction to Differential Equations

Introduction to Differential Equations

R. Creighton Buck *University of Wisconsin*

in collaboration with

Ellen F. Buck

Houghton Mifflin Company BOSTON

ATLANTA DALLAS GENEVA, ILLINOIS

HOPEWELL, NEW JERSEY PALO ALTO LONDON

Some of the material in this text also appears in *Calculus of Several Variables*, by R. Creighton Buck and Alfred B. Wilcox. Houghton Mifflin, Boston, 1971.

PRINTED IN THE UNITED STATES OF AMERICA
Library of Congress Catalog Card Number: 75-25009
ISBN: 0-395-20654-5

Contents

Preface

As the title indicates, this text is intended to be used in a first course in ordinary differential equations, following a standard course in elementary calculus. It is based on a one-semester three-hour-a-week course given at the University of Wisconsin; however, the book is adaptable also for a one-quarter course or a two-quarter course.

We have made a conscious effort to present two aspects of differential equations that we believe are of particular importance for today's students, namely modeling and computers. It is obvious that the advent of the programmable computer has had a major impact on trends in applied mathematics. We have therefore introduced computer awareness into our approach to the solution of differential equations without slighting the essential mathematical ideas of the subject and without turning the book into a text on numerical analysis or computing science.

A similar effort has been made to present "modeling" as an essential step in the application of mathematics to a real problem without requiring the reader to have a depth of experience in physics, economics, engineering, biology, and a host of other disciplines. We long ago learned that mathematics does not become easier if it is explained in terms of something else which is even more unfamiliar! Most students, however, find it extremely gratifying to discover that familiar mathematical techniques can be used to produce quantitative answers to practical questions dealing with the behavior of a rocket or to predict the future actions of two competing biological populations.

On the mathematical side, this text is not oriented toward abstract theory and proof, but rather toward concrete techniques and understandings. This does not mean that we have been afraid to bring in some modern concepts when we felt that they were useful or helped to tie ideas together into a more understandable package. We have also provided occasional glimpses of certain more advanced aspects of the subject, trying to convey a correct picture of the facts but omitting complete details and proofs.

The organization of the book is shown in the following chart:

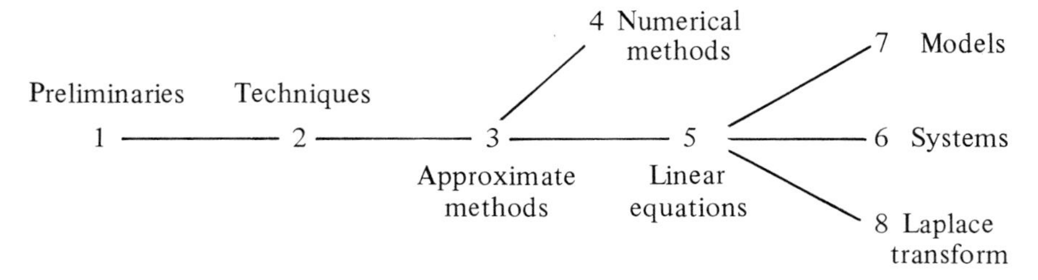

Starred sections can be omitted safely without serious impairment of later material; these include the brief treatment of the existence and uniqueness theorems in Section 3.6, the convergence of the Euler method in Section 4.2, a discussion of more advanced numerical methods in Section 4.4, and Section 5.5 (our favorite!), which contains a comparison of the physical behavior and the mathematical behavior of the damped harmonic oscillator. We have also used the symbol ⊆ in the margin to warn the reader when the water is getting deep.

If desired, the treatment of linear systems in Chapter 6 can be cut short after Section 6.3. The remainder of the chapter is devoted to the use of matrices and matrix-valued functions, particularly in the solution of systems with constant coefficients. Note that we do not assume previous knowledge of linear algebra; the treatment is self-contained and the small amount of matrix theory that is needed is developed here. (Of course, students who have already been exposed to a course in linear algebra will be able to move through this chapter more rapidly.)

We have included an appendix on Laplace transforms containing the usual applications of this method to the solution of single linear equations and linear systems.

We have included a section on infinite series (Section 3.2) which may be review for many. We also include a short section (Section 3.1) dealing with numerical sequences which is very elementary in nature but which we regard as very important; it contains much that is apt to be unfamiliar to students coming directly from calculus.

The work on numerical solution of differential equations is largely confined to Chapter 4. The emphasis is upon the description of certain standard algorithms and the rationale behind them, not upon computer programming or error analysis. In particular, we do not assume that readers have access to a computer or are able to write programs themselves. In order that students may appreciate better some of the problems that may arise, we have provided vicarious experience with computers by showing the results of certain experiments with specific algorithms and specific equations.

The book contains more than 500 exercises of varying difficulty, weighted toward the easier end rather than the honors level. Some of these carry the symbol PC to indicate that they involve arithmetic calculations that are more efficiently done with an inexpensive pocket calculator.

We also call especial attention to Chapter 7, which contains a variety of separate topics, including a discussion of Kepler's Laws, a treatment of the motion of a vibrating string and a vibrating circular drumhead, and a solution of the brachistochrone problem. Each topic illustrates some aspect of the way in which differential equations have been used to answer an interesting question. Each also serves to point the reader toward other areas of mathematics (for example, Fourier series, calculus of variations). This chapter, together with the list of miscellaneous word problems that follows it, can serve as a basis for open-ended class discussions, group projects, or individualized instruction.

If this text is to be used for a one-quarter course, we recommend that Chapters 1–5 be covered, omitting all starred sections. This can be covered in about 25 class meetings. If there is more time, the first three sections of Chapter 6 should be added, as well as Section 5.5.

In a full semester course, Chapters 1–6 should be covered, together with at least two of the following: Section 4.4, Section 7.1, Section 7.2, Section 7.3, and the appendix on Laplace transforms.

In a two-quarter course, the entire text should be covered.

If students have had some material on differential equations in the calculus course, it may be possible to cover Chapter 2 more rapidly, since portions of it will be review.

The approach we have used in this book was first tried by one of the authors some years ago at Wisconsin in an experimental course, funded in part by the National Science Foundation. However, the viewpoint was first explored while we were teaching ordinary differential equations at Brown University in 1948, and owes much to the frequent discussions with our colleagues George Carrier and J. B. Diaz on the philosophy of applied mathematics.

It is difficult to single out all the individuals whose influence is visible in the book. Much appreciation must certainly go to those with whom we worked for so long, starting in 1961, in connection with the CUPM panels on computing and on applied mathematics; it is evident that we have benefited much from conversations with Henry Pollak, Bob Walker, Bert Colvin, Ben Noble, Carl de Boor, Fred Brauer, John Nohel, and many, many others, both here at Wisconsin and elsewhere. We would also like to thank those who reviewed the manuscript: Philip Crooke, Vanderbilt University; Francis J. Murray, Duke University; Noboru Suzuki, University of California, Irvine; Thomas L. Sherman, Arizona State University; and James F. Hurley, University of Connecticut.

R. C. Buck
E. F. Buck

Introduction to Differential Equations

Preliminaries

1.1 Models and Reality

Most people study mathematics because they are interested in using mathematics as a tool in some other discipline. Indeed, one of the rewards for many research mathematicians is seeing how a newly discovered mathematical theory makes it possible to better understand and deal with a portion of the world that was formerly a mystery—for example, the changing structure of the economy, or the strange nature of the electromagnetic winds that sweep across the solar system, or the mixture of human decisions that lies behind the play of a poker game.

In this respect, mathematics seems to have a unique versatility, and many scientists with a philosophical turn of mind have tried to understand why mathematics has proved to be so useful. Without attempting to answer this question, we can describe in very general terms the way a scientist uses mathematics. The key idea is given in the title of this section. One may think of the totality of mathematics thus far discovered or invented as contained in a large warehouse. Each mathematical concept or structure has a definite pattern and a definite collection of associated rules and techniques. A scientist who is interested in studying some aspect of reality comes to the warehouse to find a mathematical system which he thinks will be a model of that aspect of reality. He uses mathematical techniques to explore the behavior of the model and to answer questions about the model which reflect questions about nature.

Perhaps the diagram in Figure 1.1 will help to describe this process. Starting from reality (whatever that may be), we first pass to an idealized situation and then construct (or choose) a mathematical model. Both of these steps can be of great difficulty, and in many cases represent major scientific advances. The model itself can be as simple as a set of algebraic equations or so complex that an entire book is required to describe it. It should be possible to translate each question concerning reality into a parallel question about the model. The next step is to use specialized mathematical techniques to answer the questions about the model. Sometimes the questions are so difficult that no known techniques will answer them, and in this case, new techniques must be found or a new model chosen. When the questions can be answered, one is then ready to translate the answers back into the language of reality and if possible test them by experiment.

As a first illustration, let us reexamine briefly the way in which the basic ideas of calculus arise as models for aspects of the physical concept we call "motion." The first step was taken by Descartes, who realized that the idea of spatial position, or location, could be made concrete by the use of coordinate geometry; the second step was the realization that the subjective experience of time could be modeled by the real numbers. From these steps follows the basic model for the motion of a point along a line—that is, straight-line motion—as a function f from the time axis to one-dimensional space:

$$x = f(t) \tag{1.1}$$

Similarly, the model for the motion of a point in space becomes a function F

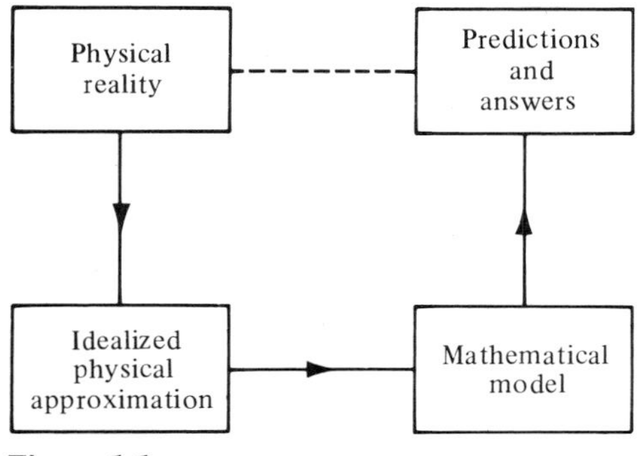

Figure 1.1

from the time axis to three-dimensional space

$$(x, y, z) = F(t) = (f(t), g(t), h(t)) \tag{1.2}$$

which specifies the position of the moving point at each instant of time by giving its coordinates. We note that the path, or track, of the moving point described by (1.2) is a curve in space; thus, as a dividend (1.2) becomes the equation in parametric form for a general space curve.

The simplest motions are the straight-line motions in which the point moves back and forth along a fixed line; the model for this is a single function of t, as in (1.1). Suppose we look for something associated with this model which can be calculated, and which can be connected with the intuitive concept of "speed" or "velocity." This is a familiar situation, and it leads to the standard definition

$$\begin{aligned}\text{Velocity at time } t &= \lim_{h \to 0} \frac{f(t+h) - f(t)}{h} \\ &= f'(t) = \frac{dx}{dt}\end{aligned}$$

Historically, it happened that the same mathematical formula also arose in an entirely different context. In the study of curves, people wanted to know how to construct tangent lines to arbitrary curves, and this led to the study of the graphs of functions and then to the development of a formula for the slope of a curve at an arbitrary point. This turned out to be the same as the preceding formula for $f'(t)$.

It helps to think of three different contexts in which all this is going on: subjective reality (motion), geometry (curves in the plane), and analysis (functions). We can summarize the situation by means of the accompanying table.

Context	*Concept*	*Concept*
Reality	Straight-line motion	Velocity
Geometry	Curve	Slope
Analysis	Function	Derivative

All the concepts represented in the table can be displayed on a single diagram. (See Figure 1.2.) The graph of the function which describes a straight-line motion is a curve in the two-dimensional space-time plane, and the velocity of the moving point at a particular moment of time can be either read off the

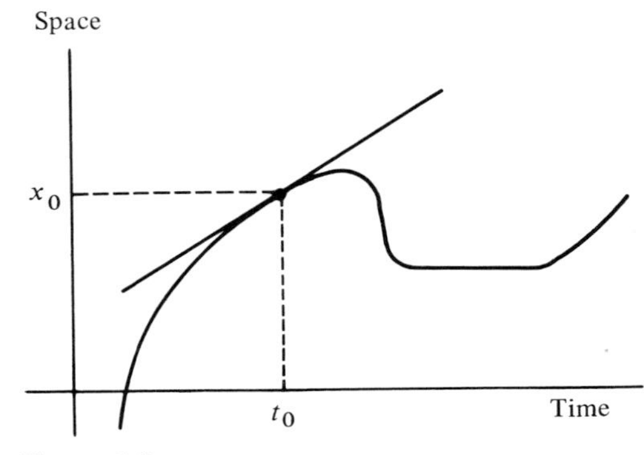

Figure 1.2

diagram as the slope of the curve at the corresponding point, or calculated as the value of the derivative function f' at t.

In the context of mathematical analysis divorced from all applications, differentiation becomes a process which is applied to one function f to yield a new function f'. It is then natural to examine the results of applying this process again, obtaining a new function f''. Returning to our table, it is logical to wonder if f'' has some obvious interpretation in either of the other two contexts—in geometry or in the real world. As it happens, in the geometry of curves, f'' does not correspond directly to any immediately relevant geometric concept (although it is related to the concept of curvature). However, in the study of motion, f'' fits at once into the experimental facts: it can be identified with the physical concept of acceleration, so that one may give precision to the intuitive notion of "force."

Things like this have happened often enough in the interplay between mathematics and science that some people believe that any mathematical concept that fits in the mathematical context must also correspond to some significant physical concept. Sample: Maxwell's equations were formulated partly for mathematical reasons; they predicted that certain experiments should produce "waves" that could be detected, and Hertz verified this in the laboratory.

Almost every branch of mathematics has served as a source for models of some aspect of reality. However, the most common models that have led to the most striking advances in science have been those associated with the calculus and, in particular, with differential equations. Without attempting to give an all-inclusive formal definition at this time, we may state simply that a *differential equation* is any equality relation involving a function and one or more of its derivatives.

Starting with the next section, we shall present a systematic treatment of some of the most useful techniques for analyzing a simple ordinary differential equation. Thus, in terms of Figure 1.1, we shall concentrate upon the right-hand side of the diagram, the passage from the mathematical model to the box called "predictions and answers." Before doing so, however, we shall examine a number of simple illustrations of the modeling process. We shall not be able to carry through to the final step until we have obtained the needed mathematical techniques, but we hope to be able to show how such models arise and how some of the translations are made.

The Pendulum Problem

Suppose that "physical reality" is a pendulum consisting of a spherical bob mounted at the end of a rod and swung from a pivot. We say that we "understand" the physics of this pendulum if we can predict the behavior of this pendulum. Can we answer such questions as the following: (1) How does the nature of the motion depend on the mass of the spherical weight? Or on the length of the rod, or the material of which it is made? (2) If I hold the rod out to form an angle θ_0 and release it, what will be the speed at the bottom of the swing and how long will it take the pendulum to reach the bottom? (3) Will the pendulum eventually stop swinging, and how long will that take?

Experience and experimentation suggest that we had better start with a simplified situation. (Example: Try swinging a weight on the end of a rubber band!) We restrict ourselves to the following physical approximation. We assume that the rod is completely rigid and of zero mass, that the bob is a point of mass M, that the pivot has no friction, that the force of gravity is constant, that there is no air resistance, and that Newton's laws are sufficient to describe the situation. (Note that the last assumption is equivalent to saying that we are willing to buy the Newtonian model for motion.) With

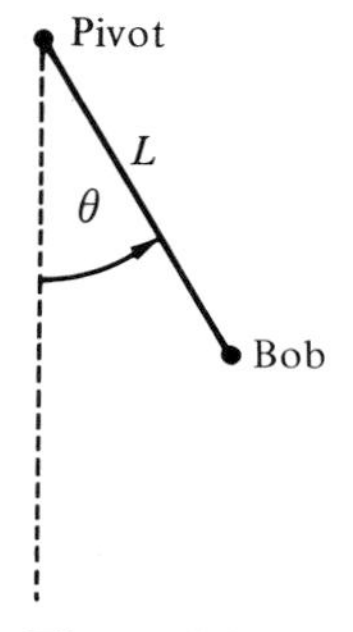

Figure 1.3

these assumptions, we can now move to the mathematical model. (See Figure 1.3.) The position of the pendulum swinging in a plane can be given at any moment of time by the angle θ. Thus, the motion of the pendulum can be translated into a function $\phi(t)$ defined for all $t \geq 0$ with $\theta = \phi(t)$ giving us the value of the angle at time t. We regard $t = 0$ as the instant at which we released the pendulum, so that we start with $\phi(0) = \theta_0$. Since we released the pendulum from rest, so that its velocity was zero at $t = 0$, we also have $\phi'(0) = 0$.

If we use Newton's laws to further restrict the nature of the function ϕ, we can derive a differential equation that must hold. In Figure 1.4, we show the force of gravity F acting on the bob of the pendulum in a vertical direction. The magnitude of F is Mg, where g is the acceleration of gravity and M is the mass of the bob. F can be resolved into two components, F_1 and F_2, as shown. Since the rod is absolutely rigid, F_1 has no effect. The motion of the pendulum bob is on a circle of radius L. Newton's Second Law often appears in the form

$$\begin{aligned} \text{Force} &= \text{rate of change of momentum} \\ &= \frac{d}{dt}\{mv\} \end{aligned}$$

where m is the mass at time t, and v the velocity at time t. When the mass does not change with time, as in our problem, this becomes

$$\begin{aligned} \text{Force} &= m\frac{dv}{dt} = ma \\ &= \text{mass} \times \text{acceleration} \end{aligned}$$

which is a more familiar form of Newton's Law.

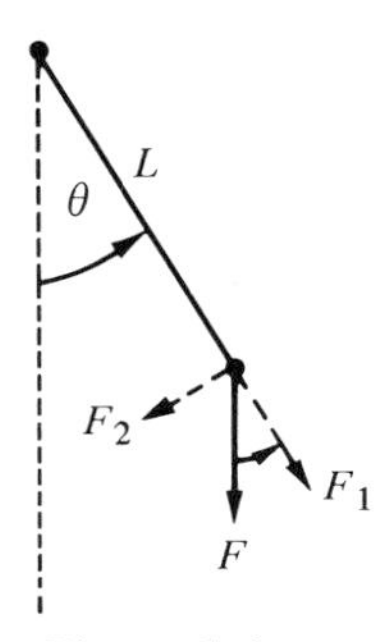

Figure 1.4

In our pendulum problem, the mass is M (which is constant) and the velocity of the moving bob is $L\,d\theta/dt$. The only force to be considered is F_2, so Newton's Law yields the equation

$$F_2 = M\frac{d}{dt}\left\{L\frac{d\theta}{dt}\right\} = ML\frac{d^2\theta}{dt^2}$$

By trigonometry, we find from Figure 1.4 that $F_2 = -Mg\sin\theta$, so that, after canceling M, we arrive at the desired differential equation for a swinging pendulum:

$$\frac{d^2\theta}{dt^2} = -\frac{g}{L}\sin\theta \tag{1.3}$$

Restating in terms of the function ϕ, we obtain the following:

A mathematical model for the motion of this pendulum is a function ϕ, defined and continuous on the interval $t \geq 0$, which has a continuous second derivative and is such that

$$\phi''(t) = -\frac{g}{L}\sin\phi(t) \qquad \textit{for all } t \geq 0 \tag{1.4}$$

$$\begin{aligned} \phi(0) &= \theta_0 \\ \phi'(0) &= 0 \end{aligned} \tag{1.5}$$

How do the questions we asked at the beginning of our problem translate into questions about this model? The first, which was

How does the nature of the motion depend on the mass of the bob or the length of the rod?

becomes

How does the graph of the function ϕ depend on the number M or the number L?

Since the number M does not appear in (1.4), we see that the motion is independent of M. We are not ready yet to explore the nature of the dependence upon L.

The second question about reality,

What will be the speed at the bottom of the swing and how long will it take the pendulum to reach the bottom?

becomes

For what value of t, say t_0, is it true that

(1.6) $$\phi(t_0) = 0$$

and what is the value of $L\phi'(t_0)$?

Finally, the last question, which was

Will the pendulum eventually stop swinging, and if so, how long will this take?

becomes

Is there a number T such that

(1.7) $$\phi'(t) = \phi(t) = 0$$

for all $t > T$?

We have not yet presented the mathematical techniques needed to answer the questions about the function ϕ. However, here are several of the results that one can obtain.

There is a number $t_0 > 0$ such that

(1.8) $$0 < \phi(t) \qquad \textit{for all } t, \ 0 < t < t_0$$

and

$$\phi(t_0) = 0$$

Moreover,

(1.9) $$t_0 = \sqrt{\frac{L}{2g}} \int_0^{\theta_0} \frac{d\theta}{\sqrt{\cos\theta - \cos\theta_0}}$$

and the value of $L\phi'$ at t_0 is given by

(1.10) $$V_0 = 2\sqrt{gL}\sin(\tfrac{1}{2}\theta_0)$$

Furthermore, for all $t > 0$,

(1.11) $$\phi(t + 4t_0) = \phi(t)$$

and $\lim_{t\to\infty} \phi(t)$ does not exist.

These statements are next translated back into statements about reality, or rather into statements about the idealized physical pendulum from which our problem arose. (In each case that follows, you should check back to see exactly how the mathematical statement answers one of the questions asked about the mathematical model, and how it in turn relates back to the original physical reality.)

Thus, from (1.8) we see that the length of time it takes the pendulum bob to reach the bottom of its first swing is t_0, the number given in formula (1.9). Moreover, the velocity which it has reached at that moment is the number V_0 given in formula (1.10). Finally, the motion of the pendulum is periodic, repeating exactly after the time interval $4t_0$. It never stops moving nor does it ever slow down.

These are the results which the mathematical model predicts about the behavior of the real pendulum. The next step is to compare them with the results of experimentation with real pendulums. For example, we can test (1.8) and (1.9) by measuring the length of time it takes the pendulum bob to descend to the bottom of its swing and comparing this with the value of t_0 predicted from (1.9) for various choices of the initial angle θ_0 and the pendulum length L. We could also do the same for formula (1.10) for the velocity V_0. The degree of agreement is some sort of measure of the degree to which the model is a true reflection of reality. It is also a measure of the confidence one could place in the predictions of the model in areas where its agreement with reality has not been tested by experiment.

Regarding the latter, one must never lose sight of the fact that a model is not the same as reality. Although a model may behave in many ways like the real situation which gave rise to it, it will also predict behavior which does not occur. For example, (1.11) says that once set moving, a pendulum will never stop, and any experiment will show that this is false. Furthermore, the values of t_0 and V_0 given in formulas (1.9) and (1.10) will not match exactly the values obtained from measurement. A partial answer is to say that our model was constructed for an "ideal" pendulum, and that requirements such as a weightless, perfectly rigid rod and a frictionless pivot cannot be satisfied in practice. It is, of course, possible to start from a less ideal physical picture, permitting the rod to have mass and to be slightly elastic, the bob to have nonzero radius, and the pivot to have frictional forces, etc. When all this is done, the resulting mathematical model will be immensely complicated, possibly beyond the limits of presently understood techniques of mathematical solution. Nevertheless, it would still be our contention that this last model would predict kinds of behavior not exhibited by real pendulums and that in no sense can we identify such a model with reality, whatever indeed that is.

This discussion suggests that it is often useful to consider several different models for the same physical situation. Perhaps you have noticed that the formula (1.9) for the quarter-period t_0 of a pendulum depends in a complicated way upon the initial angle θ_0. If you recall the frequently quoted statement that the period of a pendulum does not depend upon the amplitude (which is determined by the angle θ_0), then you may wonder where the error lies. The facts are easily checked, and a simple experiment will show you that the length of the period of a pendulum is visibly different for $\theta_0 = 10°$ and for $\theta_0 = 80°$.

The explanation is that in many discussions of the motion of a pendulum, a much cruder mathematical model is used than that set forth in (1.4) and (1.5). In the cruder model, the differential equation (1.4) is replaced by the equation

$$\phi''(t) = -\frac{g}{L}\phi(t) \tag{1.12}$$

or, in more familiar notation,

$$\frac{d^2\theta}{dt^2} = -\frac{g}{L}\theta$$

The predictions which result from this model are different from those of the previous model, but are more easily obtained. We will find out that

$$t_0 = \frac{\pi}{2}\sqrt{\frac{L}{g}} \tag{1.13}$$

$$V_0 = \sqrt{gL}\,\theta_0 \tag{1.14}$$

and it is evident that the value of t_0 does not involve θ_0 at all.

The connection between these two models for the motion of a pendulum is a mathematical one and not a physical one, although it can be given a physical interpretation. When θ is small, $\sin\theta$ is very nearly the same number as θ. Thus, if the motion of the pendulum is confined to a small neighborhood of the bottom of its arc, then the term $\sin\phi(t)$ which occurs in (1.4) is nearly equal to $\phi(t)$; this replacement turns (1.4) into (1.12). For the same reason, one should expect that the predicted values of t_0 and V_0 given in formulas (1.9) and (1.10) turn into those in (1.13) and (1.14) as θ_0 tends toward zero.

Another simple illustration may help here. Physicists speak of light sometimes as a wave and sometimes as a particle. Which is it? The answer, of course, is that light is neither. Both words describe specific mathematical models for light, but neither *is* light; one must not confuse a model with reality itself. Both of these models successfully predict ("explain") some of the observed phenomena of light; both predict behavior that light does not exhibit. A model is most useful when it imitates nature closely, but there will always be aspects of reality it does not reproduce, and it will always predict events that do not in fact occur. The skill of a scientist lies in knowing how far and in which contexts to rely on a particular mathematical model. The final moral is clear; there is never a unique *correct* mathematical model for an aspect of reality. Those "laws of nature" that are not merely tautologies, akin to the statement that there are three feet to every yard, are models, and as such they are subject to revision and replacement by others. None is "correct" or "true" in some eternal sense. (At this time, we have moved far onto the quicksands of metaphysics; those interested in further speculation are encouraged to read H. Poincaré, *Science and Hypothesis*. Dover Publications, New York, 1952.)

The Rocket Flight Problem

The force of gravity is not constant, but depends upon the distance between objects; in the case of the earth, it also depends upon local rock densities and other factors. In dealing with objects at large distances from the earth, it is sufficient to regard the earth as a sphere, and to assume that the force on an object of mass M far above the atmosphere is given by

$$F = \frac{kM}{x^2} \tag{1.15}$$

where k is a negative constant and x is the distance from the object to the center of the earth. This relationship is usually described by saying that the force of gravity varies inversely as the square of the distance. The constant k is negative because the force is an attraction and acts to decrease the distance x.

During the initial powered-flight period, the motion of a rocket obeys rather complicated laws, since both the force of the exhaust and the mass of the rocket are changing. After burnout, things are much simpler: since the mass is then constant, the rocket is coasting, and its velocity is slowly being decreased by the effects of the earth's gravity.

If we assume that the rocket is moving directly away from the earth and we know its height and velocity at burnout, what will its future motion be? Will it continue moving upward and never stop, or will it reach a maximum height and then fall back?

Under the conditions we have described, an appropriate mathematical model for the motion of the rocket (now considered to be a particle of mass M) will be a function f such that $f(t) = x$ is the distance from the center of the earth to the rocket at time t. Let $t = 0$ correspond to the instant of burnout. Then, the distance from the earth at this moment is $R_0 = f(0)$, and the vertical velocity of the rocket at this moment is $V_0 = f'(0)$. We know both R_0

and V_0, and we want to predict the shape of the graph of f for $t > 0$. The final step in constructing the model is to produce a differential equation which the function f must satisfy. Again, this comes by the use of Newton's Second Law. From Figure 1.5 and formula (1.15), we obtain first

$$M\frac{d^2x}{dt^2} = \frac{kM}{x^2}$$

We can find the value of k by using the fact that on the surface of the earth, the acceleration of gravity is $-g$, which is 32 ft/sec^2. Thus, we put $d^2x/dt^2 = -g$ and $x = R$ (the radius of the earth), cancel M, and find $k = -gR^2$. This gives us the final form of the rocket flight equation:

$$\frac{d^2x}{dt^2} = -\frac{gR^2}{x^2} \tag{1.16}$$

We combine all the preceding discussion to formulate the mathematical model for our rocket flight problem.

The function f is defined and has a continuous second derivative f'' (that is, f is of class $\mathcal{C}^2$) for all $t \geq 0$, and obeys the differential equation

$$f''(t) = -\frac{gR^2}{(f(t))^2} \qquad \text{for all } t > 0$$

In addition, $f(0) = R_0$ and $f'(0) = V_0$.

The questions which we raised about the motion of the rocket are now translated into the following questions about the function f in our model.

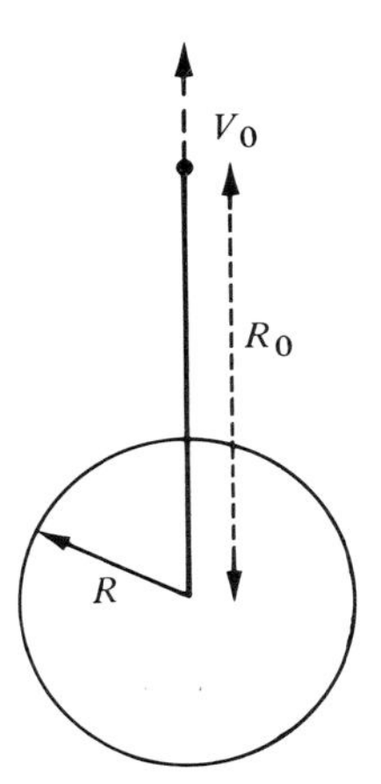

Figure 1.5

Is the function f monotonic increasing for all $t > 0$ and unbounded, or does f have a maximum value; and if so, what is it, and for what value of t is it achieved? More generally, what is the shape of the graph of f?

Without entering into the methods used to answer these questions, which will be discussed in Section 2.3, it may be proved that the graph of f has one of two typical shapes. The first, shown in Figure 1.6, occurs whenever $V_0 \geq R\sqrt{2g/R_0}$, and the second, shown in Figure 1.7, occurs when $V_0 < R\sqrt{2g/R_0}$.

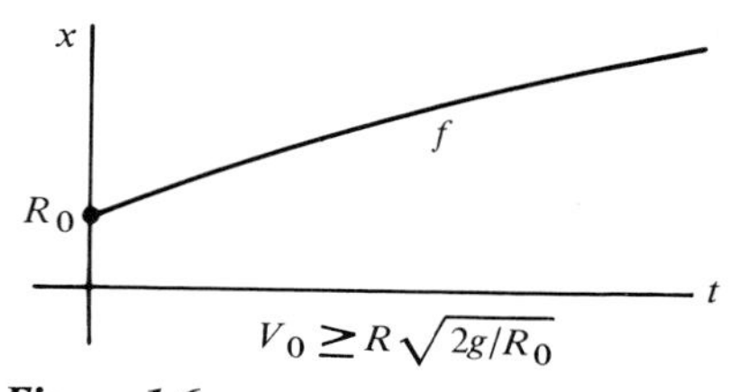

Figure 1.6

These facts about the function f can be translated back to yield useful statements about the expected behavior of the rocket. Thus, if the distance and velocity of the rocket at burnout are such that $V_0 \geq R\sqrt{2g/R_0}$, then the model predicts that the rocket will continue to move upward forever; it will recede indefinitely from the earth, eventually reaching any assigned distance. On the other hand, if $V_0 < R\sqrt{2g/R_0}$, the rocket will reach a maximum height and then fall back toward the earth, reaching it after a specific length of time. The most frequently examined case occurs when R_0, the distance at burnout, is approximately the same as R, the radius of the earth. In this case, the condition for nonreturn becomes $V_0 \geq \sqrt{2gR}$. Taking $g = 32$ ft/sec^2 and $R = 4000$ miles, computation gives the critical escape velocity as $V_0 = 40/\sqrt{33} = 6.9$ mi/sec. Thus, a rocket that at burnout is traveling vertically at about 7 miles per second will never fall back to the earth.

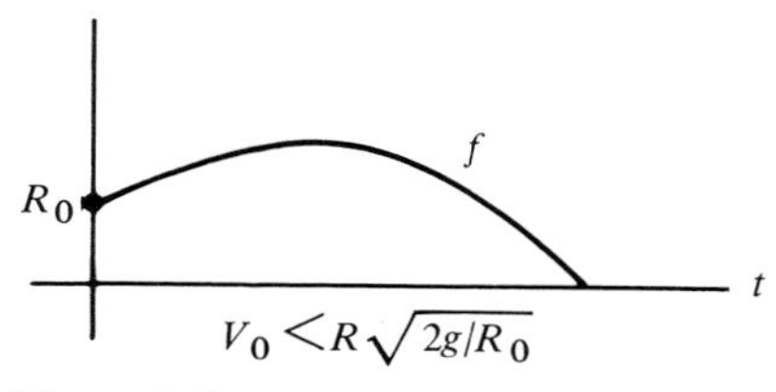

Figure 1.7

The Two-Tank Problem

Suppose that tanks I and II are placed as shown in Figure 1.8 and provided with inlet and outlet pipes as indicated. Each tank contains 100 gallons of liquid, and the flow capacity of each pipe is as shown on the diagram. The left-hand pipe admits pure water to tank I. Suppose that at the moment that all pipes are opened, tank II contains 10 grams of a toxic substance such as rat poison (Warfarin), and tank I contains only pure water. How long will it be until the liquid flowing out of the final outlet pipe on the right is virtually free of the poison (that is, until the level of concentration decreases to 0.001 g/gal)?

As a starting point in building a model for this situation, let x be the amount of the toxic substance in tank I at time t, and y the amount in tank II at time t. For example, we know that at $t = 0$, when the system starts, $x = 0$ and $y = 10$. The rates of change of x and y are governed by a pair of related differential equations which we shall derive shortly:

$$\begin{aligned} \frac{dx}{dt} &= \frac{2}{100}y - \frac{6}{100}x \\ \frac{dy}{dt} &= \frac{6}{100}x - \frac{6}{100}y \end{aligned} \tag{1.17}$$

Thus a model for the two-tank situation is provided by a pair of continuous functions $\phi(t)$ and $\psi(t)$ such that $x = \phi(t)$ and $y = \psi(t)$ satisfy the equations (1.17), and in addition, are such that $\phi(0) = 0$, $\psi(0) = 10$. The basic question which we must ask of the model is

For what value of t is it true that $\psi(t) < 0.1$?

Other questions about the model are suggested in the exercises. Again, we have not yet discussed the techniques needed to obtain an answer to questions about this model, and we therefore postpone further discussion until later.

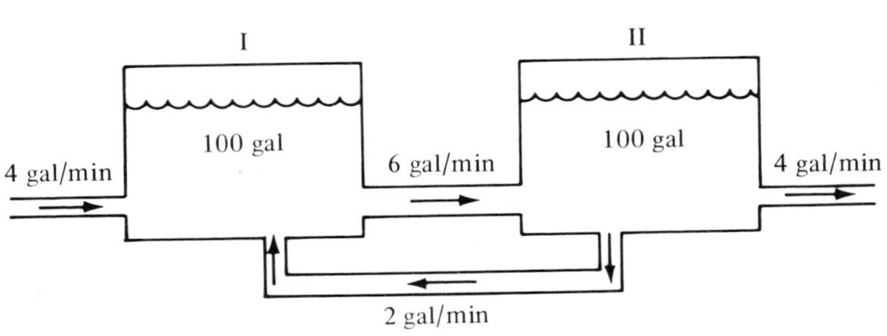

Figure 1.8

Some general comments about the differential equations (1.17) are, however, in order. It should first be noted that the process of going from reality to the mathematical model is not a logical exercise like the proof of a theorem in geometry. The concept of deduction that is involved is of a different sort. We must present an argument that suggests that equations (1.17) are an appropriate mathematical image of the process of mixing that takes place in the pair of tanks. The procedure is often a combination of precise mathematical reasoning and arguments based on physical intuition. In the creation of a model, there is room for wild guesses and even luck. The role of rigorous mathematics is the deduction of consequences of the model, and the ultimate test of a model is how well its predictions match what is seen in reality. Quantum mechanics is accepted not because it is a rigorous mathematical deduction from observed phenomena, but because it enables one to calculate the results of measurements which can be verified.

We can arrive at formula (1.17) as follows. Looking at the diagram in Figure 1.8, we suppose that liquid is flowing as indicated. At a particular moment of time, we suppose that x and y are the amounts of the toxic substance in tanks I and II in grams. Its concentration in each tank in grams per gallon is $x/100$ and $y/100$. During a short interval of time Δt, what change will take place? Looking at each pipe in turn, we find

$4\,\Delta t$ gallons of pure water enters tank I,
$2\,\Delta t$ gallons enters tank I carrying the toxic substance at concentration $y/100$,
$6\,\Delta t$ gallons leaves tank I at concentration $x/100$,
$6\,\Delta t$ gallons enters tank II at concentration $x/100$,
$2\,\Delta t$ gallons leaves tank II at concentration $y/100$,
$4\,\Delta t$ gallons leaves tank II at concentration $y/100$.

From these findings, we can calculate the total change in the amount of the toxic substance in each tank at the end of the Δt-long interval.

$$\Delta x = 0 + 2\,\Delta t(y/100) - 6\,\Delta t(x/100)$$

$$\Delta y = 6\,\Delta t(x/100) - (2+4)(\Delta t)(y/100)$$

Dividing by Δt, we are immediately led to impose the relations given in (1.17) upon the functions that govern the values of x and y.

This is an example of one of the methods used in building a model. It may be noted that in building a model, we have assumed that mixing is slow enough to allow us to use $x/100$ as the concentration of the liquid leaving tank I while we are adding the new toxic substance, but fast enough that we need not worry about how long it takes for the new material to diffuse throughout the tank for the next step. Perhaps these assumptions seem inconsistent. One might therefore desire a different model which takes account of the true speed of diffusion and the unequal concentration in different parts of either tank. Such a model, although a better reflection of reality, is likely to be so difficult to work with that we may not be able to learn much from it.

In the exercises that follow, we ask you to examine a number of models and to attempt to build others. There are few guidelines that can be given. Remember that there is no uniquely correct model of any portion of nature, although some models are more useful than others.

Exercises

1. Each of the space-time diagrams in Figure 1.9 describes the straight-line motion of one or more points. Give an equivalent verbal description of the events.

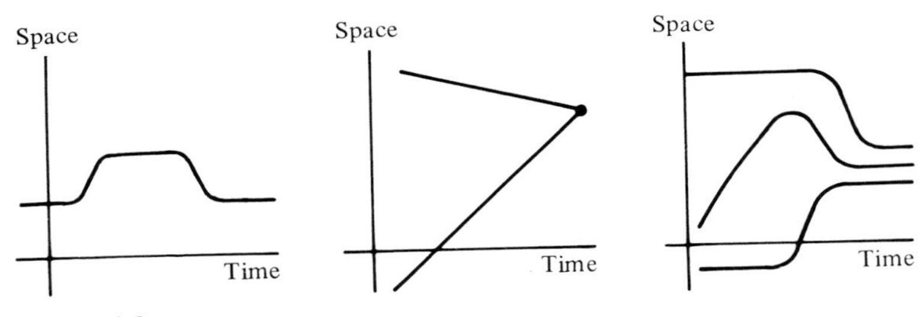

Figure 1.9

2. The graph of f' is shown in Figure 1.10. Given that $f(0) = 1$, draw the graph of f'' and f.

3. If the function $F(t) = (3t, t^3 - t^2, t^2 - 4t)$ describes a motion in space, find the velocity, speed, and acceleration at the points $(-3, -2, 5)$ and $(6, 4, -4)$.

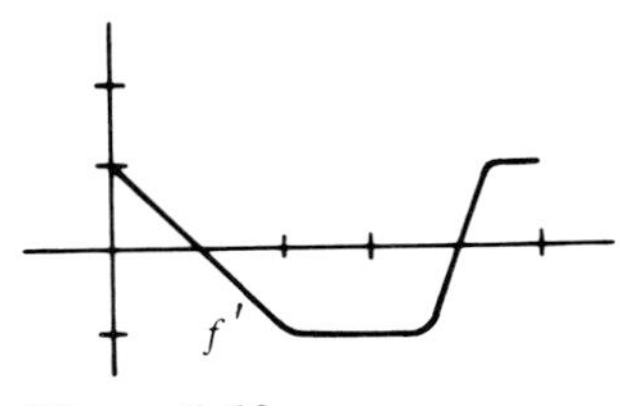

Figure 1.10

4. Can a space-time diagram of a straight-line motion turn out to be a circle?

5. Verify analytically that for values of θ_0 near zero, formula (1.10) is approximately the same as formula (1.14).

6. According to either of the models for a simple pendulum, what is the effect of lengthening the rod? What happens if you make it twice or four times as long? How could you test your answer?

7. When $\theta_0 = \frac{1}{2}\pi$, calculate the predicted value of V_0 by each of the models for a simple pendulum. Can you design an experiment that would tell you which better represents reality? Is this easier to do if $\theta_0 = \pi$?

8. Construct simple pendulums in which the rod is replaced by (a) a string, (b) a long rubber band, (c) a heavy rod whose mass is comparable to that of the bob. Can you decide if either of the models in the text represents any of these real pendulums?

9. (If you have enough physics experience.) Can you devise a mathematical model for the pendulum shown in Figure 1.11 similar to either of those in the text?

10. A company which makes radios has two factories and five warehouses, all of which are in different cities. The various costs of shipping a radio from either factory to any of the warehouses are known. We are given the available storage room at each warehouse, and the maximum production capacity of each factory. Devise a model which could enable you to answer a question such as "What plan for producing and shipping radios will be the cheapest?" [*Note:* The simplest model will be an algebraic one, not involving functions or differential equations at all.]

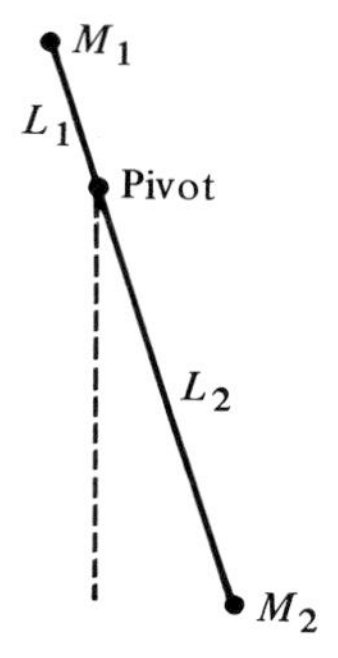

Figure 1.11

11. A lake is being polluted constantly by the addition of a fixed amount of a certain chemical per day. At one end of the lake, pure water is entering at a constant rate, and at the other end, water is flowing out at the same rate. Can you construct a model for this lake which will enable you (after you have learned to solve differential equations) to discuss the future behavior of the lake and to predict when the pollution level will exceed a certain threshold?

12. How would you modify the model for Exercise 11 if the amount of chemical being added per day varies in a known way? If the amount of water flowing in and out per day is not a constant, but also changes with time?

13. It is observed that a quantity of material A left by itself emits radiation and after a while is found to be a mixture of substance A and substance B. It is conjectured that the atoms of substance A are "unstable," and that they randomly turn into atoms of substance B, emitting radiation as they do so. If the quantity of material is large, so that there are many, many atoms involved, it is conjectured that the rate at which substance A turns into substance B can be regarded as proportional to the amount of A still present. Construct a mathematical model which would enable you to answer questions such as the following: "If it takes 10^5 years for 1 gram of A to turn into 0.5 gram A and 0.5 gram B, then how long will it take 1 gram of A to become 0.8 gram A and 0.2 gram B?" What is the connection here with carbon dating of human artifacts?

14. Construct a model for human population growth. (Since no answer to this question can be either simple or unique, it is an ideal topic for class discussion; an optimal answer may be a list of the factors that should be taken into account in building such a model and some of the functions which behave in the required way.)

15. (a) Use your experience and physical intuition to guess the shape of the functions ϕ and ψ in the Two-Tank Problem.
    (b) How do you think these graphs would change if we had started with 10 grams of the toxic substance in tank II and 50 grams in tank I? (Everything else stays the same.)
    (c) Under the circumstances holding in part (b), what would be the effect of changing the flow in the two pipes connecting tank I and tank II from 2 gal/min and 6 gal/min to 12 gal/min and 16 gal/min?

16. Can you devise a model for a system consisting of three tanks and a number of interconnecting pipes, and one or more pollutants?

1.2 First-Order Equations

Having seen how physical problems can give rise to differential equations, we shall begin a systematic study of these equations. From the mathematical point of view, a differential equation is one in which the unknown is a function, rather than a number as in the case of an algebraic equation; the unknown function is to be determined from the identities and relations which it and its derivatives are known to satisfy. When we are dealing with functions of one (independent) variable, the equations are called **ordinary differential equations**; when we are dealing with functions of several variables and the identities involve partial differentiation, then they are called **partial differential equations**. As an example, the study of heat distribution leads to the partial differential equation

$$\frac{\partial^2 U}{\partial x^2} + \frac{\partial^2 U}{\partial y^2} + \frac{\partial^2 U}{\partial z^2} = 0$$

Functions U that satisfy this identity are called *potential* or *harmonic* functions. Until Chapter 7, we shall be concerned only with ordinary differential equations.

Differential equations are classified in various ways. The term **order** is used to refer to the highest-order derivative of the unknown function that appears in the identity to be solved. Thus, a first-order equation may involve the unknown function and its first derivative, but will not involve its second derivative or higher derivatives. Since time is so often the independent variable in physical applications, we choose to label points in the plane as (t, y) rather than (x, y). Then, the standard form of a first-order differential equation is

$$\frac{dy}{dt} = f(t, y) \tag{1.18}$$

where f is a function of two variables defined on a region D in the TY-plane.

A solution of (1.18) on an interval I is a function ϕ such that (see Figure 1.12)

1. $\phi'(t)$ exists for all $t \in I$.
2. The graph of ϕ lies in the region D.
3. For each $t \in I$, $y = \phi(t)$ satisfies the equation (1.18), meaning that

$$\phi'(t) = f(t, \phi(t)) \qquad \text{for all } t \in I$$

As an example, the function $y = 3e^{t^2}$ is a solution of the equation

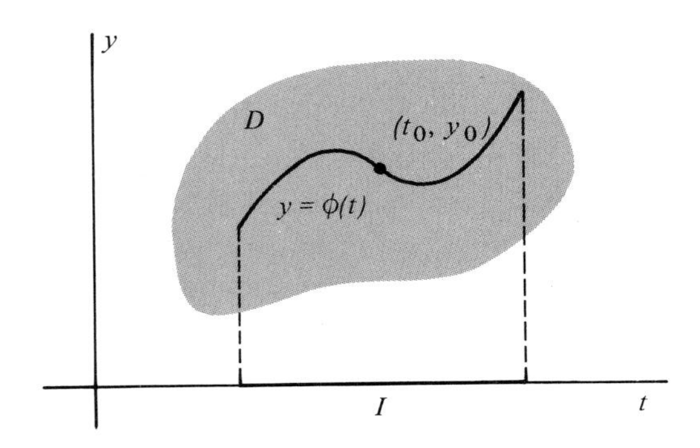

Figure 1.12

$dy/dt = 2ty$ on the interval $-5 < t < 5$ because for every such t,

$$\frac{d}{dt}(3e^{t^2}) = (3e^{t^2})(2t) = 6te^{t^2}$$

and

$$2ty = (2t)(3e^{t^2}) = 6te^{t^2}$$

In this case, there is nothing special about the choice of the interval I, and in fact $3e^{t^2}$ is a solution of the given equation on the whole T-axis, $-\infty < t < \infty$.

A more typical example is provided by the equation

$$\frac{dy}{dt} = -\frac{t}{y} \tag{1.19}$$

Here, the function f is defined for any choice of t and y, with $y \neq 0$. Thus, D could be taken to be any region in the upper half-plane. The function given by $y = \sqrt{1-t^2}$ is defined only on the interval $-1 < t < 1$; but on that interval, it is a solution of the equation, for

$$\begin{aligned}\frac{dy}{dt} + \frac{t}{y} &= \tfrac{1}{2}(1-t^2)^{-1/2}(-2t) + \frac{t}{\sqrt{1-t^2}} \\ &= 0\end{aligned}$$

It is also easily checked that $y = \sqrt{2-t^2}$ is also a solution of this same equation—this time on the larger interval $(-\sqrt{2}, \sqrt{2})$. Indeed, for any positive value of A, the function $y = \sqrt{A-t^2}$ is a solution of this equation on the interval $-\sqrt{A} < t < \sqrt{A}$.

There is a simple geometric relationship between a first-order differential equation $dy/dt = f(t, y)$ and the functions that are its solutions. Moreover, this relationship is the basis for a simple and effective method for finding out something about the solutions of an equation; and it also lies at the heart of the procedures by which high-speed computers obtain approximate solutions of differential equations.

Since dy/dt is associated with the concept of slope, we can use the given differential equation to construct a direction field or tangent field in the region D where f is defined. We proceed as follows. At each point $p = (t, y)$ in D, we evaluate f and use this value $f(p) = f(t, y)$ as the slope of a short line segment which we draw centered on p. The totality of such line segments for all points p in D is called the **direction field** or tangent field for the differential equation. In Figure 1.13, we have drawn a portion of the tangent field defined by the equation $dy/dt = -y$, and in Figure 1.14, the field defined by $dy/dt = y - t$. (A special trick which makes it easier to construct a tangent field, called the method of isoclines, is given in the exercises—see Exercise 4.)

What does the direction field for an equation have to do with solutions of the equation? The fact that $y = \phi(t)$ satisfies the equation $dy/dt = f(t, y)$ means that at any point p_0 on the curve that is the graph of ϕ, the slope of the curve must coincide with the slope of the line segment in the direction field passing through p_0. In other words, the curve must be tangent to each of the line segments whose midpoints it passes through. In Figure 1.14, we have drawn a number of the solution curves to show this tangency property.

It is now evident that this procedure can be used in reverse to find out something about the solutions of a given differential equation. In fact, given a carefully drawn tangent field, a skilled person can often produce reasonably

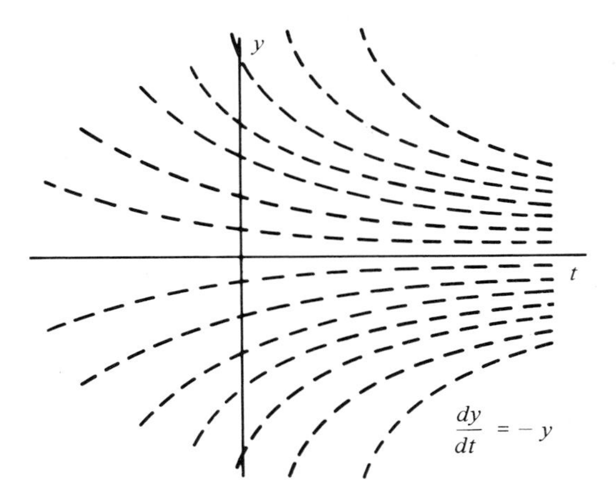

Figure 1.13

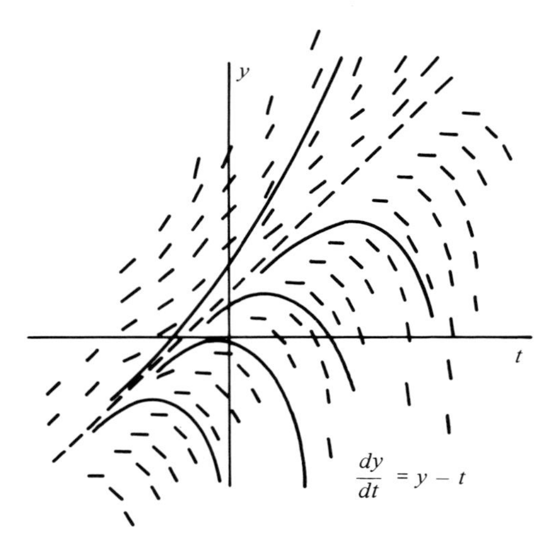

Figure 1.14

accurate sketches of many of the solutions by drawing curves that have this property of being tangent to, and therefore of flowing along with, the directions of the field. Such freehand sketches often reveal something of the general qualitative nature of a solution—indicating, for example, whether it is increasing, oscillating, tending to a limit, etc. (For accuracy, this eyeball method must be combined with the calculating ability of a digital computer.)

Although it is possible to invent a differential equation that has only one (or even no) real solution, in general a differential equation will have an infinite number of solutions. For example, a differential equation associated with a swinging pendulum will have as many solutions as there are ways of starting the motion, one for each initial angle θ_0 and each initial velocity. In a physical situation, we are usually looking for only one of the possible solutions of the equation, one which satisfies certain additional restrictions associated with the real problem from which the model came.

In first-order equations, these normally take the simple form of requiring that y be some specific value y_0 when t is a particular moment t_0 (often $t_0 = 0$). This condition is called an **initial condition**. We therefore seek a function ϕ which is a solution of the given equation $dy/dt = f(t, y)$ and which in addition obeys $\phi(t_0) = y_0$. Translated into geometric language, this merely means that the graph of ϕ must go through the point (t_0, y_0). Of course, this point (t_0, y_0) must be in the region D. Such a problem is usually called an *initial-value problem.*

We can summarize this geometric approach to the approximate solution of first-order equations as follows.

To solve the differential equation $dy/dt = f(t, y)$ with given initial condition $y = y_0$ when $t = t_0$, first construct the tangent field in the region D where f is defined. Then, construct a curve in D which passes through the point (t_0, y_0) and is "tangent" to the field everywhere. This curve will be the graph of a solution.

In most cases, and certainly in all the most useful ones, there will be one and only one solution curve passing through an interior point of the given region D. However, as we shall see, this depends upon the nature of the defining function f and may fail if the function f isn't sufficiently smooth.

A physical problem may involve two or more interrelated quantities that depend upon time. In this case, a standard model for the situation may be a system of differential equations. This was the case, for example, in the Two-Tank Problem in Section 1.1 [see formula (1.17)]. Systems of equations are also classified by order, and the standard form for a system of two first-order differential equations would be

$$\frac{dx}{dt} = f(t, x, y)$$
$$\frac{dy}{dt} = g(t, x, y) \tag{1.20}$$

The general approach is quite similar to that of a single equation. We assume that both f and g are defined in a region D in TXY-space. A solution of the system (1.20) is a pair of functions ϕ and ψ, each defined on the same T-interval I, such that ϕ' and ψ' are also defined there. Furthermore, when $t \in I$, the points $(t, \phi(t), \psi(t))$ must all lie in D, and $x = \phi(t)$, $y = \psi(t)$ must satisfy the equations as an identity; that is,

$$\phi'(t) = f(t, \phi(t), \psi(t))$$
$$\psi'(t) = g(t, \phi(t), \psi(t))$$

for all $t \in I$.

It is possible to give a geometric interpretation here. The pair of functions f and g can be used to create a direction field on the region D, and the set of points $(t, \phi(t), \psi(t))$ trace a curve in 3-space which must be tangent to the direction field in order for $x = \phi(t)$ and $y = \psi(t)$ to be a solution.

Exercises

1. Verify that for $C > 0$, each of the functions $y = \sqrt{C - t^2}$ is a solution of the differential equation (1.19). What about the functions $y = -\sqrt{C - t^2}$? Sketch some of these solutions for several choices of C.

2. Show that the functions $y = 1/(t - C)$ are solutions of the equation $dy/dt = -y^2$ on certain intervals which depend upon the choice of C. Sketch these solutions for $C = 0, \pm 1, \pm 2$.

3. Verify that for each C, the function $y = (1 + Ce^t)/(1 - Ce^t)$ is a solution on an appropriate interval of the equation $dy/dt = \frac{1}{2}(y^2 - 1)$. Sketch the graphs of these solutions for $C = 0$, $C = 1$, and $C = -1$.

4. The **isocline** method for constructing the direction field for the equation $dy/dt = f(t, y)$ consists in drawing line segments of slope m at all the points on the level curve of f with equation $m = f(t, y)$. Apply this method to the equation $dy/dt = 2y$ and sketch several of the solution curves for the differential equation.

5. By the isocline method sketch the direction field for the equation $dy/dt = t + y$ and draw several of the solutions.

6. Do the same as in Exercise 5 for the equation $dy/dt = t/(t - y)$.

7. Do the same as in Exercise 5 for the equation $dy/dt = t^2 + y^2$. Do you think that the solution passing through the point $(0, 1)$ can be extended to arbitrarily large values of t?

8. Among the solutions discussed in Exercise 1, can you find one which satisfies the condition $y = 2$ when $t = 1$? for which $y = -1$ when $t = 3$?

9. Among the solutions discussed in Exercise 3, is there one which passes through the point $t = 1$, $y = 5$? Is there one such that $y = 2$ when $t = 0$?

10. A class of functions is described by the formula $y = Ct/\sqrt{C^2 - t^2}$ for each real number C. Is there a function in this family such that $y = 2$ when $t = 1$? such that $y = 1$ when $t = 2$?

11. (a) Verify that the functions given by $x = e^{4t} + e^{-6t}$, $y = e^{4t} - e^{-6t}$ are solutions of the system of equations

$$\frac{dx}{dt} = 5y - x, \qquad \frac{dy}{dt} = 5x - y$$

(b) Verify that all functions of the form $x = Ae^{4t} + Be^{-6t}$ and $y = Ae^{4t} - Be^{-6t}$ are solutions of the same system.

12. (a) Find a solution of the system in Exercise 11 which obeys the initial condition $t = 0$, $x = 1$, $y = 3$.
    (b) Do the same for the initial condition $t = 1$, $x = 1$, $y = 3$.

13. Show that $x = -t^3/3$ and $y = 3/t$ are solutions of the system of first-order equations $dx/dt = xy$, $dy/dt = 3/(xy)$. Do you think there are others?

14. Show that for every choice of A, the functions $x = t + At^2$, $y = 1 + 2At - At^2$ are solutions of the system $dx/dt = x + y - t$, $dy/dt = 2y/t - 2x/t^2$. Do you think there are other solutions?

1.3 Higher-Order Equations

A differential equation may involve higher-order derivatives, as in the case of the first two illustrations of mathematical models given in Section 1.1. One obvious reason is that many models involve Newton's Laws of Motion and therefore forces and accelerations which involve second derivatives with respect to time.

Let f be a function defined in a region D in 3-space. Then, with f we can associate a general second-order differential equation whose standard form is

$$\frac{d^2y}{dt^2} = f(t, y, y') \tag{1.21}$$

(Here, we have written y' for dy/dt. Another common notation is $\dot{y}$.)

A function ϕ is a solution of the equation (1.21) on an interval I of the T-axis if the following conditions hold:

1. $\phi(t)$, $\phi'(t)$, and $\phi''(t)$ are defined for all $t \in I$.
2. For each $t \in I$, the point $(t, \phi(t), \phi'(t))$ lies in the region D.
3. $y = \phi(t)$ satisfies the equation (1.21) on I, meaning that the identity

$$\phi''(t) = f(t, \phi(t), \phi'(t))$$

holds for all $t \in I$.

As an example, consider the equation

$$y'' - 2y' + y = 0$$

which is an equivalent way to write

$$\frac{d^2y}{dt^2} - 2\frac{dy}{dt} + y = 0$$

This can be rewritten as

$$\frac{d^2y}{dt^2} = 2\frac{dy}{dt} - y$$

which has the form (1.21) with $f(t, y, y') = 2y' - y$. The set D can be taken to be all of 3-space. Then, the function $\phi(t) = e^t$ is easily seen to be a solution of the equation on the interval $-\infty < t < \infty$. For, $y' = e^t$ and $y'' = e^t$, so that $y'' - 2y' + y = e^t - 2e^t + e^t = 0$ for all values of t.

It is possible with second-order differential equations, as with first-order equations, to attach a geometric interpretation to the relationship between an equation and one of its solutions, but it is far less useful either as a practical guide in finding solutions or as a theoretical tool. If we look at a particular solution function ϕ which satisfies the equation (1.21), then the knowledge of the coordinates $t = t_0$, $y = y_0$ of a point on the graph of ϕ and the knowledge of the slope y_0' of this graph at (t_0, y_0) enables us to use f to compute the value $f(t_0, y_0, y_0')$, which is the value of $\phi''(t_0)$. Thus, we can determine the curvature of the graph of ϕ at any point where we know its slope. This does not lead to a practical scheme for obtaining freehand sketches of solutions to (1.21).

A technique to single out one among the infinite number of solutions of a second-order equation is to specify initial conditions; this means that the desired solution $y = \phi(t)$ must have prescribed values A and B for y and $y' = dy/dt$ corresponding to $t = t_0$. In terms of the geometric interpretation given above, this means that we have asked for a solution of the given equation which passes through a specified point (t_0, A), having there the assigned slope B.

For example, the differential equation

$$\frac{d^2y}{dt^2} = -\frac{t}{1+t}y' + \frac{1}{1+t}y$$

has for solutions the functions

$$y = Ae^{-t} + Bt$$

for any choice of the numbers A and B. Suppose we want a solution such that at $t = 1$, $y = 3$ and $y' = 2$. We have

$$y' = \frac{dy}{dt} = -Ae^{-t} + B$$

Setting $t = 1$, $y = 3$, and $y' = 2$, we obtain the algebraic equations

$$3 = Ae^{-1} + B$$

$$2 = -Ae^{-1} + B$$

These can be solved to find $B = \frac{5}{2}$ and $A = \frac{1}{2}e$.

A second type of condition which picks one of the solutions of an equation from all the rest occurs in the study of higher-order equations; it has no analogy in the study of first-order equations. Sometimes, the nature of the problem that gave rise to a particular second-order differential equation leads to the requirement that the unknown function ϕ must satisfy a condition like $\phi(t_1) = a_1$, $\phi(t_2) = a_2$, where t_1 and t_2 are different moments of time. (For example, we may ask that the moving object be at specified locations at the start and at the end of its motion.) This type of condition is called a **two-point boundary-value problem** and is of great significance in the theory and application of differential equations. As a statement about the graph of a solution, it merely says that the graph of ϕ passes through two specified points, but says nothing about the slope at either point.

For example, if we return to the solutions $Ae^{-t} + Bt$ of the previous example and ask that $y = 2$ when $t = 0$, and $y = 3$ when $t = 1$, then substitution again gives a set of algebraic equations:

$$2 = Ae^0 + B(0) = A$$

$$3 = Ae^{-1} + B$$

from which we find $A = 2$ and $B = 3 - 2/e$. Thus the desired solution is $y = 2e^{-t} + (3 - 2/e)t$.

Finally, we point out a mathematical fact about differential equations of higher order that is of great importance for creating a general theory. It simplifies the general viewpoint and even helps in the discussion of an individual equation. The first special case of the general principle is that any second-order equation can be regarded as a system of two first-order equations. In general, a single nth-order equation can be regarded as a special system of n first-order equations.

To see how this is possible, suppose we have the second-order equation

$$y'' = \frac{d^2y}{dt^2} = f(t, y, y') \tag{1.22}$$

Now, put $x = dy/dt$ so that $dx/dt = d^2y/dt^2$. Then, equation (1.22) is clearly equivalent to the system

$$\frac{dx}{dt} = f(t, y, x)$$

$$\frac{dy}{dt} = x$$

As another example, the system of first-order equations

$$\frac{dx}{dt} = f(t, y, z, x)$$

$$\frac{dz}{dt} = x$$

$$\frac{dy}{dt} = z$$

is equivalent to the general third-order equation

$$\frac{d^3y}{dt^3} = f(t, y, y', y'') \tag{1.23}$$

Exercises

1. (a) Verify that the functions $y = Ae^{4t} + Be^{-6t}$ are solutions of the equation $d^2y/dt^2 = 24y - 2\,dy/dt$ for every choice of A and B.
   (b) Find a solution such that $y = 2$ and $y' = 3$ when $t = 0$.

2. (a) Show that the functions $y = A/(t+B)$ are solutions of the equation $d^2y/dt^2 = 2(y')^2/y$ for every A and B with $A \neq 0$. Can you guess some other solutions?
   (b) Can you find a solution which satisfies the initial conditions $y = 2$, $y' = 3$, when $t = 0$? Are there any admissible initial conditions you cannot satisfy?

3. (a) The equation $y'' = (5/2t)y' - (3/2t^2)y$ has solutions of the form $y = t^\alpha$. Find two of these, ϕ and ψ.
   (b) Show that every function $y = A\phi + B\psi$ is also a solution, and find such a solution that satisfies the initial conditions $y = 3$, $y' = 2$, when $t = 1$.

4. Can you find a solution of the equation discussed in Exercise 1 which satisfies the following two-point boundary condition: when $t = 0$, $y = 0$ and when $t = 1$, $y = 1$?

5. Can you find a solution of the equation discussed in Exercise 2 which satisfies the following two-point boundary condition: when $t = 0$, $y = 1$ and when $t = 1$, $y = 3$?

6. Can you find a solution of the equation in Exercise 3 which obeys the two-point condition $t = 1, y = 2; t = 4, y = 1$?

7. Check the equivalence of the equation (1.23) to the system of three first-order equations given.

8. Write D for d/dt so that, for example,

$$\frac{d^2x}{dt^2} - 4\frac{dx}{dt} + 5x = D^2x - 4Dx + 5x$$

$$= (D^2 - 4D + 5)x$$

Show that the system $dx/dt = 3x + 2y$, $dy/dt = 7x - 5y$ can be written in the form

$$(D-3)x = 2y$$

$$(D+5)y = 7x$$

9. Using the previous exercise, show that the functions x and y in this particular system must satisfy the identity

$$\phi'' + 2\phi' - 29\phi = 0$$

10. If x and y are functions satisfying the system $dx/dt = x + y$, $dy/dt = x - y$, then show that x and y each must satisfy the relation $\phi'' = 2\phi$.

11. Show that a general nth-order differential equation is equivalent to a special system of n first-order equations.

12. Show that the equation $t^2y'' + Aty' + By = 0$ has $y = t^\alpha$ and $y = t^\beta$ as solutions whenever $A = 1 - \alpha - \beta$ and $B = \alpha\beta$.

1.4 Differential Equations and Families of Functions

The collection of solutions of a differential equation is a parametrized family of functions or curves. Conversely, given such a family, it is often possible to find a differential equation which characterizes this particular family. For example, consider the family of all straight lines in the plane. If we exclude the vertical lines, the family can be described by the general formula $y = Ax + B$, where A and B can be any pair of real numbers. (Such a family is sometimes said to have two **degrees of freedom**, or to be a two-parameter family.) It should be evident that this same family can be described as the set of all solutions of the second-order differential equation $d^2y/dx^2 = 0$.

A second illustration will show better how one can go from a formula for a family of curves to a differential equation describing the family. Suppose we are given the curves whose equations have the form

$$y = \frac{x^2+A}{x+A} \tag{1.24}$$

This is a one-parameter family of curves, and for each choice of A, a new curve results. (In Figure 1.15, we have shown a number of these curves for various choices of A.) We pose the problem: Is there a single differential equation which has all of these curves among its solutions? Such an equation will be a single relationship composed of x, y, dy/dx, and perhaps higher derivatives, which is satisfied by functions of the form (1.24), no matter what value A has.

In effect, we want to eliminate A from equation (1.24) by means of differentiation. One approach is to solve for A, first writing $y(x+A) = x^2+A$, and then writing

$$A = \frac{x^2-xy}{y-1}$$

Now, differentiate this with respect to x, remembering that y is a function

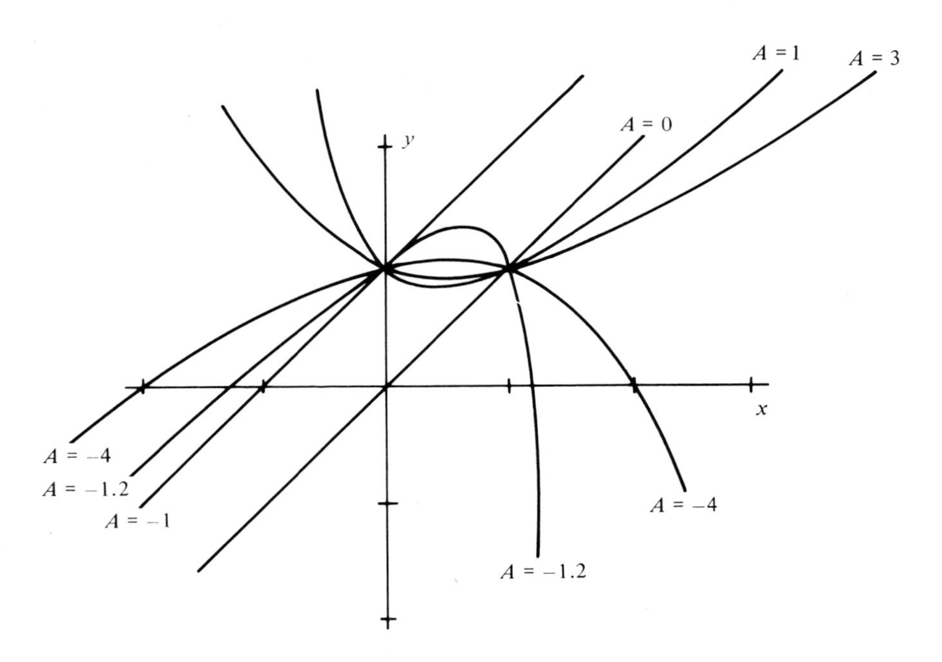

Figure 1.15

of x and that A is merely a number. We obtain

$$0 = \frac{(y-1)\left(2x - x\dfrac{dy}{dx} - y\right) - (x^2 - xy)\dfrac{dy}{dx}}{(y-1)^2}$$

If we solve this for dy/dx, we obtain

$$\frac{dy}{dx} = \frac{(2x-y)(y-1)}{x^2-x} \tag{1.25}$$

This is the desired differential equation. It can be verified directly that each of the functions given by (1.24) is a solution of this equation; one need only substitute for y and compare both sides of (1.25).

Another example may be helpful. Consider the family of all circles that pass through the origin and have their centers on the X-axis. The equation of such a circle is

$$(x-c)^2 + y^2 = c^2$$

We seek a differential equation that is satisfied by each of these curves. First, transform the circle equation to the form

$$x^2 + y^2 = 2cx$$

and then differentiate with respect to x, obtaining

$$2x + 2y\frac{dy}{dx} = 2c$$

This enables us to eliminate c, obtaining first

$$x^2 + y^2 = x\left(2x + 2y\frac{dy}{dx}\right)$$

and then

$$\frac{dy}{dx} = \frac{y^2 - x^2}{2xy}$$

which is the desired differential equation.

When a family of curves is defined by a formula that contains n independent parameters, one expects that the associated differential equation will be of the nth order. For example, let us find a differential equation for the two-parameter family of parabolas

$$y = A(x-B)^2$$

If we differentiate this twice, we obtain

$$y' = 2A(x-B)$$

$$y'' = 2A$$

We can then use these to eliminate both A and B, obtaining first

$$A = \tfrac{1}{2}y''$$

$$x - B = \frac{y'}{y''}$$

and then

$$y = \tfrac{1}{2}y''\left(\frac{y'}{y''}\right)^2$$

which finally yields the desired equation

$$y'' = \frac{(y')^2}{2y}$$

Since the family of curves was a two-parameter family, we are not surprised to find that the appropriate equation is of second order.

A very important application of this process is in the study of **orthogonal families** of curves. Two families are called orthogonal if each curve in one family is perpendicular to every curve in the other family at every crossing point. As an illustration, consider the following two families of hyperbolas:

$$\mathcal{F}_1: \quad x^2 - y^2 = A$$

$$\mathcal{F}_2: \quad xy = B$$

These have been sketched in Figure 1.16; the dashed curves belong to family $\mathcal{F}_1$, and each is orthogonal to every curve of $\mathcal{F}_2$ which it crosses.

To prove this, we find the appropriate differential equation for each family. Using the techniques explained in the preceding examples, we first work with $\mathcal{F}_1$. If we differentiate the formula for a curve of the family, we immediately arrive at the desired equation:

$$\mathcal{F}_1: \quad y' = \frac{x}{y} \tag{1.26}$$

If we apply the same process to $\mathcal{F}_2$, we arrive at a different differential equation:

$$\mathcal{F}_2: \quad y' = -\frac{y}{x} \tag{1.27}$$

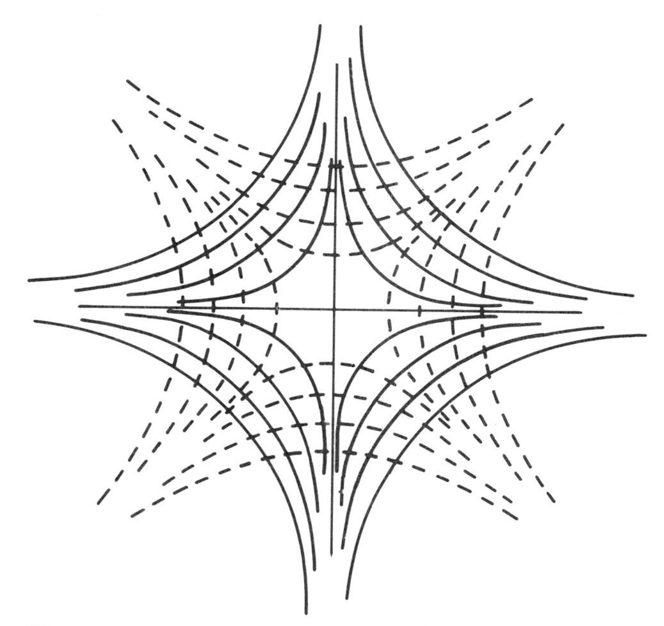

Figure 1.16

The mutual orthogonality property of the two families is at once evident from (1.26) and (1.27), for we observe that the product of the expressions for each of the slopes is -1: using m_1 for the slope of any curve in $\mathcal{F}_1$ and m_2 for the slope of any curve in $\mathcal{F}_2$, we have

$$m_1 m_2 = \left(\frac{x}{y}\right)\left(-\frac{y}{x}\right) = -1$$

(Recall that two lines are perpendicular if the product of their slopes is -1.)

In many physical problems, it is useful to be able to find the family of curves that is orthogonal to a given family of curves. For example, in a force field, the flow lines along which a free particle will move will often be the curves that are orthogonal to the family of equipotential curves of the force field. In perhaps a more familiar context, the paths of steepest ascent of a mountain will be the curves that are orthogonal to the contour lines on its topographic map.

To see how such a problem is solved, let us find the orthogonal family (also called the **orthogonal trajectories**) associated with the following family of ellipses:

$$\tfrac{1}{4}x^2 + y^2 = A$$

(These could be the contour lines for a single mountain peak, or the force field for a single charged particle.) If we ask for the differential equation for this family, we are led to the first-order equation

$$y' = -\frac{x}{4y} \tag{1.28}$$

We can immediately write down the differential equation for the orthogonal family, merely by taking the negative reciprocal of the right-hand side, obtaining

$$y' = \frac{4y}{x} \tag{1.29}$$

While we are not yet in a position to solve this equation to find the formula for the associated family of curves, we can verify directly (Exercise 12) that it is given by the simple equation

$$y = Cx^4 \tag{1.30}$$

We give a number of additional illustrations of this among the exercises.

Exercises

1. Find a differential equation satisfied by all cubic curves $y = Ax^3 + Bx^2 + Cx + D$.
2. What differential equation is satisfied by the family $y = e^{Ct}$?
3. Find a differential equation for the family of all lines that go through the point $(-\frac{1}{2}, -1)$.
4. Find a differential equation for the family of cubic curves given by $y = (t-c)^3$.
5. What differential equation is satisfied by the two-parameter family $w = Ae^{By}$?

8. What differential equation is satisfied by $y = a \tan(at+b)$?

7. Find a differential equation for the family of all circles which are tangent to $y = x$ at $(0, 0)$.

6. What is a differential equation for the family of all circles going through the origin?

9. Find a differential equation for the family of all lines that are tangent to the parabola $y = x^2$. Does $y = x^2$ also satisfy that differential equation? Draw a picture showing various members of the family.

10. Verify equations (1.26) and (1.27) for the families $\mathcal{F}_1$ and $\mathcal{F}_2$.

11. Verify equation (1.28).

12. Show that the differential equation for the family (1.30) is indeed that in (1.29).

13. Show that the family of cubic curves with general formula $y = Ax^3$ is orthogonal to the family of ellipses with formula $\frac{1}{3}x^2 + y^2 = B$.

14. (a) Find the differential equation for the family of curves with formula $1/x + 1/y = A$.
    (b) Write the differential equation for the orthogonal family.
    (c) Verify that this family has the formula $x^3 - y^3 = B$.

15. (a) Find the differential equation for the family of all circles with centers at $(0, 2)$.
    (b) Find the differential equation for the orthogonal trajectories.
    (c) Use your geometric intuition to identify this family, and verify that it leads to the equation found in part (b).

1.5 Theory versus Experiment

We have seen that the task of formulating a mathematical model for a physical phenomenon (the motion of a pendulum, for example) can lead to a differential equation. Different ways of approaching the phenomenon, different physical approximations, may result in different models. In the pendulum example of Section 1.1, we arrived at the equation

$$\frac{d^2\theta}{dt^2} = -\frac{g}{L}\sin\theta \tag{1.31}$$

In order to answer some of the questions raised about the motion of the pendulum, we must find a solution of this equation satisfying the initial conditions $\theta = \theta_0$, $d\theta/dt = 0$, when $t = 0$.

It would certainly be disconcerting if there were to be no such solution! Since an actual pendulum really moves, this would mean that our model is quite useless, and we would have to construct a new model. This means that the only useful models are those which have solutions; an important area of mathematical study has to do with proving that certain very general classes of differential equations do have solutions, even when we do not have any techniques for finding these solutions.

Nor is the existence of solutions the only reasonable requirement that we might wish for in constructing a differential equation that is to lead to a useful model. Suppose that we take an actual pendulum, displace it an angle θ_0, and release it from rest. Experience suggests that if we could repeat this experiment exactly, we would obtain precisely the same motion. The time of descent of the bob and the speed attained would be the same each time. This is really a philosophical assumption, connected with the hypothesis of determinism. If we accept this assumption, with regard to a mathematical model for the motion of a pendulum it means that a differential equation such as (1.31) should have exactly one solution satisfying a given set of initial conditions. If ϕ and ψ are solutions of (1.31) for which $\phi(0) = \psi(0) = \theta_0$ and $\phi'(0) = \psi'(0) = 0$, then we must have $\phi(t) = \psi(t)$ for all $t > 0$. The model must also predict that repetitions of the same experiment have the same results. The corresponding mathematical study is concerned with the uniqueness of solutions of differential equations which have assigned initial conditions or boundary condition.

Finally, experience also suggests a third requirement for models. We know that in practice we cannot repeat an experiment in exactly the same way. However, if all the initial conditions are almost the same, then we expect the outcomes to be almost the same. If the pendulum is released at an angle very nearly θ_0, then the time of descent ought to be very nearly that obtained when the angle was exactly θ_0. The solutions of our model ought to have the same behavior. Stating this in mathematical language, we say that the solutions of a differential equation for a physical situation ought to depend in a continuous way upon the numerical values imposed in the initial conditions. For brevity, we refer to this as "continuity of the solution."

We digress for a moment to observe that there are areas of modern physics where it has seemed necessary to relax some of these assumptions in order to build adequate models. Especially at the atomic and nuclear level, there seem to be phenomena that may violate both the determinism and the continuity hypotheses.

Summarizing the three mathematical requirements that seem desirable for a differential equation used to model a physical situation, we have

1. *Existence:* The equation has a solution satisfying the given set of initial conditions.

2. *Uniqueness:* Two solutions of the equation that satisfy the same initial conditions are identical.
3. *Continuity:* Solutions depend continuously on the initial conditions.

Much of the research of mathematicians who study the theory of differential equations has dealt with the effort to show that wider and wider classes of equations satisfy these three requirements. For example, they have succeeded in showing that equations so complicated that there is no known method for finding formulas for their solutions satisfy all three requirements. This accomplishment is particularly important for applications; faced with such an equation, an applied mathematician can turn to a high-speed computer and calculate an approximation to the solution, knowing by requirements 2 and 3 that there is only one exact solution and that his values must be close to the true values.

We shall explain the meaning of certain of the basic theorems of this type without attempting to prove them now. The first, dealing with first-order equations, is the following:

THEOREM 1.1 *Let D be an open convex set in the TY-plane. Let $f(t, y)$ be defined for all (t, y) in D, and suppose that f and its partial derivative f_y are continuous in D. Let $p_0 = (t_0, y_0)$ be any interior point of D. Then the differential equation $dy/dt = f(t, y)$ has a unique solution $y = \phi(t)$ which satisfies the initial condition $\phi(t_0) = y_0$. Its graph passes through p_0, and ϕ is defined on an interval I containing t_0. Moreover, if t is in I, then the value of $\phi(t)$ is a continuous function of the initial conditions (that is, of the point p_0.)*

A **convex** set in the plane is one that contains the line segment joining any two points of the set. For example, the set of points inside an ellipse is convex. An **interior point** of D is one that is the center of a disk consisting entirely of points of D; this will not be true of a point located on the boundary of a set.

To give a simple illustration of how Theorem 1.1 can be used, consider the equation

$$\frac{dy}{dt} = t^2 + te^{-y} \tag{1.32}$$

Here, we have $f(t, y) = t^2 + te^{-y}$, which is defined and continuous in the whole TY-plane. Since $f_y(t, y) = 0 - te^{-y}$, which is also continuous everywhere, we conclude that (1.32) can be solved uniquely to satisfy any given initial condition. For example, there is exactly one solution of (1.32) that satisfies the condition that y shall be 103 when $t = 1.71$.

When would an equation fail to satisfy the hypotheses of this theorem? The standard example is the equation

$$\frac{dy}{dt} = 3y^{2/3} \tag{1.33}$$

Here, $f(t, y) = 3y^{2/3}$, which is continuous for all t and y. However, $f_y(t, y) = 2y^{-1/3} = 2/\sqrt[3]{y}$. This is not continuous or even defined at points (t, y) with $y = 0$. Thus, the theorem would not apply if we were trying to discuss solutions of (1.33) on a region D that contained any points of the T-axis. What makes this equation interesting is that although it has solutions for any given initial condition, the uniqueness condition breaks down completely. If the point $p_0 = (t_0, y_0)$ is not on the horizontal T-axis, then there is exactly one solution curve passing through p_0. However, if p_0 is of the form $(t_0, 0)$, then there are infinitely many solutions of the equation that pass through p_0.

For example, if $p_0 = (0, 0)$, then two solutions are easily found. Both $y = 0$ for all t and $y = t^3$ for all t are solutions, and both fulfill the requirement $y = 0$ when $t = 0$. Less obvious is that each of the functions defined by

$$y = \begin{cases} 0 & -\infty < t < b \\ (t-b)^3 & b \le t < \infty \end{cases} \tag{1.34}$$

is a continuous function having a continuous derivative satisfying the equation (1.34) and the assigned initial condition. (See Figure 1.17.)

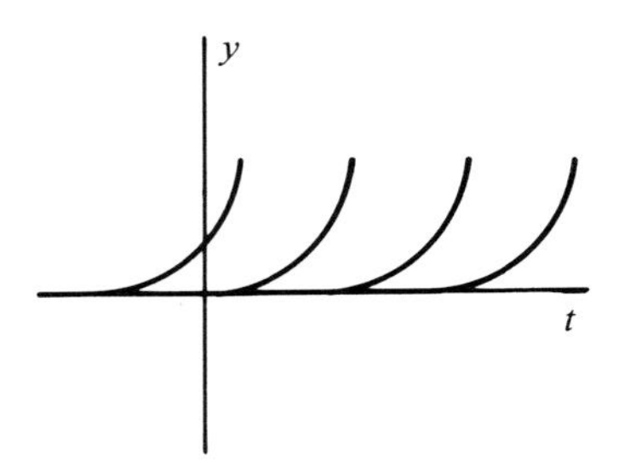

Figure 1.17

This example shows that uniqueness is not automatic, and that an adequate theory of differential equations is needed to enable us to know conditions which will ensure that solutions for assigned initial conditions are unique.

A similar result can be stated for second-order equations.

THEOREM 1.2 *Let D be a convex set in 3-space. Let f and its partial derivatives be defined and continous in D. Let (t_0, A, B) be an interior point of D. Then, the* *differential equation* $d^2y/dt^2 = f(t, y, y')$ *has a unique solution* $y = \phi(t)$ *which satisfies the initial conditions* $y = \phi(t_0) = A$, $y' = \phi'(t_0) = B$. *The function* ϕ *is defined on an interval* I *containing* t_0, *and if* $t \in I$, *then the value of* $\phi(t)$ *depends continuously on the initial values* A *and* B.

To illustrate this, consider the following equation, which is a model for a pendulum swinging in a resisting medium such as oil:

$$\frac{d^2\theta}{dt^2} = -\frac{g}{L}\sin\theta - k\left(\frac{d\theta}{dt}\right)^2 \tag{1.35}$$

If we set $y = \theta$ and $y' = d\theta/dt$, then this takes the standard form $y'' = f(t, y, y') = -(g/L)\sin y - k(y')^2$. To check the hypotheses, we compute the partial derivatives of f. Since we are dealing with a complicated situation, it is convenient to introduce another notation for partial derivatives that has many advantages, and use f_1, f_2, and f_3 to indicate the function obtained by differentiating f with respect to the first variable, the second variable, or the third variable, respectively. In our particular case, we have

$$\begin{aligned} f_1(t, y, y') &= 0 \\ f_2(t, y, y') &= -\frac{g}{L}\cos y \\ f_3(t, y, y') &= -2ky' \end{aligned} \tag{1.36}$$

Since all these are continuous everywhere, we conclude that equation (1.35) has a unique solution corresponding to any assignment of initial conditions.

Theorems 1.1 and 1.2 are commonly called the **Existence and Uniqueness theorems**. Their proofs are not particularly difficult but do involve some ideas that are often taken up in advanced calculus, and so we have postponed their treatment to Section 3.6. It should be remarked that the hypotheses placed on the function f are unnecessarily strong, and better theorems have been discovered which show that existence and uniqueness hold even under weaker requirements on f. (The example $y' = 3y^{2/3}$ should remind you that *some* restrictions are needed.)

As has been mentioned earlier, once it is known that a particular differential equation has exactly one solution obeying a given set of initial conditions, then one may have much more confidence in using approximate numerical methods or other techniques that describe general properties of the solution, especially in those cases where there is no known method for obtaining formulas for the exact solutions.

It should also be stated that there are interesting physical problems which give rise to differential equations that do not satisfy the hypotheses of Theorems 1.1 and 1.2, and which in fact do not have unique solutions.

Exercises

1. Apply Theorem 1.1 to the equation $dy/dt = 1+y^2$. What is the region D in which one might expect to find solutions? Are there any points in this region which should not be used as initial points?

2. Do the same as in Exercise 1 for the equation $dy/dt = \sqrt{y^2-1}$. (Note that $y = \phi(t) \equiv 1$ is a solution.)

3. Check that the equations (1.34) do provide solutions of the differential equation $dy/dt = 3y^{2/3}$ and that Figure 1.17 is correct.

4. Use Theorem 1.2 to discuss the existence of solutions to the equation $d^2y/dt^2 = a(t)+b(t)\,y+c(t)(dy/dt)$.

5. Do the same as in Exercise 4 for the equation

$$y'' = \frac{y}{4-(y')^2}$$

   Are there any initial conditions that cannot be expected to be usable for solutions?

6. Do the same as in Exercise 4 for the equation $y'' = (t-y)/(t-y')$.

7. Speculative question: Suppose a differential equation is a model for a certain type of chemical reaction. Could the fact that the equation does not have a solution indicate that the reaction cannot take place? Would the fact that the equation has a solution guarantee that the reaction *does* take place?

8. Why can't the family of curves shown in Figure 1.18 be the solution curves for the differential equation $y' = f(t, y)$, where f is a polynomial in t and y?

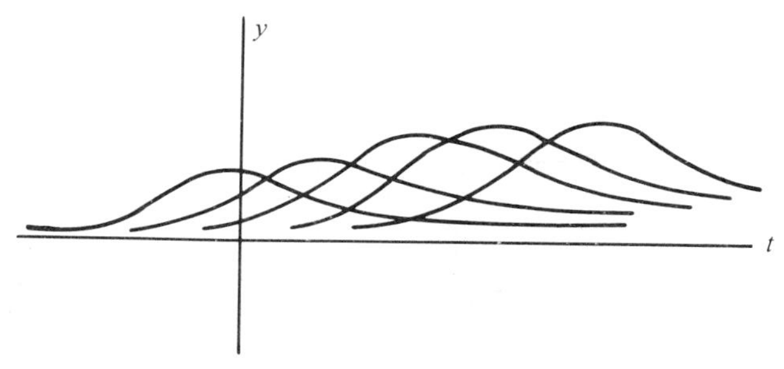

Figure 1.18

9. The parabola $y = x^2$ and the line $y = 2x-1$ are both solutions of the equation $y' = 2x-2\sqrt{x^2-y}$ and obey the initial conditions $x = 1$, $y = 1$. Does this contradict Theorem 1.1?

2

Techniques of Solution

Introduction

The general existence and uniqueness theory discussed in the preceding chapter tells us when to expect solutions for a particular differential equation with given initial conditions. The next step is to learn some methods for finding these solutions. It must first be said that there are no general techniques that work for all equations, even for all first-order equations. The best that can be expected is to discover procedures for obtaining a sequence of better and better approximations to the true solution. We would prefer to have a formula that expresses the solution in terms of polynomials or other relatively simple functions which we regard as elementary. As an analogy, we may be able to express the solution of an algebraic equation as the real number defined by a converging sequence of rational numbers (for example, in its decimal expansion), but we often prefer to have an answer that is expressed in terms of rational numbers and certain familiar real numbers like π, e, and $\sqrt{3}$.

Lacking a general universal method, it is useful to have a number of special methods that have been discovered over the years which apply only to certain equations. Note that *any* device for obtaining a solution is legitimate, provided that the function obtained satisfies the equation. The reason is that the theory tells us that the equation has only one solution for a given set of initial conditions, so the function we have found must be this unique solution.

2.1 Separable Equations

Recall the general form for a first-order equation:

$$\frac{dy}{dt} = f(t, y) \tag{2.1}$$

Two special cases can be disposed of at once, namely when the right side contains only one of the variables t, y. For example, if $f(t, y) = g(t)$, then (2.1) merely tells us that the unknown function $y = \phi(t)$ is an antiderivative of g. For example, the solutions of the equation

$$\frac{dy}{dt} = te^{-t^2}$$

consist of the functions $y = -\frac{1}{2}e^{-t^2} + A$ for any choice of the number A.

The case in which the right-hand side of (2.1) has the form $h(y)$ is almost as simple. Make the assumption that we are dealing with a point where $h(y) \neq 0$. Since this implies $dy/dt \neq 0$, we can regard the equation $y = \phi(t)$ as one that can be solved for t, in the form $t = \psi(y)$. Recall that we then have $dt/dy = 1/(dy/dt)$. We have now converted the equation $dy/dt = h(y)$ into the equation

$$\frac{dt}{dy} = \frac{1}{h(y)}$$

and again we can solve this directly by integrating

$$t = \int \frac{dy}{h(y)}$$

We then regard the result as an equation defining y as a function of t.

For example, to solve the equation

$$\frac{dy}{dt} = \frac{y}{y+1}$$

we first rewrite it as

$$\frac{dt}{dy} = \frac{y+1}{y}$$

We then have, by integration,

$$t = \int \frac{y+1}{y}\, dy = \int (1 + y^{-1})\, dy$$

$$= y + \log y + C$$

This equation, for any C, is taken as a relation that implicitly defines the desired function $y = \phi(t)$.

We inject the comment that a more usual procedure is to go from the equation

$$\frac{dy}{dt} = \frac{y}{y+1} \tag{2.2}$$

to the expression

$$\frac{y+1}{y}\, dy = dt \tag{2.3}$$

and then to the integral form

$$\int \frac{y+1}{y}\, dy = \int dt = t \tag{2.4}$$

Although it is possible to become very technical about the meaning of the expression in (2.3), it is much easier to regard it as nothing more than a notational device which leads one conveniently from (2.2) to (2.4).

A more general case, which includes both of the above, is of wide usefulness. A differential equation is said to have **variables separable** if it has the form

$$\frac{dy}{dt} = g(t)h(y) \tag{2.5}$$

The corresponding solution method is called **separation of variables.** Proceeding as in the previous case, we go from (2.5) to the expression

$$\frac{dy}{h(y)} = g(t)\, dt$$

(This step explains the name of the method, since the variables are now separated.) We then obtain the integrated equation

$$\int \frac{dy}{h(y)} = \int g(t)\, dt \tag{2.6}$$

Each side yields an indefinite integral (or antiderivative)—the left a function of y, and the right a function of t. The resulting equation is again regarded as implicitly defining $y = \phi(t)$, and we obtain a solution of (2.5). We note that each side of (2.6) will include constants of integration, since antiderivatives are not unique. However, we do not need to add a constant on *both* sides, since we can achieve the same result by adding a constant on only one side. In general, for each choice of a numerical value for this constant, we obtain a solution of the equation (2.5). It is possible that this process does not yield a full set of solutions of the original equations, so that we may not be able to satisfy an arbitrary set of initial conditions among these solutions, but must look elsewhere.

An illustration may help to clarify the last remarks. Suppose we consider the first-order equation $dy/dt = 4y^2t$. This is clearly of separable type, so that we arrive at

$$\int \frac{dy}{y^2} = \int 4t\,dt$$

or

$$-\frac{1}{y} = 2t^2 + C$$

from which we obtain the desired family of solutions of the given equations:

$$y = \frac{-1}{2t^2 + C} \qquad \text{for any } C \tag{2.7}$$

(This can be checked by direct substitution into the equation.) Since the right-hand side of the equation, $4y^2t$, is continuous and has continuous partial derivatives everywhere, we can be sure that the equation has a unique solution passing through any given point (t_0, y_0). If, for instance, $t_0 = 1$ and $y_0 = 3$, then we must choose C so that $3 = -1/(2+C)$, or $C = -\frac{7}{3}$. This works for every assignment of initial condition except an assignment such as $(t_0, 0)$, for the equation $0 = -1/(2t_0{}^2 + C)$ has no solution. Does this mean that the given equation has no solution through the point $(1, 0)$ or $(0, 0)$? This would contradict Theorem 1.1, which we cannot accept. The answer lies in the simple fact that there is another solution of the equation, $y = \phi(t) \equiv 0$, which is not described by the formula (2.7) but which fulfills all initial conditions corresponding to points $(t_0, 0)$.

How did the above method happen to overlook this missing solution? The answer lies in the first step of the solution. We divided by $h(y)$, which in this case was y^2, and thus we inadvertently limited ourselves to solutions other than $y = 0$. It is exactly the solution $y = 0$ that was omitted.

The same thing happens in the general case. If a separable equation of the form $dy/dt = g(t)h(y)$ is solved by the method described above, then the final solutions will not take into account solutions which satisfy initial conditions (t_0, y_0) where $h(y_0) = 0$. However, they will be covered by straight-line solutions of the form $y \equiv y_0$, and the combination usually gives a solution of the given equation going through the admissible point (t_0, y_0).

Another illustration will show how the method of separated variables is used in a more complicated case. Let

$$\frac{dy}{dt} = \frac{y^2+y}{t} \tag{2.8}$$

We first write

$$\frac{dy}{y(y+1)} = \frac{dt}{t} \tag{2.9}$$

Noting that

$$\frac{1}{y(y+1)} = \frac{1}{y} - \frac{1}{y+1}$$

we integrate (2.9), obtaining

$$\int \left(\frac{1}{y} - \frac{1}{y+1}\right) dy = \int \frac{1}{t}\, dt$$

or

$$\log y - \log(y+1) = \log t + C$$

In order to make the work of simplifying easier, we replace C by $\log A$ and then combine the terms on each side, obtaining

$$\log\left(\frac{y}{y+1}\right) = \log(At)$$

$$\frac{y}{y+1} = At$$

This can now be solved for y, yielding the desired family of solutions

$$y = \frac{At}{1-At} \tag{2.10}$$

Can we now satisfy any admissible initial condition by choosing one of these solutions of (2.8)? Since the right side of (2.8) is not defined or continuous when $t = 0$, the region D for which the equation is meaningful cannot contain any points on the line $t = 0$. However, there are no other restrictions, so we might wish to consider any initial point (t_0, y_0), where $t_0 \neq 0$ but y_0 is arbitrary. Putting $t = t_0$ and $y = y_0$ in (2.10), we can solve for A, obtaining the (unique as expected) solution

$$A = \frac{y_0}{(1+y_0)\,t_0}$$

However, this solution breaks down when $y_0 = -1$. This means that no number can be chosen for A to give a solution of this form which passes through a point such as $(2, -1)$ or $(1, -1)$. Return to (2.8) and observe that the right side factors as $h(y)g(t)$, $h(y) = y^2 + y$, which vanishes when $y = -1$. As explained in the general discussion preceding this example, we can conclude that we must adjoin the special solution $y = \phi(t) \equiv -1$ to the set described by (2.10). We then will have a complete set of solutions, one passing through every admissible initial point (t_0, y_0).

There is one class of differential equations which can be reduced to the separable form by a simple change of variables, and which is sufficiently common in practice to deserve emphasis. We first need a useful definition.

DEFINITION **2.1** A function f of two variables is called **homogeneous** if it has the property that $f(u, v) = f(cu, cv)$ for all u, v, and c for which these are defined. A differential equation of first order is called **homogeneous** if it has the form $dy/dt = f(t, y)$ and the function f is homogeneous.

For example, each of the following expressions defines a homogeneous function

$$\frac{2x^2 - y^2}{3xy}, \qquad \sin\left(\frac{5u}{v}\right), \qquad \frac{\sqrt{x^2 + 2y^2 - xy}}{4x + y}$$

(The last of these satisfies the required identity only for positive c, but this is often enough for most purposes.) In contrast, none of the following functions is homogeneous.

$$\frac{xy}{x + y}, \qquad \sin(x + y), \qquad \frac{u + uv + v}{u^2 + v^2}$$

An alternative definition is to say that $f(u, v)$ is homogeneous if it can be expressed simply as $g(u/v)$, that is, as a function of the quotient u/v alone. (See Exercise 9.)

THEOREM **2.1** *Any homogeneous differential equation can be changed into a separable equation by the substitution $y = ut$.*

Proof If $y = ut$, then $dy/dt = t\, du/dt + u$. Thus, the differential equation $dy/dt = f(t, y)$ becomes

$$t \frac{du}{dt} + u = f(t, ut)$$

However, since f is homogeneous, $f(t,ut)=f(t\cdot 1,t\cdot u)=f(1,u)$, so that the equation can now be rewritten as

$$t\frac{du}{dt} = f(1,u) - u$$

which is a separable equation.

We illustrate this by solving the equation

$$\frac{dy}{dt} = \frac{t-3y}{3t+y} \tag{2.11}$$

Putting $y = ut$, this becomes

$$t\frac{du}{dt} + u = \frac{t-3ut}{3t+ut} = \frac{1-3u}{3+u}$$

and

$$t\frac{du}{dt} = \frac{1-3u}{3+u} - u = \frac{1-6u-u^2}{3+u}$$

We rewrite this separable equation in the form

$$\int \frac{u+3}{u^2+6u-1}\,du = -\int \frac{dt}{t}$$

and integrate, obtaining

$$\tfrac{1}{2}\log(u^2+6u-1) = -\log t + \text{constant}$$

This in turn can be rewritten as

$$2\log t + \log(u^2+6u-1) = \text{constant}$$

Taking exponentials, we arrive at

$$t^2(u^2+6u-1) = \text{constant}$$

Since $y=ut$, we replace u by y/t, and finally obtain for the differential equation (2.11) the desired solution curves

$$y^2 + 6yt - t^2 = C$$

These are conic sections, and in fact hyperbolas.

Note: In the solution of this differential equation, we have used a convenient notational device which perhaps deserves more explanation. We refer to the use of the word "constant," which has represented a different arbitrary constant each time it appears in the solution. To understand this better, suppose we use the letter A instead, when we first need an arbitrary constant. When we integrate the separated equation, we would then have

$$\tfrac{1}{2}\log(u^2+6u-1) = -\log t + A$$

The next equation results from multiplying this equation by 2, so that it now becomes

$$2\log t + \log(u^2+6u-1) = 2A$$

As before, taking exponentials, we arrive at

$$t^2(u^2+6u-1) = e^{2A}$$

and the final form would be

$$y^2 + 6ty - t^2 = e^{2A}$$

But, since A stands for *any* number at all, the expression e^{2A} is itself an arbitrary number, and so we can replace it by C.

Exercises

For each of Exercises 1–8, find solutions $y = \phi(t)$ which go through the given initial points (t_0, y_0).

1. $y' = t^2y$; $(1, 2)$
2. $y' = t^2y$; $(1, 0)$
3. $y' = t/y$; $(1, 3)$, $(0, 1)$, $(1, -3)$
4. $y' = (t-t^2)/(2y+1)$; $(2, 1)$, $(1, 1)$
5. $y' = (y^2-y)/t$; $(1, 2)$ $(1, 1)$
6. $y' = e^{t+y}$; $(0, 1)$
7. $y' = t^2y - t^2$; $(1, 3)$, $(1, 2)$, $(1, 0)$, $(0, 2)$
8. $y' = (y-y^2)/(t+t^2)$; $(1, 2)$
9. Show that $f(u, v)$ is homogeneous if it can be expressed as a function $g(u/v)$, that is, as a function of the quotient u/v alone.
10. Solve the equation $y' = (y^2-x^2)/2xy$ and sketch the family of curves which are the solutions.

11. Solve the equation

$$\frac{dy}{dt} = \frac{2t^2 - y^2}{2ty}$$

Verify your solution by differentiation.

12. Solve the equation

$$\frac{du}{dv} = \tan\left(\frac{5u}{v}\right) + \frac{u}{v}$$

Verify.

13. Solve and identify the solutions of

$$\frac{dy}{dx} = \frac{y^2 + 2xy - x^2}{x^2 + 2xy - y^2}$$

14. Find a relationship between x and y satisfying the differential equation

$$\frac{dy}{dx} = \frac{xy}{x^2 - y^2}$$

15. Find the orthogonal trajectories (Section 1.4) for the family of curves

$$\frac{1}{x^2} + \frac{1}{y^2} = C$$

16. Find the orthogonal trajectories for the family of curves $y = Cx^3$. Sketch a few members of both families.

17. Find the family orthogonal to the family $4x^2 - y^2 = C$.

18. A tank contains 100 liters of water and 10 kilograms of salt, thoroughly mixed. Pure water is added at the rate of 5 liters per minute, and the mixture is drained off at the same rate. How much salt is left in the tank after 1 hour? Assume complete and instantaneous mixing.

19. Experience suggests that the rate at which people contribute in a charity drive is proportional to the difference between the current total and the announced "target" goal. A drive is announced with target set at $90,000 and an initial contribution of $10,000. After one month, $30,000 has been contributed. What does the model predict at the end of two months?

20. The mass of a spherical hailstone grows at a rate proportional to its surface area. Show that its diameter grows at a constant rate.

21. A mold grows at a rate which is proportional to the amount present. Initially the quantity is 2 grams; in two days the quantity has increased to 5 grams. Find the quantity at the end of eight days.

22. A crucible is removed from an oven whose temperature is 500 °C and placed in a room in which the air temperature is kept at 20 °C. Assume that the rate of cooling is proportional to the difference in the temperatures of the air and the crucible, and that after $\frac{1}{2}$ hour the temperature of the crucible has fallen to 260 °C. How long will it take to reach 80 °C?

23. The yeast in a vat grows at a rate proportional to its size. At 7:00 a.m. it fills one-twentieth of the vat. By 7:30 a.m. it has grown to fill one-tenth. Describe the situation at 9:00. When will it fill the vat?

24. The rate at which water empties through a hole in the bottom of a vessel is proportional to the square root of the height of the liquid above the hole. A cylindrical tin can of radius 6 cm and height 36 cm is filled with water. There is a small hole in the bottom. In $\frac{1}{2}$ hour, half the water has leaked out. How much water will be left after 1 hour?

25. The rate at which water evaporates in a container is proportional to the surface area. A conical glass has height 6 inches and diameter 4 inches at the top and is full of water. By the end of the first day the level of the water has dropped $\frac{1}{2}$ inch from the top. How long does it take to drop to a depth of 3 inches? How long will it take for all the water to evaporate?

26. In a recently used classroom of volume 6000 cubic feet, the air tested 0.16% carbon dioxide. The ventilation system brings in air containing 0.04% carbon dioxide at a rate of 1000 ft^3/min and drives out the mixture of stale air and fresh air at the same rate. What will be the percentage of carbon dioxide in the room after 6 minutes of ventilation?

27. A spherical raindrop falls through a uniform fog at a constant speed V, and grows in volume as it collects all the moisture in its path. How does its radius grow with time?

28. Assume that the rate of increase of a nation's gross national product (GNP) is proportional to the size of the GNP. The GNP's of two nations are shown below (in billions):

	1960	*1970*
Nation A	60	70
Nation B	50	60

(a) Predict the GNP's for 1980.

(b) Show that at some future time nation A and nation B will have the same GNP, and determine the date.

29. The rate at which a body changes temperature is, according to Newton, approximately proportional to the difference between its temperature and that of the surrounding medium. A hot meteorite weighing 30 pounds falls into a rapidly flowing stream whose temperature is 40°. Ten minutes later, the temperature of the meteorite is down to 200°, and in ten more minutes, it is 60°. What was the initial temperature of the meteorite, and what will be the temperature half an hour after it has landed?

30. Consider the following model for the growth of a city. The shape of the city remains always roughly circular. The maximum travel time is proportional to the diameter of the city. The population is proportional to the area of the city. The rate of increase of the city's population is inversely proportional to the maximum travel time. If the population of the city was 5000 in 1900 and 20,000 in 1940, what is the predicted population for 1980?

31. Many "curves of growth" satisfy what is called the **logistic law**, described by saying that the rate of growth is proportional to the product of the size and the difference between the size and some maximum size. Solve the corresponding differential equation, and sketch a typical logistic curve.

32. Assume that a college's enrollment follows the logistic law (Exercise 31) and that at the start there are 10,000 students. If the maximum is set at 25,000 students and in five years enrollment has reached 20,000, what will it be in five more years?

33. In a simple chemical reaction, equal parts of chemicals A and B combine to yield two parts of substance C. The rate at which C is produced is proportional to the product of the amounts of A and B which are left uncombined at that moment. If the reaction starts with 20 grams of A and 10 grams of B, 10 grams of C is obtained at the end of 5 minutes. How much C will there be after 5 more minutes?

*34. In a simple chemical reaction, 2 grams of A and 1 gram of B combine to yield 3 grams of C. The rate at which C is produced is proportional to the product of the amounts of A and B that are left uncombined at the moment. If we start with 20 grams of A, 10 grams of B, and no C, it is found that 10 grams of C is produced at the end of 5 minutes. How long will it take to obtain a total yield of 25 grams of C? Would it have saved much time to have added 10 more grams of A at the start?

2.2 Linear First-Order Equations

Sometimes the form of the function f in a differential equation

$$\frac{dy}{dt} = f(t, y)$$

makes it possible either to solve the equation directly or to change it to an equation which can be solved directly. We will illustrate two general methods for the latter.

Suppose it is possible to expand the right-hand side in powers of y, as

$$\frac{dy}{dt} = a_0(t) + a_1(t)\,y + a_2(t)\,y^2 + \cdots \tag{2.12}$$

An equation is said to be a **linear equation of first order** if only the first two terms in this expansion are present. For example,

$$\frac{dy}{dt} = t^3 - 2ty \tag{2.13}$$

is a linear equation. An equation that is not linear can sometimes be replaced by a linear equation simply by throwing away all the terms with higher powers of y. The new equation will often be a usable model for the original problem as a first approximation. This process, called **linearization,** makes sense if we expect that the values of y will always be very small. [It is the process that led from pendulum model (1.4) to pendulum model (1.10).]

Any first-order linear equation can be solved. The technique is easily learned, but in many cases the solution in its final form is not very useful because it is given in terms of integrals that can be evaluated only in special cases. We illustrate the process first for the equation (2.13), and then we discuss the general case.

Let us rewrite (2.13) as

$$y' + 2ty = t^3$$

and then multiply both sides by e^{t^2}, obtaining

$$e^{t^2}y' + 2te^{t^2}y = t^3e^{t^2} \tag{2.14}$$

(As happens with many mathematical discoveries, this technique, attributed to one of the Bernoulli brothers, was probably the result of a lucky insight. Inspiration is very difficult to motivate.) We observe that the two terms on the left arise from differentiating the expression $e^{t^2}y$. Thus, we can rewrite (2.14) as

$$\frac{d}{dt}\{e^{t^2}y\} = t^3 e^{t^2}$$

If we now integrate, we first obtain

$$e^{t^2}y = \int t^3 e^{t^2}\, dt + C$$

and then the desired solution of the equation (2.13),

$$y = e^{-t^2}\int t^3 e^{t^2}\, dt + Ce^{-t^2}$$

We can carry out the integration to obtain a simpler answer. Set $u = t^2$, so that

$$\int t^3 e^{t^2}\, dt = \tfrac{1}{2}\int ue^u\, du = \tfrac{1}{2}(u-1)\, e^u$$
$$= \tfrac{1}{2}(t^2-1)\, e^{t^2}$$

and we obtain the final form

$$y = \tfrac{1}{2}(t^2-1) + Ce^{-t^2}$$

for the set of all solutions to (2.13).

How does one generalize from this special case to obtain a general method? Since a linear first-order equation is one in which the right-hand side has the form $f(t, y) = a_0(t)+a_1(t)\, y$, we can always rewrite the equation so that it takes the form

$$y' + p(t)\, y = q(t) \tag{2.15}$$

(At this point, we insert the warning that the coefficient of y' must be 1 if the method is to work; in some cases, it will be necessary to divide the equation by some factor to bring this about.)

The trick we used to solve (2.13) was to multiply the equation by a specific function which turned the left-hand side into an expression that was exactly the derivative of a product of y and something else. Restating, then, we would like to find a function $G(t)$ that has the property that

$$G(t)\,(y'+p(t)\, y) = \frac{d}{dt}\{G(t)\, y\}$$

If we expand both sides, we have

$$G(t)\,y' + G(t)\,p(t)\,y = G(t)\,y' + G'(t)\,y$$

In order to have this satisfied, we need to choose G so that $G(t)\,p(t) = G'(t)$.

To solve this separable equation for G, we first write it as $dG/dt = Gp(t)$, and then, in the usual way, arrive at

$$\int \frac{dG}{G} = \int p(t)\,dt$$

$$\log G = \int p(t)\,dt$$

If we abbreviate the right-hand side by $P(t)$, we find that the correct choice for $G(t)$ is $e^{P(t)}$.

We now have the key to the desired general method. To solve the differential equation

$$y' + p(t)\,y = q(t) \tag{2.15}$$

we first evaluate the function $P(t) = \int p(t)\,dt$, and then calculate the function $G(t) = e^{P(t)}$. We next multiply both sides of the equation (2.15) by $G(t)$ and observe that the result can be rewritten in the form

$$\frac{d}{dt}\{G(t)\,y\} = q(t)\,G(t)$$

Integration then yields

$$G(t)\,y = \int q(t)\,G(t)\,dt + C$$

and we have the desired solution

$$y = \frac{1}{G(t)} \int q(t)\,G(t)\,dt + \frac{C}{G(t)} \tag{2.16}$$

If we replace $G(t)$ by $\exp\{P(t)\}$, this becomes

$$y = e^{-P(t)} \int q(t)\,e^{P(t)}\,dt + Ce^{-P(t)} \tag{2.16$'$}$$

We give several additional examples of this general method for solving linear first-order equations.

(a) Solve

$$ty' = -3y + t^3 - t^2$$

Rewrite the equation as $y'+(3/t)y = t^2-t$. Then, $p(t) = 3/t$, so that $P(t) = 3\log t$. Hence, $e^{P(t)} = t^3$. Thus, if we multiply both sides of the equation by t^3, obtaining

$$t^3\left(y' + \frac{3}{t}y\right) = t^3(t^2-t)$$

or

$$t^3y' + 3t^2y = t^5 - t^4$$

then it can be rewritten as

$$\frac{d}{dt}\{t^3y\} = t^5 - t^4$$

and our solution is

$$t^3y = \int (t^5-t^4)\,dt$$

$$= \frac{1}{6}t^6 - \frac{1}{5}t^5 + C$$

or

$$y = \frac{1}{6}t^3 - \frac{1}{5}t^2 + \frac{C}{t^3}$$

This example shows that there are some cases in which the method is very sucessful in obtaining solutions easily.

(b) Solve

$$y' = \frac{1}{t^2}y + t^3$$

We first rewrite the equation as

$$y' - \frac{1}{t^2}y = t^3 \tag{2.17}$$

Hence, $p(t) = -1/t^2$, so we can take $1/t$ as $P(t)$. Multiplying (2.17) by $e^{1/t}$, we obtain

$$\frac{d}{dt}\{e^{1/t}y\} = t^3e^{1/t}$$

and thus

$$e^{1/t}y = \int t^3e^{1/t}\,dt + C$$

or

$$y = e^{-1/t} \int t^3 e^{1/t}\, dt + Ce^{-1/t} \tag{2.18}$$

This example illustrates the basic handicap in the method. The formula in (2.18) for the solution is of very little use; the integral cannot be evaluated in terms of the ordinary functions of analysis. Although the result is formally correct, it cannot be used directly to find the values of the solution function $y = \phi(t)$ for a specified value of t.

(c) Solve the equation

$$\frac{dx}{ds} = \frac{x}{s} - s^2$$

[For variety—and to break the habit of using only certain names for variables—we have changed to x and s; thus, we are thinking of x as $\phi(s)$. If this causes trouble in following the solution, rewrite it with x replaced by y and s by t. But try to get used to working with different variables, since that is the way differential equations usually arise.]

The equation is linear since only the first power of x is present. First rewrite it in the form

$$\frac{dx}{ds} + \left(-\frac{1}{s}\right)x = -s^2$$

Since $p(s) = -1/s$, $P(s) = -\log s$ and $e^{P(s)} = 1/s$. Multiply *both* sides by this, and obtain

$$\frac{1}{s}\frac{dx}{ds} - \frac{1}{s^2}x = -s$$

which can also be written as

$$\frac{d}{ds}\left\{\frac{1}{s}x\right\} = -s$$

Integrating, we have

$$\frac{1}{s}x = -\tfrac{1}{2}s^2 + \text{constant}$$

so that the solution of the differential equation is

$$x = -\tfrac{1}{2}s^3 + Cs$$

Even when the final integration step cannot be carried out in terms of elementary functions, the result can be useful. Consider the equation

$$\frac{dx}{dt} = 2tx - t^2 \tag{2.19}$$

Note that this is linear since the right-hand side contains only the first power of x. Following the standard scheme, which is usually better than using the final formula (2.16′), we first have

$$\frac{dx}{dt} - 2tx = -t^2$$

Since $p(t) = -2t$, $P(t) = -t^2$ and $e^{P(t)} = e^{-t^2}$. Multiplying by this (on both sides of the equation!), we arrive at

$$e^{-t^2}\frac{dx}{dt} - 2te^{-t^2}x = -t^2e^{-t^2}$$

or

$$\frac{d}{dt}\{e^{-t^2}x\} = -t^2e^{-t^2}$$

leading to the formal solution

$$x = -e^{t^2}\int t^2e^{-t^2}\,dt + Ce^{t^2} \tag{2.20}$$

This time, again, the integral cannot be evaluated in terms of elementary functions. However, using techniques that go a little deeper than elementary calculus, it can be shown that the solutions of this equation must have the form

$$x = F(t) + Ae^{t^2}$$

where $F(t)$ is a function very much like $\frac{1}{2}t$. Indeed, by examining the integral in (2.20), it can be shown that F obeys the inequality

$$\frac{t}{2} \leq F(t) \leq \frac{t}{2} + \frac{1}{4t}, \qquad t > 2$$

Note that this gives a good estimate for F when t is large, since the two sides of the inequality are close to each other. [Proofs of this and further discussion can be found in the *American Mathematical Monthly*, **63**, 414–416 (1956).]

This example illustrates that it is possible to use the formal solution formula, even when it cannot be integrated in simple form, to obtain useful qualitative information about the solutions of a differential equation. In

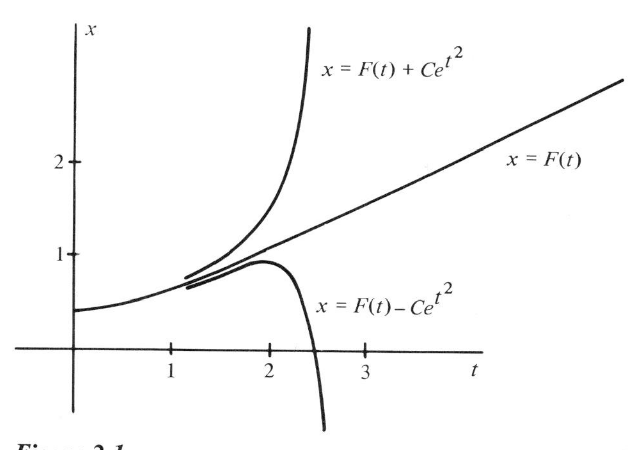

Figure 2.1

Figure 2.1 we have shown the graph of F and several of the other solution curves for equation (2.19). (We will discuss this equation again in the next chapter when we investigate numerical techniques for solving differential equations.)

Sometimes a first-order equation that is nonlinear can be converted into a linear equation by a simple change of variable. As a first illustration, consider the equation

$$y' = \frac{4}{t}y - 6ty^2 \tag{2.21}$$

This is nonlinear because of the y^2 on the right. Suppose we set $y = 1/u$. Then, $y' = (-u^{-2})u'$ and the equation becomes

$$-\frac{1}{u^2}\frac{du}{dt} = \frac{4}{tu} - \frac{6t}{u^2}$$

Multiply through by $-u^2$, and obtain

$$\frac{du}{dt} = -\frac{4}{t}u + 6t$$

which is now a linear equation for u. The integrating factor $e^{P(t)}$ turns out to be t^4, and one arrives at

$$\frac{d}{dt}\{t^4u\} = 6t^5$$

so that

$$u = t^2 + Ct^{-4}$$

Returning to y, we find

$$y = \frac{1}{t^2+Ct^{-4}} = \frac{t^4}{t^6+C}$$

This general formula does not include one special solution of the original equation, namely, the solution $y \equiv 0$. Note that this solution is the limit of those given by the formula if we allow C to increase without bound. The discovery of such special solutions for a differential equation, falling outside the general formula supplied by solution procedures, can be frustratingly difficult. Most can be found by checking the original equation for solutions of the form $y \equiv$ const, or by checking back over the algebraic work to see if at any time a factor has been discarded that could have been identically zero.

This example is one of a class of differential equations covered by the following general result, first observed by Leibniz, although the equation is named after James Bernoulli.

THEOREM 2.2 *The Bernoulli differential equation*

$$\frac{dy}{dt} = a(t)\,y + b(t)\,y^m \tag{2.22}$$

can be converted into a linear equation by the special substitution $y = u^{1/(1-m)}$.

Proof Suppose we try a substitution $y = u^r$, where we have not yet decided upon the choice of r. Then, $y' = ru^{r-1}u'$, so that the differential equation becomes

$$ru^{r-1}u' = a(t)\,u^r + b(t)\,u^{mr}$$

Dividing by u^{r-1}, we obtain

$$ru' = a(t)\,u + b(t)\,u^{mr-r+1}$$

This will be a linear equation in u if the last term is merely a function of t—in other words, if the exponent of u is zero. Setting $mr-r+1 = 0$, we find $r = 1/(1-m)$. [We remark that if $m > 0$, then $y \equiv 0$ is always a solution of (2.22), which may not be found by the substitution method.]

To see how this general result applies to the first illustration, return to equation (2.21); there, we had $m = 2$ so that $r = 1/(1-2) = -1$ and the substitution to try was $y = 1/u$. For a second illustration, consider the equation

$$\frac{dx}{dt} = \frac{1}{t}x + \sqrt{x}$$

Suppose we want the solution that obeys the initial condition $x = 0$ when $t = 1$. Since this is of the form required by the theorem with $m = \frac{1}{2}$, we choose $r = 2$ and make the substitution $x = u^2$. The result is

$$2u\frac{du}{dt} = \frac{1}{t}u^2 + u \tag{2.23}$$

so that dividing by $2u$ (and assuming $u \neq 0$), we arrive at the linear equation

$$\frac{du}{dt} - \frac{1}{2t}u = \frac{1}{2}$$

Solving this in the usual way, we multiply by $e^{P(t)} = t^{-1/2}$ and have

$$\frac{d}{dt}\{t^{-1/2}u\} = \tfrac{1}{2}t^{-1/2}$$

$$\frac{u}{\sqrt{t}} = \sqrt{t} + \text{const}$$

$$u = t + C\sqrt{t}$$

$$x = u^2 = t^2 + 2Ct^{3/2} + C^2t$$

In addition to these solutions, we must include the solution $x \equiv 0$, which comes from the possibility that u is zero. If we want the solution of the original equation that obeys the initial condition $x(1) = 0$, we must choose $C = -1$.

Two standard pitfalls that could be overlooked in solving linear equations are present in this example. The first occurred in (2.23) when we divided by $2u$. If we had divided by u alone, we would have obtained an equation that was clearly linear in u but that did not have 1 as the coefficient of du/dt. The method explained in this section for solving linear equations depends upon having exactly y' or u' on the left, and not some multiple of it. Of course, as noted above, it is important to ask if u can be zero identically, since this may also lead to a solution. The second pitfall has been mentioned before: Be sure to multiply *both* sides of the equation by $e^{P(t)}$.

Exercises

For Exercises 1–5, find solutions $y = \phi(t)$ passing through the given initial points (t_0, y_0).

1. $y' = -2y + e^t$; $(0, 1)$
2. $y' = 2y + e^t$; $(0, 1)$
3. $y' = y + e^t$; $(0, 1)$
4. $y' = 2y/t + t^4$; $(1, 1)$
5. $t^2y' = 1 - 2ty$; $(1, 1)$, $(1, 0)$

For Exercises 6–13, find an expression for the general solution. Note any special solutions.

6. $y' = 2ty + ty^2$
7. $ty' = 2y + 4t^3\sqrt{y}$
8. $xy' + 2y = e^x$
9. $y' + y = xy^2$
10. $y' - y\tan x = x^2$
11. $y' + y = y^3e^x$
12. $xy' + y = y^2x^2$
13. $yy' + y^2 = t$
14. Find a solution for $y' = 1/(y-x)$. [*Hint:* This is linear if $x = \phi(y)$.]
15. Solve $y' = y^2 - 2t^{-2}$.
16. A government agency has a current staff of 6000, of whom 25% are women. Employees are quitting randomly at the rate of 100 per week. Replacements are being hired at the rate of 50 per week, with the requirement that half be women. (a) What is the size of the agency staff in 40 weeks, and what percentage is then female? (b) What would have been the percentage if it had been required that *all* new employees were women?
17. A curve in the plane has the property that at each point P on the curve in the first quadrant, the trapezoid formed by the tangent at P, the Y-axis, the X-axis, and the vertical line through P has area 12. If the curve goes through the point $(8, 1)$, where does it cross the line $x = 1$?
18. A tank with a capacity of 100 gallons is half full of fresh water. A pipe is opened which admits treated sewage into the tank at 4 gal/min; at the same time, a drain is opened to allow 3 gal/min of the mixture to leave the tank. If the treated sewage contains 10 g/gal of usable potassium, what is the concentration of potassium in the tank when it is full?
19. A 200-gallon tank is full of fresh water initially. A drain is opened that removes 3 gal/sec from the tank, and, at the same moment, a valve is

opened that admits a 1% solution of chlorine at 2 gal/sec. (a) When is the tank half full, and what is the concentration of chlorine then? (b) If the drain is closed when the tank is half full and the tank is allowed to fill, what will be the final concentration of chlorine in the tank?

2.3 Reduction of Certain Second-Order Equations

It is possible to use the techniques already explained to solve some second-order differential equations if they happen to be of the right form. For example, consider the equation

$$\begin{aligned} \frac{d^2y}{dt^2} &= -2\frac{dy}{dt} + 4t \\ &= -2y' + 4t \end{aligned} \tag{2.24}$$

and suppose we wish a solution obeying the initial conditions $y = 1$, $y' = 2$, at $t = 0$. This equation is a special case of the general second-order equation $y'' = f(t, y, y')$ in that there is no appearance of y itself in the equation. Let us put $u = y'$ in (2.24) and then rewrite it in terms of u and t, thus

$$\frac{du}{dt} = -2u + 4t$$

This is now a first-order equation which happens to be linear. The method of the previous section applies, so that we have

$$\begin{aligned} e^{2t}(u' + 2u) &= 4te^{2t} \\ \frac{d}{dt}\{e^{2t}u\} &= 4te^{2t} \\ e^{2t}u &= \int 4te^{2t}\,dt + C_1 \\ &= (2t-1)e^{2t} + C_1 \end{aligned}$$

and we arrive at

$$u = 2t - 1 + C_1 e^{-2t}$$

We next replace u by y' and find that we have another first-order equation to solve:

$$y' = 2t - 1 + C_1 e^{-2t} \tag{2.25}$$

The solution of this is immediate:

$$y = t^2 - t - \tfrac{1}{2}C_1 e^{-2t} + C_2 \tag{2.26}$$

To satisfy the initial conditions, put $t = 0$ in (2.25) and (2.26), and require that we obtain $y = 1$ and $y' = 2$. This leads to the algebraic equations

$$2 = 0 - 1 + C_1$$

$$1 = 0 - 0 - \tfrac{1}{2}C_1 + C_2$$

from which we find $C_1 = 3$, $C_2 = \tfrac{5}{2}$. The desired solution of the original equation (2.25) is then

$$y = t^2 - t - \tfrac{3}{2}e^{-2t} + \tfrac{5}{2}$$

Another example which yields to the same technique is the second-order equation

$$y'' = e^{-t}(y')^2 \tag{2.27}$$

If one sets $u = y'$, then this becomes the separable first-order equation $u' = e^{-t}u^2$, whose solution is

$$u = \frac{e^t}{1 + Ae^t}$$

Replacing u by dy/dt, one arrives at the solution of (2.27) as

$$y = \frac{\log(1 + Ae^t)}{A} + B$$

There is another extremely important class of second-order equations for which there is a general procedure that converts them into first-order equations. Since this class happens to include many equations that frequently occur in physical problems, we devote the rest of the section to the study of this technique.

The general second-order equation has the form

$$y'' = f(t, y, y')$$

The characteristic property of the special class to be studied is that the independent variable is missing. Thus, in the present notation, such an equation would have the form

$$y'' = g(y, y')$$

Such equations are often called **autonomous**; they are independent of time, and describe motions which do not depend upon any absolute notion of the time elapsed from some fixed universal starting point. For such an equation, we can always take the initial conditions to be at $t = 0$. Physical problems in which there is no time-dependent external force lead to autonomous differential equations.

The following pair of equations belong to this class:

$$\frac{d^2y}{dt^2} = y + \frac{dy}{dt}$$

$$y'' = 3yy'$$

The following pair do not:

$$\frac{d^2y}{dt^2} = ye^t + \frac{dy}{dt}$$

$$y'' = (t+y)\,y'$$

When the variables involved in the differential equation are not y and t, as in the illustrations above, it may be harder to recognize the autonomous form; many of the equations that have been mentioned in earlier sections belonged to this class. For example, the following equations were cited in (1.4), (1.12), and (1.35) as possible models for a swinging pendulum:

$$\frac{d^2\theta}{dt^2} = -\frac{g}{L}\sin\theta$$

$$\frac{d^2\theta}{dt^2} = -\frac{g}{L}\theta$$

$$\frac{d^2\theta}{dt^2} = -\frac{g}{L}\sin\theta - k\left\{\frac{d\theta}{dt}\right\}^2$$

Each of these is autonomous since t is missing. This is also true for the equation (1.16) used for the rocket problem,

$$\frac{d^2x}{dt^2} = -\frac{gR^2}{x^2}$$

Finally, the differential equation

$$\frac{d^2T}{dx^2} = 2T^2 - 3\frac{dT}{dx}$$

is autonomous since here the independent variable is x, which is missing from the right-hand side.

The special trick by which such equations can be changed into first-order equations was probably discovered by John Bernoulli in about 1720. Let us recapture his method in modern notation. Suppose we are interested in solving the equation

$$y'' = g(y, y') \tag{2.28}$$

A solution of this would be a function ϕ such that the substitution $y = \phi(t)$ satisfies (2.28). Suppose we could solve the equation $y = \phi(t)$ for t, obtaining a formula $t = \psi(y)$. Next, set $u = y' = \phi'(t)$. Then, we also have $u = \phi'(\psi(y))$; in words, we propose to regard y as the independent variable, and think of t and u (which is y') as being functions of y. With this speculation in mind, Bernoulli might have written the following application of the chain rule for differentiation:

$$y'' = \frac{d}{dt}\{y'\} = \frac{d}{dt}\{u\} = \frac{du}{dt} = \frac{du}{dy}\frac{dy}{dt} = \frac{du}{dy}u$$

Using this result to replace y'', and u to replace y', (2.28) becomes

$$u\frac{du}{dy} = g(y, u)$$

which is a first-order equation in u and y.

If we are lucky, this equation is one that we can solve in closed form, obtaining a solution of the form

$$u = F(y)$$

But, this equation is equivalent to the separable first-order equation

$$\frac{dy}{dt} = F(y)$$

To see how this general method works in a specific case, consider the equation

$$y'' = \frac{2(y')^2}{y}$$

Following the general pattern, we put $u = y'$, and then write

$$y'' = \frac{du}{dt} = \frac{du}{dy}\frac{dy}{dt} = u\frac{du}{dy}$$

The original equation then becomes

$$u\frac{du}{dy} = \frac{2u^2}{y}$$

which is separable. If $u \neq 0$, we can rewrite this equation as

$$\frac{du}{u} = \frac{2\,dy}{y}$$

and integrate as usual, getting

$$\log u = 2 \log y + \text{const}$$

or

$$u = Ay^2$$

Replacing u by dy/dt, we have another first-order equation

$$\frac{dy}{dt} = Ay^2$$

whose solution is seen to be

$$y = \frac{-1}{At+B}$$

We next apply the general method to a less artificial problem, namely, the rocket problem discussed in Chapter 1. Recall that according to equation (1.16), the motion of the rocket after burnout is modeled by

$$\frac{d^2x}{dt^2} = -gR^2\left(\frac{1}{x^2}\right) \tag{2.29}$$

and that the initial conditions are $x = R_0$, $dx/dt = V_0$ when $t = 0$ (the rocket is at distance R_0 from the center of the earth and has velocity V_0 at the moment of burnout). As in the general method, we set $u = dx/dt$ and assume that u can be taken to be a function of x. Then by the chain rule,

$$\frac{d^2x}{dt^2} = \frac{du}{dt} = \frac{du}{dx}\frac{dx}{dt} = u\frac{du}{dx}$$

Equation (2.29) becomes

$$u\frac{du}{dx} = -gR^2\left(\frac{1}{x^2}\right)$$

This is a separable equation, so we write

$$u\,du = -gR^2\frac{dx}{x^2}$$

Integrating, we have

$$\frac{u^2}{2} = gR^2\left(\frac{1}{x}\right) + C \tag{2.30}$$

We can evaluate the number C by using the initial conditions. When $t = 0$, we have $x = R_0$ and $u = dx/dt = V_0$. Hence, making these substitutions,

$$\frac{V_0{}^2}{2} = \frac{gR^2}{R_0} + C$$

Solving this for C, we use in (2.30) the value thus obtained and derive the formula

$$\left(\frac{dx}{dt}\right)^2 = u^2 = \frac{2gR^2}{x} + V_0{}^2 - \frac{2gR^2}{R_0} \tag{2.31}$$

Already we have enough information to answer one of the questions asked about the flight of an unpowered rocket. How far up will it go? As with a thrown rock, the vertical speed will be zero at the moment (if any) that it reaches its maximum height and starts to return toward the earth. If we set $dx/dt = 0$ in (2.31) and solve for x, we obtain

$$x = \frac{2gR^2}{(2gR^2/R_0) - V_0{}^2}$$

If the number in the denominator is positive, then this formula gives us the maximum height from the center of the earth achieved by the rocket.

What interpretation should we give to the formula when the denominator is zero? (Or negative!) The answer is clear if we return to equation (2.31). Suppose that $(V_0)^2$ is larger than $(2gR^2)/R_0$. Then, the right-hand side of (2.31) can never be zero, and in fact is bounded below by some positive quantity. This means that $(dx/dt)^2$ can never be zero, and in fact always exceeds some positive quantity. Hence, the rocket will recede indefinitely far while the velocity constantly decreases toward this limiting value, which from (2.31) we see to be

$$\text{Velocity at infinity} = \sqrt{(V_0)^2 - 2gR^2/R_0}$$

The case of equality is also treated easily. If $(V_0)^2 = 2gR^2/R_0$, then (2.31) becomes

$$\frac{dx}{dt} = \frac{\sqrt{2gR^2}}{\sqrt{x}}$$

which is a separable equation with a solution

$$\tfrac{2}{3}x^{3/2} = \sqrt{2g}\,Rt + \tfrac{2}{3}R_0^{3/2}$$

or

$$x = \left(\tfrac{3}{2}\sqrt{2g}\,Rt + R_0^{3/2}\right)^{2/3} \tag{2.32}$$

which shows that x becomes arbitrarily large as t increases, growing roughly like $t^{2/3}$.

This solution shows that the critical value of the initial velocity V_0 is $\sqrt{2gR^2/R_0} = V_E$, usually called the escape velocity. If the velocity at burnout is V_E or larger, then the rocket will never fall back but will continue to move away from the earth forever. (This of course ignores the effects of the sun and other planets.) Since the height R_0 at burnout is usually about the same as R, the radius of the earth, we can obtain a simpler formula for computing V_E by setting $R_0 = R$ and obtaining $V_E = \sqrt{2gR}$. If we set $R = 4000$ miles and $g = 32$ ft/sec^2 (and change everything to miles), we find that V_E is about 7 mi/sec.

As our second illustration, consider the first model for a pendulum

$$\frac{d^2\theta}{dt^2} = -\frac{g}{L}\sin\theta \tag{2.33}$$

where we have the initial conditions $t = 0$, $\theta = \theta_0$, and $d\theta/dt = 0$ (we release the pendulum from rest at an angle θ_0). Using Bernoulli's technique, we set $u = d\theta/dt$, regard u as a function of θ, and use the chain rule

$$\frac{d^2\theta}{dt^2} = \frac{du}{dt} = \frac{du}{d\theta}\frac{d\theta}{dt} = u\frac{du}{d\theta}$$

to replace (2.33) by the equation in u and θ

$$u\frac{du}{d\theta} = -\frac{g}{L}\sin\theta$$

This is separable and has the solution

$$\tfrac{1}{2}u^2 = \frac{g}{L}\cos\theta + C$$

Checking the initial conditions, we have $d\theta/dt = 0$ when $\theta = \theta_0$, which corresponds to $u = 0$ when $\theta = \theta_0$. This gives

$$C = -\frac{g}{L}\cos\theta_0$$

and we have a partial solution

$$\left(\frac{d\theta}{dt}\right)^2 = u^2 = \frac{2g}{L}(\cos\theta - \cos\theta_0)$$

From this equation, we obtain a formula for the angular speed of the swinging pendulum at any position:

$$\left|\frac{d\theta}{dt}\right| = \sqrt{\frac{2g}{L}}\sqrt{\cos\theta - \cos\theta_0} \tag{2.34}$$

In particular, at the bottom of its swing $\theta = 0$ and the speed of the bob of the pendulum is

$$V_0 = L\left|\frac{d\theta}{dt}\right| = \sqrt{2gL}\sqrt{1-\cos\theta_0}$$
$$= 2\sqrt{gL}\sin(\tfrac{1}{2}\theta_0)$$

as given earlier in formula (1.10).

Can we carry on beyond this point to determine the actual motion of the pendulum by solving for the function $\theta = \phi(t)$? Let us take a part of the motion where $d\theta/dt$ has constant sign, for example during the first downward swing when θ is decreasing from θ_0 to zero. As in Section 1.1, we suppose that t_0 is the time at which the pendulum first reaches the vertical position. During the interval $0 < t < t_0$, θ is decreasing and $d\theta/dt$ is negative. Hence, $|d\theta/dt| = -d\theta/dt$, and from (2.34) we find

$$\frac{d\theta}{dt} = -\sqrt{\frac{2g}{L}}\sqrt{\cos\theta - \cos\theta_0}$$

This is a separable equation, and we obtain the integrated solution

$$\int \frac{d\theta}{\sqrt{\cos\theta - \cos\theta_0}} = -\sqrt{\frac{2g}{L}}\int dt \tag{2.35}$$

Unfortunately, it is not possible to express the indefinite integral on the left in terms of the elementary functions of analysis. (It *can* be expressed in terms of what are called elliptic functions.) However, we can answer one of the questions we have asked about the motion by finding an exact expression for the time t_0. We use the device of inserting corresponding appropriate limits in the indefinite integrals in (2.35). We know that $\theta = \theta_0$ when $t = 0$; we also know that $t = t_0$ when $\theta = 0$, since at that moment the pendulum is at the bottom of its swing. Take $\theta = \theta_0$, $t = 0$ for the bottom limits of integration and $\theta = 0$, $t = t_0$ for the upper limits. Then, (2.35) becomes

$$\int_{\theta_0}^{0} \frac{d\theta}{\sqrt{\cos\theta - \cos\theta_0}} = -\sqrt{\frac{2g}{L}} \int_0^{t_0} dt$$

Evaluating the right-hand side, we end with the relation

$$t_0 = \sqrt{\frac{L}{2g}} \int_0^{\theta_0} \frac{d\theta}{\sqrt{\cos\theta - \cos\theta_0}}$$

as stated in formula (1.9), Section 1.1.

Exercises

Find the complete solutions of the differential equations for each of Exercises 1–12.

1. $y'' = -2y' + e^t$
2. $y'' = (2/t)\, y' + t^4$
3. $xy'' = y'$
4. $t^2 x'' = (x')^2$
5. $y'' + 2y' = x$
6. $x'' + 3x' = t^2$
7. $y''(e^x + 1) + y' = 0$
8. $(1 + t^2)\, y'' + 1 + (y')^2 = 0$
9. $y'' = 2yy'$
10. $x'' = 2x'$
11. $x'' = (x')^2/2x$
12. $y'' = \sqrt{1 + (y')^2}$
13. If $y'' = -4/y^3$, find a particular solution if $y(2) = 4$ and $y'(2) = 0$. What is $y(4)$?
14. Use Bernoulli's method to solve the equation used for the second model for the pendulum—see formula (1.12), Section 1.1.
15. Combine Bernoulli's method with Theorem 2.2 to show that the solution of the equation $y'' = Ay + B(y')^2$, with A and B constants, can be reduced to a problem in integration.
16. Combine Bernoulli's method with Theorem 2.2 to show that it is possible to solve equation (1.35) in Section 1.5, which was used to model the motion of a pendulum with a special nonlinear resistance law.
17. Show that if a rocket is fired at escape velocity, it will take about 54 hours to reach a distance of 64 earth radii, approximately the distance to the moon.

18. Show that if gravitational attraction varied inversely as the cube of the distance, the formula for escape velocity would be $V_E = \sqrt{gR}$.

19. Show that there is no escape velocity for a force of attraction that is proportional to the reciprocal of the distance, by showing that the height reached with initial velocity V is $R \exp(V^2/2gR)$ and is thus always finite no matter how large V may be.

20. A special braking system for a racing car supplies a frictional force proportional to the square root of the velocity of the car. In a test, the moving car was brought to a stop in 3 seconds after traveling 100 feet. How fast was it going when the brakes were applied?

21. An object moving at 300 ft/sec falls vertically into a deep lake and slows to a stop in 10 seconds. Assuming that its density is 1 so that the effect of gravity can be ignored, and that the resistance of the water is proportional to the square root of the velocity, find the depth reached.

22. Assume that the air resistance of a large falling object is proportional to its velocity. Show that such a body, falling from a great height, approaches a maximum terminal velocity. (Treat gravitational attraction as constant.)

23. A bullet of mass M is fired straight upward, with a muzzle velocity 800 ft/sec. If the air resistance is $0.001\ v^2 M$, what height does the bullet reach? (Assume constant gravitational acceleration $g = 32$ ft/sec^2.)

24. A repulsive force on a moving object is described as proportional to the velocity but inversely proportional to the square of the distance from the object to a point P. If the object starts from rest at a distance R from P and under this force moves away from P in a straight line, will its velocity become arbitrarily large, or will it approach a finite maximum?

2.4 Handbooks

The special differential equation $dy/dt = g(t)$ requires no special theory, since we have the solution $y = \int g(t)\, dt$. Thus, finding the indefinite integral of a given function is a special case of solving a differential equation. Generalizing this result, we often speak of integrating a differential equation, rather than solving it.

It is clear that a table of indefinite integrals is a handy thing to have, since it makes it unnecessary to remember all the specialized substitutions that have been discovered to be useful in carrying out a complicated integration.

One of the standard references is the *Handbook of Chemistry and Physics*, published by the Chemical Rubber Company, which also puts out a separate volume containing only the mathematical tables. The book has a list of almost 400 definite and indefinite integrals, covering many of the complicated integrals that turn up frequently in applied work. Another handbook, prepared by the U.S. National Bureau of Standards* contains a great deal of information about infinite series and special functions that are important in science.

In the past 300 years, people have discovered a great many special tricks for solving certain differential equations; in this chapter, we have described only a very few of these, such as the use of the substitution $y = ut$ in solving a homogeneous equation, or the Bernoulli trick for certain second-order equations. Many of these special methods have been collected in a handbook which is available in most mathematical libraries: *Differentialgleichungen, Lösungsmethoden und Lösungen*, by E. Kamke.†

One needs to know only a very small amount of German to use this handbook; however, as with a table of integrals, it does help to know something about mathematical theory. With an adequate theoretical background, you will be able to find in this book the explicit solution to a great many of the classes of differential equations for which specialized substitution techniques are known. For this reason, we have not included many of these techniques in the present book, but urge readers to become acquainted with Kamke. You will not use it often, but it is always there if you should need it.

There are other reasons for spending less time on learning specialized techniques for solving differential equations. In the 30 years since Kamke prepared his handbook, the high-speed computer has come upon the scene. An expert in the use of a modern computer finds much of Kamke unnecessary, except for certain directions of theoretical interest. If one is mainly interested in obtaining numerical information about the solution of a certain differential equation, it is often much easier to use well-chosen numerical techniques for finding approximate solutions than to pore through Kamke for the particular equation and then employ the specialized techniques described there. Indeed, at times these techniques merely convert the given equation into another which is not much better or easier to solve. Finally, it is very often the case that the equation that interests you cannot be found

* U.S. Department of Commerce, National Bureau of Standards. *Handbook of Mathematical Functions*, Applied Mathematics Series 55.

† This book was made available in the United States by Edwards Brothers Publishers, Ann Arbor, Michigan, in 1946, and reissued in an expanded version in Leipzig in 1959, and issued by Chelsea Publishers (New York) in 1971.

in the handbook at all. (Just as the integral that you are really interested in doesn't turn up in the list of sample integrals.)

For these reasons, we devote the next two chapters to a discussion of approximatation methods for solving differential equations.

2.5 Review Problems

Each of the exercises that follow can be done by one of more of the mathematical techniques discussed in this chapter. The order in which the exercises are listed is not the same as the order in which the techniques appear in the chapter. Thus, an essential part of each exercise is the problem of deciding which technique is appropriate; several may have to be tried before the right one is discovered. This, of course, is the way things are outside of the artificial world of the classroom. In solving problems that actually arise in science, sometimes one does not even know which book to look in—or if that book has been written yet.

Exercises

Find solutions for the following differential equations;

1. $xy' + y = xy^2 \log x$
2. $y' = 2ty^2$
3. $y' = (3y^2 - x^2)/2xy$
4. $2yy' + y^2 = 5$
5. $y'' = y' + 2$
6. $y' + e^x y = e^x$
7. $y' + 2y/x = 2x^3y^3$
8. $y' = (x^2 - 6xy)/(3x^2 + y^2)$
9. $y' = (5/t)\,y + 3t^2$
10. $y'' = (y')^2/y$
11. $y' = y - (2t/y)$
12. $y' + y = t^2/\sqrt{y}$
13. $(y')^2 + (x+y)\,y' + xy = 0$
14. $(y')^2 - (x+1)\,yy' + xy^2 = 0$
15. Find a solution $y = \phi(t)$ for $y' = (y/t)\log y$ which goes through $(1, e^2)$; through $(-1, e^2)$. Sketch the graphs. Why is there no solution through $(1, -2)$?
16. Water is heated to the boiling point, 100 °C. The water is then removed from the heat and placed in a room which is at a constant temperature of 60 °C. After 3 minutes, the water temperature is 90 °C.
    (a) Find the water temperature after 6 minutes.
    (b) When will the water temperature be 75 °C?

17. A tank contains 50 gallons of fresh water. Brine at a concentration of 2 lb/gal starts to run into the tank at 3 gal/min; at the same time, the mixture of fresh water and brine runs out at 2 gal/min.
    (a) How much liquid is there in the tank after 50 minutes?
    (b) How many pounds of dissolved salt is in the tank then?

18. A rocket has mass M_0 when empty, and can hold an equal mass of fuel. During flight, it burns fuel at the rate of $(M_0/50)$ lb/sec, achieving an exhaust velocity of 3000 ft/sec. Assume that gravity is independent of height. (a) How high will the rocket be at the moment all the fuel is gone? (b) What will the speed be then? (c) What is the maximum height the rocket will reach?

19. A large stone is dropped from a very high bridge at night. The splash is heard 8 seconds later. How high above the surface of the water is the bridge? (Assume that sound travels at 1100 ft/sec.)

20. A block is sliding along a horizontal track. The friction force is proportional to the square root of the velocity. The block is going at 100 ft/sec as it starts down the track, and it comes to rest after going 500 feet.
    (a) How long did the block take to stop?
    (b) At the midpoint (250 feet) was it going faster than 60 ft/sec?

21. A large, square water purification tank has sides of length 100 feet and is 3 inches deep. Initially it is empty. Water is run in at 2 ft^3/sec. However, there is a hole in the bottom of radius $6/\sqrt{\pi}$ inches. How long does it take for the water to overflow? [*Hint:* Assume the velocity of outflow is given by $\sqrt{2gh}$, where h is the depth of the water and g, the gravitational constant, is 32 ft/sec^2.]

22. A small sphere made of a substance half as dense as water is dropped into a deep lake from 100 feet above the surface. How deep does the sphere go? (Ignore resistance in both air and water.)

23. A bullet was fired into a bale of cotton which slowed it to a stop by a friction force proportional to the square root of its velocity. It stopped in 0.1 second and penetrated 10 feet. How fast was it going when it hit?

24. In a large tank are 100 gallons of brine containing 75 pounds of dissolved salt. Water runs into the tank at the rate of 3 gal/min and the mixture runs out at the rate of 2 gal/min. The concentration is kept uniform by stirring. How much salt is there in the solution after 1.5 hours?

25. A swimming pool holds 10,000 gallons of water. It can be filled at the rate of 100 gal/min and emptied at the same rate. At the present moment the pool is filled, but there are 20 pounds of an impurity dissolved in the water; for safety, this must be reduced to less than 1 pound. It would take 200 minutes to empty the pool completely and refill it, but during part of this time the pool could not be used. How long will it take to restore the pool to safe condition if at all times the pool must be at least half full?

*26. A hollow sphere of weight 10 pounds is of sufficient size to displace 25 pounds of water. In moving through the water, its resistance is $\frac{1}{4}\sqrt{v}$, where v is the speed in feet per second. It is released at an unknown depth, and breaks the surface 5 seconds later. What was its velocity when it emerged? Could you find its original depth?

27. A large underground cylindrical storage tank suddenly develops a leak in the side at an unknown depth. If the liquid being held in the tank was originally at ground level and fell 10 feet the first day and 8 feet the second day, estimate the position of the leak. [*Hint:* Assume the amount leaking out is proportional to the square root of the height of liquid above the hole.]

Approximation Methods

Introduction

If "solving a differential equation" means finding a usable formula for the unknown function, then, as we have seen in the last chapter, only a few special classes of equations can be solved. However, the ultimate purpose that led to the differential equation may not demand this much. It may be sufficient to have a reasonably accurate table of values of the solution function, or a fairly good graph, or an approximate formula which serves the purpose, such as the first dozen terms of a series expansion of the solution.

The advent of the high-speed computer has made numerical methods of great importance and usefulness, and we have devoted all of Chapter 4 to this topic. In the present chapter, we discuss some older approximation methods which are still very important.

The first three sections deal with some necessary preliminaries on sequences and series. The presentation of series is probably review for most readers since the topics are very standard, but the material on sequences and difference equations may be new to those coming directly from a calculus course.

Section 3.4 develops the method of repeated differentiation, which leads to the calculation of polynomial approximations to the solution of a differential equation, using Taylor's Theorem. Section 3.5 introduces the related, useful method of power-series solution. Section 3.6 deals with the method of Picard iteration, which produces a sequence of functions that converges to a solution.

All of these are very important for theoretical work but have limited practical application.

The chapter ends with a proof of one version of the basic existence and uniqueness theorems for initial-value problems, based on Picard iteration. Since this is more advanced in content and viewpoint than the rest of the chapter, and is not used in later sections, it may be regarded as optional.

3.1 Recursions and Difference Equations

Many of the methods used to obtain approximate solutions of differential equations involve the determination of a sequence of numbers associated with the solution. Recall the formal definition of a sequence.

DEFINITION **3.1** A *sequence* of numbers is a function on the set $\{0, 1, 2, \ldots\}$ to the set $\mathbf{R}$ of real numbers.

If f is the function and we denote the values of f by $f(n) = a_n$, then it is customary to write $\{a_n\}$ for the sequence, and to display the terms of the sequence in order as $\langle a_0, a_1, a_2, \ldots \rangle$. Thus, $\{1/2^n\}$ is the sequence whose terms are $\langle 1, \frac{1}{2}, \frac{1}{4}, \frac{1}{8}, \ldots \rangle$, given by the formula $f(n) = 1/2^n$.

It is also possible to describe a sequence of numbers by a recursive process which does not present a formula for the nth term of the sequence, but rather enables one to calculate any desired term. For example, the pair of statements

$$\begin{cases} x_0 = 1 \\ x_{n+1} = \frac{1}{2}x_n \end{cases}$$

defines the same sequence as the one above, namely, $\{1/2^n\}$, for we clearly have $x_1 = \frac{1}{2}x_0 = \frac{1}{2}$, $x_2 = \frac{1}{2}x_1 = \frac{1}{4}$, etc. The formal verification that these two sequences are the same is a trivial exercise in mathematical induction. If we have already shown that $x_n = 1/2^n$ for $n = 0, 1, \ldots, N$, then we show that it is also true for $n = N+1$ by the calculation

$$x_{N+1} = \frac{1}{2}x_N = \frac{1}{2}\left(\frac{1}{2^N}\right) = \frac{1}{2^{N+1}}$$

We say that a sequence $\{x_n\}$ is defined **recursively** if its terms are specified by a pair of equations such as

$$\begin{cases} x_0 = a \\ x_{n+1} = F(x_n) \end{cases}$$

where F is a specific function. It is intuitively clear that such a formula enables one to compute the values of x_n one after the other, starting from the value of x_0. For example, the pair

$$\begin{cases} x_0 = 1 \\ x_{n+1} = 1 + 2x_n \end{cases}$$

generates the sequence $\langle 1, 3, 7, 15, 31, \dots \rangle$. In this case, it is easy to verify by induction that the terms of the sequence are also given by the simple formula $x_n = 2^{n+1} - 1$. However, we can just as readily give a recursive definition for a sequence for which it is very hard to find a simple formula for the nth term:

$$\begin{cases} x_0 = 1 \\ x_{n+1} = x_n + (x_n)^2 \end{cases} \tag{3.1}$$

yields the sequence $\langle 1, 2, 6, 42, 1806, 1638042, \dots \rangle$.

More generally, the formula for calculating x_{n+1} might involve the value of the index itself, so that one can often have a recursion of the more general form

$$\begin{cases} x_0 = a \\ x_{n+1} = F(n, x_n) \end{cases} \tag{3.2}$$

An example of this would be

$$\begin{cases} x_0 = 3 \\ x_{n+1} = x_n + n + 2 \end{cases} \tag{3.3}$$

We have $x_1 = x_0 + 0 + 2 = 3 + 2 = 5$, $x_2 = 5 + 1 + 2 = 8$, $x_3 = 8 + 2 + 2 = 12$, and so on. The first few terms of the sequence $\{x_n\}$ defined by (3.3) are therefore

$$\langle 3, 5, 8, 12, 17, 23, 30, 38, 47, 57, 68, \dots \rangle$$

While it is clearly possible to calculate as many terms of such a sequence as one wishes, it is not clear how one might go about obtaining a general formula for the nth term, $x_n = f(n)$. In many problems, it is not at all necessary to find such a formula. As we will see in the next chapter, the numerical solution of differential equations leads to recursions of the form (3.2), and all that is desired is to find the numerical value of the terms of the sequence for some range of values of n, say $n = 1$ to $n = N$.

However, there are also times when it is desirable to have a formula that enables one to calculate a term x_N for some choice of N without having to calculate all the intervening values x_n; if we have a general formula $x_n = f(n)$, then this can be done. At other times, one may not be interested in the precise numerical value of the terms $\{x_n\}$ but rather in their behavior for large values of n. For example, one may ask if a particular sequence $\{x_n\}$ is **bounded**, that is, if there is a number B such that $|x_n| < B$ for all n. Or if a sequence $\{x_n\}$ is not bounded, one may be interested in how fast it grows. For example, it seems clear from the first few terms that the sequence defined by (3.1) grows much faster than the sequence defined by (3.3).

There is a strong analogy between first-order differential equations and the theory of recursively defined sequences defined as in (3.2). The analogy is more visible if we write $x(n)$ in place of x_n, for then (3.2) becomes

$$\begin{cases} x(0) = a \\ x(n+1) = F(n, x(n)) \end{cases}$$

which looks much like the standard initial-value problem

$$\begin{cases} y(0) = a \\ y'(t) = f(t, y(t)) \end{cases}$$

Because of such analogies, the term "difference equation" is often used instead of "recursively defined sequences," and there are books that develop the theory in a manner quite similar to that for differential equations. We need only the most elementary aspects of this theory for our applications.

The simplest difference equation that arises is the following:

$$\begin{cases} x_0 = a \\ x_{n+1} = Ax_n + B \end{cases} \tag{3.4}$$

where A and B are constants.

THEOREM 3.1 *The solution of the difference equation* (3.4) *is given by*

$$x_n = A^n x_0 + (1 + A + A^2 + \cdots + A^{n-1})B$$

When $A \neq 1$, this can also be written as

$$x_n = A^n x_0 + B\left(\frac{A^n - 1}{A - 1}\right)$$

Proof From (3.4) we have $x_1 = Ax_0 + B$, $x_2 = Ax_1 + B$, and $x_3 = Ax_2 + B$. Substituting for x_1 and then for x_2, we obtain

$$x_2 = A(Ax_0 + B) + B = A^2 x_0 + AB + B$$

$$x_3 = A(A^2 x_0 + AB + B) + B = A^3 x_0 + A^2 B + AB + B$$

From these cases, it is easy to conjecture the form of the general case:

$$x_n = A^n x_0 + (A^{n-1} + A^{n-2} + \cdots + A^2 + A + 1)B \tag{3.5}$$

To test this by induction, we see what this would predict for the form of x_{n+1}, using (3.4). We obtain

$$\begin{aligned} x_{n+1} &= Ax_n + B \\ &= A[A^n x_0 + (A^{n-1} + A^{n-2} + \cdots + A + 1)B] + B \\ &= A^{n+1}x_0 + (A^n + A^{n-1} + \cdots + A^2 + A)B + B \\ &= A^{n+1}x_0 + (A^n + A^{n-1} + \cdots + A^2 + A + 1)B \end{aligned}$$

Since this agrees with the expression that would have been obtained by using (3.5) with n replaced by $n+1$, it confirms that (3.5) is the correct formula for the general term of the sequences $\{x_n\}$ and we have proved Theorem 3.1.

As an application, we see that the sequence defined by $x_{n+1} = 3x_n + 5$, with $x_0 = 2$, has the formula $x_n = 2(3^n) + \frac{5}{2}(3^n - 1) = \frac{9}{2}(3^n) - \frac{5}{2}$.

The analog of differentiation, for sequences, is known as **differencing**. By this, we mean the calculation of the sequence of differences between successive terms of a given sequence. For example, let us consider the sequence determined by the recursion (3.3), namely,

$$\langle 3, 5, 8, 12, 17, 23, 30, 38, 47, \ldots \rangle$$

From this, we can generate a new sequence by computing the differences; it is customary to write the differences below the pair being differenced.

```
3   5   8   12   17   23   30   38   ...
  2   3   4    5    6    7    8    ...
```

Thus, the difference operation applied to a sequence yields a new sequence, just as the derivative operation applied to a function yields a new function. The notation for the difference operation is Δ, and the formula for it is simply

$$\Delta x_n = x_{n+1} - x_n$$

The difference operator can be applied repeatedly, so that one speaks, for example, of the second difference

$$\begin{aligned} \Delta^2 x_n &= \Delta(\Delta x_n) = \Delta(x_{n+1} - x_n) \\ &= (x_{n+2} - x_{n+1}) - (x_{n+1} - x_n) \\ &= x_{n+2} - 2x_{n+1} + x_n \end{aligned}$$

The first three differences of the sequence arising from (3.3) are as follows:

$$\begin{array}{lcccccccccccccccc}
x_n: & 3 & & 5 & & 8 & & 12 & & 17 & & 23 & & 30 & & 38 & \cdots \\
\Delta x_n: & & 2 & & 3 & & 4 & & 5 & & 6 & & 7 & & 8 & & \cdots \\
\Delta^2 x_n: & & & 1 & & 1 & & 1 & & 1 & & 1 & & 1 & & \cdots \\
\Delta^3 x_n: & & & & 0 & & 0 & & 0 & & 0 & & 0 & & \cdots
\end{array}$$

Obviously, there is something special about this sequence that makes its third difference identically zero. The explanation lies in the effect the difference operator has on polynomial sequences, an effect that is the analog of the fact that the derivative of a polynomial of degree p is a polynomial of degree $p-1$.

THEOREM 3.2 *If $\{x_n\}$ is described by a polynomial function $P(n)$ of degree p, then $\{\Delta x_n\}$ is described by a polynomial of degree $p-1$.*

Proof If $x_n = n^2$, then $\Delta x_n = (n+1)^2 - n^2 = 2n+1$. If $x_n = n^3$, then $\Delta x_n = 3n^2+3n+1$. Generally, if $x_n = n^p$ then $\Delta x_n = pn^{p-1}+\cdots+pn+1$, a polynomial sequence of degree $p-1$. Now, suppose $x_n = P(n)$, where $P(n) = a_0+a_1n+a_2n^2+\cdots+a_pn^p$. Then,

$$\begin{aligned}
\Delta x_n &= a_0 - a_0 + a_1\,\Delta n + a_2\,\Delta n^2 + \cdots + a_p\,\Delta n^p \\
&= 0 + a_1 + a_2(2n+1) + a_3(3n^2+3n+1) + \cdots + a_p(pn^{p-1}+\cdots+pn+1)
\end{aligned}$$

If we were to collect like powers of n, it is clear that $\{\Delta x_n\}$ would be expressed by a polynomial in n of degree $p-1$.

The following result follows at once from this.

COROLLARY *If $\{x_n\}$ is a polynomial sequence of degree p, then $\Delta^{p+1}x_n$ is zero for every n.*

Returning to the sequence arising from (3.3), one would conjecture that since $\Delta^3 x_n = 0$, there must be a quadratic polynomial P with $x_n = P(n)$. Setting $P(t) = a_0+a_1t+a_2t^2$, we want to choose the a_i so that

$$\begin{aligned}
3 &= P(0) = a_0 \\
5 &= P(1) = a_0 + a_1 + a_2 \\
8 &= P(2) = a_0 + 2a_1 + 4a_2
\end{aligned}$$

Solving these, we find $P(n) = 3+\frac{3}{2}n+\frac{1}{2}n^2$. Checking, we have $P(3) = 3+\frac{9}{2}+\frac{9}{2} = 12$, $P(4) = 3+6+8 = 17$, and $P(5) = 3+\frac{15}{2}+\frac{25}{2} = 23$, each of which agrees with the corresponding x_n. Does this hold for all values of n? The only way to prove this conclusively is to return to the difference equation that defined the original sequence, and verify by induction that $x_n = P(n)$ satisfies it. We have to show that

$$P(n+1) \stackrel{?}{=} P(n) + n + 2$$

Calculating each side separately, we have

$$3 + \tfrac{3}{2}(n+1) + \tfrac{1}{2}(n+1)^2 \stackrel{?}{=} 3 + \tfrac{3}{2}n + \tfrac{1}{2}n^2 + n + 2$$

$$3 + \tfrac{3}{2}n + \tfrac{3}{2} + \tfrac{1}{2}(n^2+2n+1) \stackrel{?}{=} 3 + \tfrac{3}{2}n + \tfrac{1}{2}n^2 + n + 2$$

Since both sides agree, we have confirmed the conjecture.

The general converse of the corollary is true, as the following theorem states.

THEOREM 3.3 *If $\{x_n\}$ is a sequence with the property that some sufficiently high difference of the sequence is identically zero, then $\{x_n\}$ is itself a polynomial sequence—that is, there is a polynomial function P such that $x_n = P(n)$ for $n = 0, 1, 2, \ldots$. Explicitly, if $\Delta^{p+1}x_n = 0$ for all n, then $x_n = P(n)$, where P is a polynomial of degree at most p.*

Proof It is easy to check this for $p = 0$ and for $p = 1$. If $\Delta x_n = 0$ for all n, then we merely have $x_{n+1} = x_n$ for all n, so that $x_0 = x_1 = x_2 = \cdots$. The sequence $\{x_n\}$ is constant, and we have $x_n = C$ for all n; such a sequence is given by a polynomial of degree 0. Suppose now that we had a sequence $\{x_n\}$ whose second differences were identically zero. Since $\Delta^2 x_n = \Delta(\Delta x_n)$, we could apply the case just completed to conclude that the sequence Δx_n is constant. This is equivalent to the recursion $x_{n+1} = x_n + C$, and the solution of this is easily seen to be $x_n = x_0 + nC$. Thus, $\{x_n\}$ is again a polynomial sequence, given by the polynomial function $P(t) = \alpha + \beta t$, where $\alpha = x_0$ and $\beta = C$.

The general case can be established by induction. The easiest method is to base the argument on the following property of polynomial sequences.

LEMMA *If $\{\Delta x_n\}$ is a polynomial sequence of degree p, then $\{x_n\}$ is a polynomial sequence of degree $p+1$.*

Proof We have verified this above for the case $p = 0$. Suppose that we have verified it for all choices of p less than some integer k and we want to show that it must then hold for $p = k$. Suppose that $\{x_n\}$ is a sequence such that $\{\Delta x_n\}$ is a polynomial sequence of degree k. We can therefore write

$$\Delta x_n = \beta n^k + Q(n)$$

where Q is a polynomial of degree less than k. Choose the special polynomial sequence

$$y_n = \frac{\beta}{k+1} n^{k+1}$$

Then,

$$\begin{aligned}\Delta y_n &= \frac{\beta}{k+1}(n+1)^{k+1} - \frac{\beta}{k+1} n^{k+1} \\ &= \frac{\beta}{k+1}[n^{k+1} + (k+1)n^k + \cdots - n^{k+1}] \\ &= \beta n^k + H(n)\end{aligned}$$

where H is a polynomial of degree less than k.

Accordingly, we have

$$\Delta x_n - \Delta y_n = \beta n^k + Q(n) - [\beta n^k + H(n)]$$

and

$$\Delta(x_n - y_n) = Q(n) - H(n) = R(n)$$

where R is a polynomial of degree less than k. By the inductive assumption made above, we may conclude that $\{x_n - y_n\}$ is a polynomial sequence of degree at most k. Explicitly, we may write $x_n - y_n = P(n)$, where P has degree at most k. But this gives us a formula for x_n, namely,

$$\begin{aligned}x_n &= y_n + P(n) \\ &= \frac{\beta}{k+1} n^{k+1} + P(n)\end{aligned}$$

which is a polynomial in n of degree $k+1$, completing the proof of the lemma.

To see how this result leads at once to a proof of Theorem 3.3, suppose that $\{x_n\}$ is a sequence such that $\Delta^{p+1} x_n = 0$ for $n = 0, 1, 2, \ldots$. Writing this as $\Delta(\Delta^p x_n) = 0$, we use the lemma to conclude that the sequence $\{\Delta^p x_n\}$ is constant (that is, given by a polynomial of degree 0). Writing $\Delta^p x_n$ as $\Delta(\Delta^{p-1} x_n)$, and applying the lemma, we conclude that $\{\Delta^{p-1} x_n\}$ is a polynomial sequence of degree 1. Continuing in the same way, we find that $\{\Delta^{p-2} x_n\}$ is polynomial of degree 2, $\{\Delta^{p-3} x_n\}$ is polynomial of degree 3, etc., eventually arriving at the desired conclusion that $\{x_n\}$ itself is polynomial of degree p.

A recursion relation may involve more than the immediate predecessor of a given term. For example, consider the equation

$$x_{n+1} = x_n + x_{n-1}$$

If we are given the values of x_0 and x_1, then this relation will serve to compute any of the later terms in the sequence. If $x_0 = 2$ and $x_1 = 5$, then $x_2 = 7$, $x_3 = 12$, $x_4 = 19$, etc. The initial choice $x_0 = 1$, $x_1 = 2$ determines a different sequence called the Fibonacci numbers:

$$1, 2, 3, 5, 8, 13, 21, 34, 55, 89, \ldots$$

(See also Exercise 23.) It is possible to find a formula which describes this sequence; the complexity of the formula may seem surprising.

$$x_n = \frac{\sqrt{5}+3}{2\sqrt{5}}\left(\frac{1+\sqrt{5}}{2}\right)^n + \frac{\sqrt{5}-3}{2\sqrt{5}}\left(\frac{1-\sqrt{5}}{2}\right)^n \tag{3.6}$$

However, it is not difficult to see where this formula came from. It is based upon a general theorem about difference equations.

THEOREM 3.4 *The difference equation*

$$x_{n+1} = Ax_n + Bx_{n-1} \tag{3.7}$$

has solution $x_n = C_1\alpha^n + C_2\beta^n$, *where* α *and* β *are the roots of the quadratic equation* $\gamma^2 = A\gamma + B$, *and* C_1 *and* C_2 *are arbitrary constants. If the quadratic has repeated roots, so that* $\alpha = \beta$, *then the sequence* $\{x_n\}$ *has the form* $x_n = C_1\beta^n + C_2 n\beta^n$.

Proof Try a solution of the form $x_n = \gamma^n$, where γ is to be determined. Substituting this into the recursion, we need to have $\gamma^{n+1} = A\gamma^n + B\gamma^{n-1}$. Dividing by γ^{n-1}, we need to have $\gamma^2 = A\gamma + B$. Thus, if the roots of this quadratic equation are α and β, we may take γ to be either of them. It is then easily checked that $x_n = C_1\alpha^n + C_2\beta^n$ is a solution of (3.7) for any choice of the constants C_1 and C_2. If the quadratic equation $\gamma^2 - A\gamma - B = 0$ has repeated roots, so that $\alpha = \beta$, then it must be true that $A = 2\beta$ and $B = -\beta^2$. Thus the difference equation in fact has the form

$$x_{n+1} = 2\beta x_n - \beta^2 x_{n-1}$$

It can then be checked that $x_n = n\beta^n$ is a solution, so that $x_n = C_1\beta^n + C_2 n\beta^n$ is also. To see that there can be no other solutions, all that has to be done is to show that these solutions allow you to obtain any pair of starting values x_0 and x_1, for once these have been obtained, the recursion (3.7) determines all succeeding terms. In the normal case when $\alpha \neq \beta$, we want to be able to choose C_1 and C_2 so that

$$\begin{aligned} x_0 &= C_1\alpha^0 + C_2\beta^0 \\ &= C_1 + C_2 \end{aligned}$$

and so that

$$x_1 = C_1\alpha + C_2\beta$$

These equations have the solution

$$C_1 = \frac{\beta x_0 - x_1}{\beta - \alpha}, \qquad C_2 = \frac{x_1 - \alpha x_0}{\beta - \alpha}$$

When we have the case of equal roots and $\alpha = \beta$, then the second form of the solution is used. We again want values of C_1 and C_2 so that

$$\begin{aligned} x_0 &= C_1\beta^0 + C_2(0)\beta^0 \\ &= C_1 \end{aligned}$$

and

$$x_1 = C_1\beta + C_2(1)\beta$$

whose solutions are $C_1 = x_0$ and $C_2 = \beta^{-1}x_1 - x_0$.

To see how this theorem gives a formula for the Fibonacci sequence, consider again the difference equation

$$\begin{cases} x_0 = 1 \\ x_1 = 2 \\ x_{n+1} = x_n + x_{n-1} \end{cases}$$

The quadratic equation is $\gamma^2 = \gamma + 1$, whose solutions are $\frac{1}{2}(1+\sqrt{5})$ and $\frac{1}{2}(1-\sqrt{5})$. Thus, the general solution is given by the formula

$$x_n = C_1\left(\frac{1+\sqrt{5}}{2}\right)^n + C_2\left(\frac{1-\sqrt{5}}{2}\right)^n$$

and to match the initial terms x_0 and x_1, we must choose

$$C_1 = \frac{\frac{1}{2}(1-\sqrt{5})(1) - 2}{\frac{1}{2}(1-\sqrt{5}) - \frac{1}{2}(1+\sqrt{5})} = \frac{1-4-\sqrt{5}}{-2\sqrt{5}} = \frac{\sqrt{5}+3}{2\sqrt{5}}$$

$$C_2 = \frac{2 - \frac{1}{2}(1+\sqrt{5})(1)}{-\sqrt{5}} = \frac{\sqrt{5}-3}{2\sqrt{5}}$$

This yields the formula (3.6) given earlier.

The general form of a recursion in which the next term is determined by knowledge of the previous two terms and the index is the following:

$$\begin{cases} x_0 = a \\ x_1 = b \\ x_{n+1} = F(n, x_n, x_{n-1}) \end{cases} \tag{3.8}$$

Notice that this is analogous to the general form of a second-order differential equation. As with the latter, it is only in very special cases that it is possible to find a simple formula for the sequence determined by a recursion such as (3.8). Such recursions do arise in many different areas of applied mathematics, and the computer makes it possible to compute many values of such sequences sufficient in many cases to answer the required questions. However, more sophisticated mathematical techniques must be used if one is interested in finding out the behavior of such a sequence for very large values of n. Such is the case if one is interested in proving that $\{x_n\}$ is a bounded sequence, or in calculating $\lim_{n\to\infty} x_n$.

Exercises

Find the first 10 terms of the sequences defined by the recursions in Exercises 1–4.

1. $x_0 = 2, \quad x_{n+1} = 2x_n + 3$
2. $x_0 = 1, \quad x_{n+1} = 2x_n - n$
3. $y_0 = 1, \quad y_{n+1} = (n+1)\, y_n$
4. $x_0 = 1, \quad x_{n+1} = x_n + n$

If you have access to a pocket calculator, examine the behavior of the sequences defined by the following two recursions.

PC5. $x_{n+1} = \frac{1}{3}[2x_n + (x_n)^2]$, when $x_0 = 2$, and when $x_0 = \frac{1}{2}$

PC6. $y_{n+1} = \frac{1}{2}(y_n + \sqrt{y_n})$, when $y_0 = 2$, and when $y_0 = \frac{1}{2}$

7. Verify that the sequence defined by the formula $x_n = n^3 - n^2$ satisfies the recursion $x_{n+1} = [1 + (3/n)]\, x_n + 4n$.

Find a formula for the sequences defined by Exercises 8–11.

8. $x_0 = 2, \quad x_{n+1} = 3x_n - 4$
9. $y_0 = 3, \quad y_{n+1} = 2 - y_n$
10. $x_0 = 1, \quad x_{n+1} = \frac{1}{2}x_n + 1$
11. $x_0 = 1, \quad x_{n+1} = 6x_n + 7$
12. Find the first, second, and third successive differences of the following sequences.
    (a) $2, 5, 4, 7, 6, 9, 8, 11, 10, 13, \ldots$
    (b) $2, 6, 12, 20, 30, 42, 56, 72, \ldots$
    (c) $3, 6, 11, 20, 37, 70, 135, 264, \ldots$
    (d) Can you tell which of these sequences are polynomial?

13. Find a polynomial sequence $\{x_n\}$ such that
(a) $\Delta x_n = 5n^2 + 3n$
(b) $\Delta^2 x_n = 6n^2 + n$

14. Solve the recursion $y_0 = 1$, $y_{n+1} = y_n + n^2 - 1$, by finding a formula for y_n.

15. Find the first 10 terms of the following sequences by using the recursion formula given.
(a) $x_{n+1} = \frac{1}{2}(x_n + x_{n-1})$, $x_0 = 1$, $x_1 = 2$
(b) $x_{n+1} = (1 + x_n)/x_{n-1}$, $x_0 = 2$, $x_1 = 3$. Can you guess x_{19}?

16. Use Theorem 3.4 to find a formula for the sequence defined by
(a) $x_0 = 2$, $x_1 = 1$, $x_{n+1} = 3x_n - 2x_{n-1}$
(b) $x_0 = 2$, $x_1 = 2$, $x_{n+1} = x_n + 2x_{n-1}$
(c) $x_0 = 1$, $x_1 = 2$, $x_{n+1} = 6x_n - 9x_{n-1}$

17. Using the operator Δ, show that the recursions in Exercise 16 are equivalent to
(a) $\Delta^2 x_n - \Delta x_n = 0$
(b) $\Delta^2 x_n + \Delta x_n = 2x_n$
(c) $\Delta^2 x_n - 4\,\Delta x_n = -4x_n$

18. A pair of sequences $\{x_n\}$ and $\{y_n\}$ are defined by the following recursion:

$$\begin{cases} x_0 = 2 \\ y_0 = 3 \\ x_{n+1} = x_n + y_n \\ y_{n+1} = 2x_n - y_n \end{cases}$$

Find the first six terms of each sequence. Can you find a general formula for the sequences?

*19. Given that $x_{n+1} = 2x_n + 3^n$ and $x_0 = 2$, calculate six terms of $\{x_n\}$. Can you find a formula for the general case when $x_0 = A$? Check for the case $A = 2$.

PC20. One discrete model for population growth that is often used is

$$\Delta x_n = \alpha x_n\left(1 - \frac{x_n}{M}\right)$$

which is a discrete analog of the logistic growth law (Section 2.1, Exercise 31). Find values for α and M which fit the sequence starting 1, 1.36, 1.83, and then calculate the next three terms of the sequence. (You may use a pocket calculator.)

*21. The square of the matrix $\begin{bmatrix} 2 & 3 \\ 0 & 4 \end{bmatrix}$ is $\begin{bmatrix} 4 & 18 \\ 0 & 16 \end{bmatrix}$ and the cube is $\begin{bmatrix} 8 & 84 \\ 0 & 64 \end{bmatrix}$. Find a formula for the nth power of the given matrix. (There is a review of matrix multiplication in Section 6.4.)

22. Use difference methods to predict the next term of each of the following sequences.
(a) $1, 4, 8, 15, 27, 46, \ldots$
(b) $0, 1, 5, 14, 24, 43, 55, \ldots$

23. The growth of certain shrubs obeys the rule that every old twig branches each year but every new twig must wait a year before branching. Each twig grows an inch each year.
(a) Show that x_n, the number of twigs in year n, obeys the relation $x_{n+1} = 2x_{n-1} + (x_n - x_{n-1})$. If $x_1 = x_2 = 1$, find x_8.
(b) What would x_8 be if each twig had to wait two years before branching?

The next exercises may require a pocket calculator.

PC24. (a) Problems in compound interest often lead to simple difference equations. If x_n is the amount owed at the end of month n, and the monthly interest rate is r, and the borrower repays P dollars each month (which includes both interest and principal), show that $x_{n+1} = (1+r)x_n - P$.
(b) What should P be to amortize a loan of \$10,000 in 25 payments at 7% per year?

PC25. Sometimes population models take into account delay terms; for example, children are not born until their parents have reached a certain age. Such a model might have the form

$$\Delta x_n = \alpha x_{n-1}\left(1 - \frac{x_n}{M}\right)$$

With $\alpha = 0.4$, $M = 10$, $x_1 = 1$, and $x_2 = 1.5$, calculate the predicted value of x_5.

3.2 Series

An infinite series is usually written as $\sum a_n$ or as

$$a_1 + a_2 + a_3 + \cdots$$

It consists of two sequences of numbers, the sequence of terms $\{a_n\}$ and the sequence of partial sums $\{A_n\}$, which are related by the identities

$$A_n = a_1 + a_2 + a_3 + \cdots + a_n \qquad n = 1, 2, 3, \ldots$$

$$a_1 = A_1$$

$$a_n = A_n - A_{n-1} \qquad n = 2, 3, 4, \ldots$$

For example, the first five terms of the series

$$1 + \tfrac{1}{2} + \tfrac{1}{4} + \tfrac{1}{8} + \tfrac{1}{16} + \tfrac{1}{32} + \cdots$$

are $1, \frac{1}{2}, \frac{1}{4}, \frac{1}{8}$, and $\frac{1}{16}$, whereas the first five partial sums are $1, \frac{3}{2}, \frac{7}{4}, \frac{15}{8}$, and $\frac{31}{16}$.

DEFINITION 3.2 The series $\sum a_n$ is said to be *convergent*, or to converge to the sum A, if $\lim_{n\to\infty} A_n$ exists and is A. If the sequence of partial sums $\{A_n\}$ is not a convergent sequence, the series is said to *diverge*.

The series

$$\tfrac{1}{2} + \tfrac{1}{6} + \tfrac{1}{12} + \tfrac{1}{20} + \cdots$$

which is

$$\frac{1}{(1)(2)} + \frac{1}{(2)(3)} + \frac{1}{(3)(4)} + \frac{1}{(4)(5)} + \cdots$$

has as its general term $a_n = 1/(n)(n+1)$. Calculation shows that its partial sums are given by $A_1 = \frac{1}{2}, A_2 = \frac{2}{3}, A_3 = \frac{3}{4}, A_4 = \frac{4}{5}$. One may show (for example, by mathematical induction) that $A_n = n/(n+1)$. Since $\lim_{n\to\infty} n/(n+1) = 1$, the given series converges and has sum 1.

Another familiar example is the **geometric series**

$$a + ar + ar^2 + ar^3 + \cdots = a\sum_0^\infty r^n$$

whose partial sums are $A_1 = a$, $A_2 = a(1+r) = a(1-r^2)/(1-r)$, and in general $A_n = a(1-r^n)/(1-r)$, assuming that the number r is not 1. (If $r = 1$, then $A_n = an$.) Since $\lim_{n\to\infty} r^n$ exists if and only if $-1 < r \leq 1$, we may conclude that the geometric series converges only when $-1 < r < 1$. The series then has sum $a/(1-r)$.

If two series $\sum a_n$ and $\sum b_n$ are convergent, then the series $\sum(a_n + b_n)$, whose terms are the sums of the terms of the other two series, is also convergent, and its sum is the sum of the other sums; we write

$$\sum(a_n+b_n) = \sum a_n + \sum b_n$$

For example, the series

$$(\tfrac{1}{2}+\tfrac{1}{3}) + (\tfrac{1}{4}+\tfrac{1}{9}) + (\tfrac{1}{8}+\tfrac{1}{27}) + \cdots = \tfrac{5}{6} + \tfrac{13}{36} + \tfrac{35}{216} + \cdots$$

is convergent and has sum

$$\frac{1}{2}\,\frac{1}{1-\frac{1}{2}} + \frac{1}{3}\,\frac{1}{1-\frac{1}{3}} = \frac{3}{2}$$

The discovery of tests to show that a given series is either convergent or divergent is central to the theory of series. Of such tests, some can be applied to the terms of the series; others may also involve the partial sums.

THEOREM 3.5 *If $a_n \geq 0$, then $\sum a_n$ is convergent if and only if the partial sums form a bounded sequence.*

Proof Since $a_n \geq 0$, the sequence $\{A_n\}$ is monotonic increasing. If it is bounded, then $\lim_{n\to\infty} A_n$ exists,* and the series $\sum a_n$ is convergent. If the sequence $\{A_n\}$ is unbounded, then $\{A_n\}$, and hence the series $\sum a_n$, is divergent.

THEOREM 3.6 *The series $\sum a_n$ is convergent if and only if*

$$\sum_{n+1}^{m} a_k = a_{n+1} + a_{n+2} + \cdots + a_{m-1} + a_m$$

converges to zero as n and m independently increase.

Proof The sum in question can be expressed in terms of the partial sums as $A_m - A_n$. Thus, the stated condition is equivalent to the statement that the sequence $\{A_k\}$ is a Cauchy sequence. This, in turn, is known to be a necessary and sufficient condition for any sequence of real or complex numbers to be convergent.

There is also a very useful test which can be used only to prove that a series is divergent.

THEOREM 3.7 *If it is* not *true that $\lim_{n\to\infty} a_n = 0$, then the series $\sum a_n$ is divergent.*

Proof Suppose that $\sum a_n$ is convergent, with sum A. Then, $\lim_{n\to\infty} A_n = A$. It must also be true that $\lim_{n\to\infty} A_{n-1} = A$ and therefore that $\lim_{n\to\infty}(A_n - A_{n-1}) = A - A = 0$. However, $A_n - A_{n-1} = a_n$, so that $\lim_{n\to\infty} a_n = 0$. This is therefore a necessary condition for convergence; without it, $\sum a_n$ diverges.

* This follows from the basic theorem often cited in an elementary calculus course: *Every bounded monotonic sequence of real numbers converges to a limit.*

There is a strong analogy between infinite series and integrals of the form $\int_a^\infty f(x)\,dx$, usually called "improper" integrals.

THEOREM **3.8 (integral test)** *Let $f(x)$ be a positive continuous monotonically decreasing function on the interval $a \le x < \infty$. Let $a_n = f(n)$ for all $n > a$. Then the series $\sum a_n$ is convergent if and only if the improper integral $\int_a^\infty f(x)\,dx$ is finite.*

Proof From Figure 3.1, we see that

$$a_n \le \int_{n-1}^{n} f(x)\,dx$$

and so if k is the first integer greater than a, then

$$A_m - A_k = \sum_{k+1}^{m} a_n \le \int_a^k f + \int_k^{k+1} f + \cdots + \int_{m-1}^{m} f \le \int_a^\infty f$$

If the integral on the right is finite, then the partial sums $\{A_m\}$ form a bounded sequence. Since $a_n \ge 0$, we see by Theorem 3.5 that the series $\sum a_n$ is convergent.

Also, from Figure 3.1, we observe that

$$a_n \ge \int_n^{n+1} f$$

and so

$$\sum_k^m a_n \ge \int_k^{m+1} f$$

If the integral $\int_a^\infty f$ is infinite (divergent), then the number $\int_k^{m+1} f$ will grow without bound as m increases, and so must the number $A_m = \sum_1^m a_n$. Since the partial sums of the series $\sum a_n$ are therefore unbounded, the series diverges. Hence $\sum a_n$ convergent implies that the improper integral is finite.

If we take $f(x) = 1/x^p$, we find that $\int_1^\infty f(x)\,dx$ is finite when $p > 1$ and

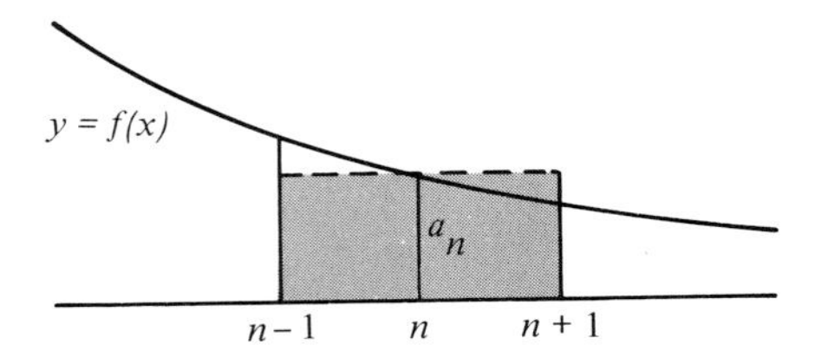

Figure 3.1

infinite if $0 \leq p \leq 1$. Accordingly, we find that the series

$$\sum_{1}^{\infty} \frac{1}{n^p} = 1 + \frac{1}{2^p} + \frac{1}{3^p} + \cdots$$

is convergent when $p > 1$, and diverges when $p \leq 1$.

It is sometimes possible to test a series with positive terms by comparing it with another positive series whose behavior is known.

THEOREM 3.9 (comparison test) *Let $0 \leq a_n \leq b_n$. Then, if $\sum b_n$ converges $\sum a_n$ converges.*

Proof If the corresponding partial sums are A_n and B_n, then we have $A_n \leq B_n$. Since $\sum b_n$ converges and has positive terms, $\{B_n\}$ is a bounded sequence. So must be $\{A_n\}$, and $\sum a_n$ must converge.

A consequence of this theorem is worth noting; if the same relation holds and $\sum a_n$ is divergent, so is $\sum b_n$.

A very similar test that is often easy to apply is the following limit form of the comparison test.

THEOREM 3.10 *Let $0 < a_n$, $0 < b_n$,* and $\lim_{n \to \infty} a_n/b_n = L$, *with $0 < L < \infty$. Then, the two series $\sum a_n$ and $\sum b_n$ are either both convergent or both divergent.*

Proof This comes immediately from Theorem 3.9. Since L lies between $2L$ and $\frac{1}{2}L$, there exists an integer N such that $\frac{1}{2}L < a_n/b_n < 2L$, for all $n > N$. Accordingly, if $n > N$, we have $(\frac{1}{2}L) b_n < a_n$ and $a_n < (2L) b_n$. By the comparison test, if $\sum a_n$ converges, so must $(\frac{1}{2}L) \sum b_n$, and if $\sum a_n$ diverges, so must $(2L) \sum b_n$, proving that $\sum a_n$ and $\sum b_n$ both converge or both diverge. (We note that Theorem 3.10 does not hold for series with positive and negative terms.)

When two sequences $\{a_n\}$ and $\{b_n\}$ of positive numbers are related as in Theorem 3.10, so that $\lim a_n/b_n = L \neq 0$ exists, then one may write $a_n \sim b_n$, read "a_n is asymptotically like b_n." For example, $n/(n^3+4) \sim 1/n^2$. Since $\sum 1/n^2$ converges, so does $\sum n/(n^3+4)$. Likewise, the series

$$\sum \frac{\sqrt{n+1}}{2+\sqrt{n^3+n+1}}$$

is divergent, since its terms behave like $1/n$.

When the terms of a series $\sum a_n$ are not all positive, the simplest indicator of convergence is the convergence of the series $\sum |a_n|$, whose terms are the absolute values of the terms of the series $\sum a_n$.

THEOREM 3.11 *A series $\sum a_n$ is convergent if the series $\sum |a_n|$ is convergent.*

Proof Suppose $\sum |a_n|$ is convergent. We give two proofs. For the first, we observe that

$$\left|\sum_{n+1}^{m} a_k\right| = |a_{n+1} + a_{n+2} + \cdots + a_m|$$

$$\leq |a_{n+1}| + |a_{n+2}| + \cdots + |a_m|$$

$$= \sum_{n+1}^{m} |a_k| \to 0$$

(by Theorem 3.6) so that (by Theorem 3.6), $\sum a_n$ converges.

For the second proof, we observe that

$$0 \leq a_n + |a_n| \leq 2|a_n|$$

Since $\sum 2|a_n|$ is convergent, so is the series $\sum(a_n + |a_n|)$ (by Theorem 3.9). Noting that

$$\sum a_n = \sum(a_n + |a_n|) - \sum |a_n|$$

the difference of two convergent series, we conclude that $\sum a_n$ is convergent.

For example, each of the following series is convergent:

$$1 - \frac{1}{4} + \frac{1}{9} - \frac{1}{16} + \frac{1}{25} + \cdots + \frac{(-1)^{n+1}}{n^2} + \cdots$$

$$\frac{1}{2} + \frac{2}{4} - \frac{3}{8} + \frac{4}{16} + \frac{5}{32} - \frac{6}{64} + \frac{7}{128} + \cdots$$

(Compare the latter series with $\sum n/2^n$.)

It is possible to have the series $\sum a_n$ convergent while the positive series $\sum |a_n|$ is divergent. This is the case for the "harmonic" series

$$1 - \frac{1}{2} + \frac{1}{3} - \frac{1}{4} + \frac{1}{5} + \cdots + \frac{(-1)^{n+1}}{n} + \cdots$$

The convergence of this series is a consequence of the following useful result.

THEOREM 3.12 (alternating series test) *Let the sequence $\{c_n\}$ be monotonic decreasing with $\lim_{n\to\infty} c_n = 0$. Then the series $\sum(-1)^{n+1} c_n$ is convergent.*

Proof The series has the form

$$c_1 - c_2 + c_3 - c_4 + c_5 - c_6 + \cdots$$

where all the numbers c_k are positive, but decrease with limit zero. The partial sums of such a series behave in a special way; they obey the inequalities

$$A_2 \le A_4 \le A_6 \le \cdots \le A_{2n} \le \cdots$$

$$\cdots \le A_{2n+1} \le \cdots \le A_5 \le A_3 \le A_1$$

$$A_{2n} \le A_{2n+1}$$

To see this, note that $A_{2n+1} = A_{2n} + c_{2n+1}$, so that $A_{2n+1} \ge A_{2n}$. Also, note that $A_{2n+2} = A_{2n} + (c_{2n+1} - c_{2n+2})$ and that $c_{2n+1} - c_{2n+2} \ge 0$, so that $A_{2n} \le A_{2n+2}$. In the same way, we see that $A_{2n+1} \le A_{2n-1}$.

Thus, the even partial sums form a bounded increasing sequence, and the odd partial sums form a bounded decreasing sequence. Both must therefore converge. Since $A_{2n+1} - A_{2n} = c_{2n+1}$, which has limit zero, we have $\lim A_{2n+1} = \lim A_{2n}$, and the original series converges.

This test can be used to show that each of the following series converges:

$$1 - \frac{1}{\sqrt{2}} + \frac{1}{\sqrt{3}} - \frac{1}{\sqrt{4}} + \cdots + \frac{(-1)^{n+1}}{\sqrt{n}} + \cdots$$

$$\frac{2}{(1)(3)} - \frac{(2)(4)}{(1)(3)(5)} + \frac{(2)(4)(6)}{(1)(3)(5)(7)} - \frac{(2)(4)(6)(8)}{(1)(3)(5)(7)(9)} + \cdots$$

Another useful test, which can be applied to series with both positive and negative terms, is the ratio test.

THEOREM 3.13 (ratio test) *If $a_n \ne 0$ and $\lim |a_{n+1}|/|a_n| = L$, $L < 1$, then the series $\sum a_n$ converges. If $L > 1$, then the series diverges. (If $L = 1$, the series might do either.)*

Proof Clearly, if $L > 1$, then for large n, $|a_{n+1}|/|a_n| > 1$ and the sequence $\{|a_k|\}$ is monotonic increasing. Certainly then, we do not have $\lim a_n = 0$, and the series diverges (by Theorem 3.7). Suppose then that $L < 1$. Choose r with $L < r < 1$. Then, for large n, we must have $|a_{n+1}|/|a_n| \le r$. This can be rewritten as

$$\frac{|a_{n+1}|}{|a_n|} \le \frac{r^{n+1}}{r^n}$$

or as

$$\frac{|a_{n+1}|}{r^{n+1}} \le \frac{|a_n|}{r^n}$$

This shows that the sequence $\{|a_n|/r^n\}$ is positive and decreasing. It must therefore be bounded, so that $|a_n|/r^n \le B$ for all n. Accordingly, we have

$|a_n| \leq Br^n$, and since $\sum Br^n$ converges when $r < 1$, the comparison test (Theorem 3.9) applies and $\sum |a_n|$ converges. By Theorem 3.11, so does $\sum a_n$.

The importance of infinite series lies largely in their use as representations for functions, especially by power series.

DEFINITION 3.3 A series of the form $\sum_0^\infty a_n x^n$, or more generally $\sum_0^\infty a_n(x-c)^n$, is called a *power series*. The convergence set is the set of all x for which such a series converges. On this set, the series defines a function $f(x)$.

$$f(x) = \sum_0^\infty a_n(x-c)^n = a_0 + a_1(x-c) + a_2(x-c)^2 + \cdots$$

THEOREM 3.14 *The set of convergence of a power series is either a bounded interval, possibly with one or both of its endpoints, or the infinite interval* $-\infty < x < \infty$. *Specifically, a power series* $\sum a_n x^n$ *will converge for all* x *such that* $-R < x < R$, *where* $R \leq \infty$ *is called the radius of convergence. It will diverge for* x *satisfying* $|x| > R$. *It may converge or it may diverge at the points* $x = R$, $x = -R$.

Proof The series $\sum_0^\infty a_n x^n$ clearly converges when $x = 0$. Suppose that it also converges for a value $x = x_0$. By Theorem 3.7, the sequence of terms $\{a_n x_0{}^n\}$ must converge to zero and must therefore be bounded: $|a_n x_0{}^n| \leq B$ for all n. Let x be any number with $|x| < |x_0|$. Then,

$$\begin{aligned} |a_n x^n| &= |a_n x_0{}^n|\, r^n \\ &\leq Br^n \end{aligned}$$

where $r = |x|/|x_0| < 1$. By the comparison test (Theorem 3.9), we conclude that $\sum |a_n x^n|$ converges, and x belongs to the convergence set for $\sum a_n x^n$. Let R be the least upper bound of the set of all numbers $x_0 > 0$ for which the series $\sum a_n x^n$ converges. If R is infinite, so that the set of x_0 has no upper bound, then the previous argument shows that the series converges for all numbers x, and the convergence set is $-\infty < x < \infty$. If R is finite, then the argument shows that the series converges for all x with $-R < x < R$ and that it diverges for all $x > R$ and for all $x < -R$. The argument gives no information about the behavior of the series for $x = R$ or for $x = -R$, where the series may converge or diverge, depending on its detailed nature.

The facts about convergence of the series $\sum a_n(x-c)^n$ follow from this special case. Putting $t = x-c$, this series becomes the power series $\sum a_n t^n$. If this series has radius of convergence R, then the series $\sum a_n(x-c)^n$ will converge for all x for which $|t| = |x-c| < R$ and will diverge for all x for which

$|t| = |x-c| > R$. Since x and c are real, the inequality $|x-c| < R$ is equivalent to $-R < x-c < R$, and thus to the inequality $c-R < x < c+R$. We therefore see that the power series $\sum a_n(x-c)^n$ converges for all points on the open interval with endpoints $c \pm R$, and diverges at all points to either side of this interval. (See Figure 3.2.) Whether or not the series converges if $x = c+R$ or if $x = c-R$ depends upon the exact nature of the coefficients $\{a_n\}$.

The ratio test is often a convenient method for finding R. For example, consider the series

$$\sum_0^\infty nx^n = x + 2x^2 + 3x^3 + 4x^4 + \cdots \tag{3.9}$$

Applying the test, we examine

$$\lim_{n\to\infty} \frac{|a_{n+1}x^{n+1}|}{|a_n x^n|} = \lim_{n\to\infty} \frac{|(n+1)x^{n+1}|}{|nx^n|}$$

$$= \lim_{n\to\infty} \frac{n+1}{n}|x| = |x|$$

With $|x| = L$ (see Theorem 3.13), the series converges for all x with $|x| < 1$ and diverges for all x with $|x| > 1$ $(R=1)$. Testing the series at $x = 1$, or at $x = -1$, we obtain the series $1+2+3+\cdots$ and $-1+2-3+4-5+\cdots$, both of which diverge. Hence, the convergence set is $-1 < x < 1$.

In contrast, the series $\sum_1^\infty (x-2)^n/n^2$ converges for exactly those x obeying the inequality $1 \le x \le 3$. To see this, replace $x-2$ by t so that the series becomes $\sum_1^\infty t^n/n^2$. Applying the ratio test, we examine

$$\lim_{n\to\infty} \frac{|a_{n+1}t^{n+1}|}{|a_n t^n|} = \lim_{n\to\infty} \frac{n^2|t^{n+1}|}{(n+1)^2|t^n|}$$

$$= \lim_{n\to\infty} \frac{n^2}{(n+1)^2}|t| = |t|$$

Hence, the series $\sum_1^\infty t^n/n^2$ converges for all t with $|t| < 1$ and diverges for all t with $|t| > 1$. Testing the series for $t = 1$ and for $t = -1$, we obtain the pair of series $\sum_1^\infty 1/n^2$ and $\sum_1^\infty (-1)^n/n^2$, which both converge. The condition

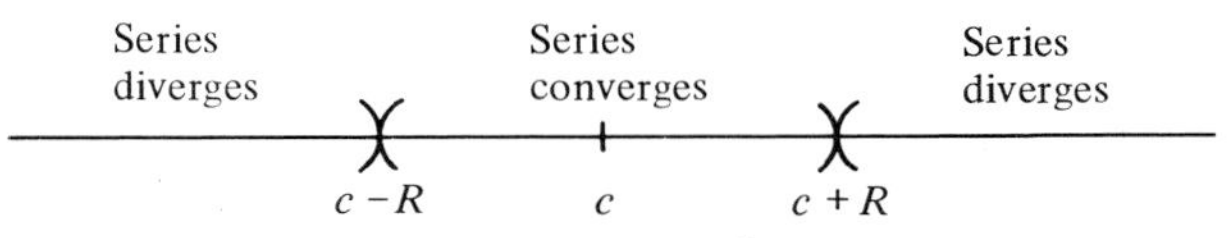

Figure 3.2 *Convergence region for* $\Sigma\, a_n(x-c)^n$

$|t| < 1$ is equivalent to $-1 < t < 1$ and thus to $-1 < x-2 < 1$, while the case $t = 1$ corresponds to $x = 3$, and $t = -1$ to $x = 1$. Thus, the series $\sum_1^\infty (x-2)^n/n^2$ converges exactly for those x with $1 \leq x \leq 3$, and diverges for all other values of x.

The following important result explains in part why power series are so useful.

THEOREM 3.15 *Power series can be differentiated termwise. That is, if*

$$f(x) = a_0 + a_1(x-c) + a_2(x-c)^2 + \cdots$$

and if this series converges for all x with $-R < x-c < R$, then the function f is continuous on this set and is of class $\mathcal{C}^2$ (that is, has continuous derivatives of all orders), and

$$f'(x) = a_1 + 2a_2(x-c) + 3a_3(x-c)^2 + \cdots$$

$$f''(x) = 2a_2 + (2)(3)a_3(x-c) + (3)(4)a_4(x-c)^2 + \cdots$$

etc. Each of these power series also converges for those x with $-R < x-c < R$.

We omit the proof of this theorem, which is to be found in any text on advanced calculus or the theory of series.

Some simple examples may show better the meaning of this result. Consider the familiar geometric series

$$\frac{1}{1-x} = 1 + x + x^2 + x^3 + \cdots + x^n + \cdots \tag{3.10}$$

We have seen that this power series converges for those x with $-1 < x < 1$. According to Theorem 3.15, we can differentiate both sides of this equation, treating the series termwise, and obtain

$$\frac{1}{(1-x)^2} = 0 + 1 + 2x + 3x^2 + 4x^3 + \cdots + nx^{n-1} + \cdots$$

Theorem 3.15 tells us that this is a correct equation, and that the series on the right converges for all x with $-1 < x < 1$. Note too that this also leads immediately to the formula $x/(1-x)^2$ for the sum of the series (3.9).

We can differentiate again, and obtain a new series identity:

$$\frac{2}{(1-x)^3} = 2 + 6x + 12x^2 + 20x^3 + 30x^4 + \cdots + n(n-1)x^{n-2} + \cdots$$

By the theorem, this too converges for $-1 < x < 1$. For example, if we set

$x = \frac{1}{3}$ we obtain a special formula for the sum of a numerical series:

$$\frac{2}{(1-\frac{1}{3})^3} = \frac{27}{4} = 2 + 2 + \frac{4}{3} + \frac{20}{27} + \frac{10}{27} + \frac{14}{81} + \cdots$$

Differentiating cannot destroy the convergence of a power series at a point that is interior to the interval of convergence of the series, but it can destroy convergence at an endpoint. For example, the series

$$1 + \frac{x}{1} + \frac{x^2}{4} + \frac{x^3}{9} + \cdots + \frac{x^n}{n^2} + \cdots$$

converges exactly for $-1 \leq x \leq 1$ [see Exercise 10(a)]. Differentiating it, we obtain the power series

$$1 + \frac{x}{2} + \frac{x^2}{3} + \frac{x^3}{4} + \cdots + \frac{x^{n-1}}{n} + \cdots$$

which converges only for $-1 \leq x < 1$ [see Exercise 10(b)]. Differentiating again, we obtain the series

$$\frac{1}{2} + \frac{2}{3}x + \frac{3}{4}x^2 + \frac{4}{5}x^3 + \cdots + \frac{n-1}{n}x^{n-2} + \cdots$$

which converges only for $-1 < x < 1$ [see Exercise 10(c)]. Thus, by differentiating we lose first one endpoint and then the other; however, no matter how many times we differentiate, the resulting series will still converge for all x with $|x| < 1$ since the radius of convergence of the original series was $R = 1$.

A function f that can be expressed as a convergent power series on a neighborhood of a point $x = c$ is said to be **analytic** at c. We can use Theorem 3.15 to find a formula for the coefficients in the power series of an analytic function; the result is usually called the Taylor Series of f.

THEOREM **3.16** *If $f(x)$ is analytic at $x = c$, then*

$$f(x) = \sum_{0}^{\infty} \frac{f^{(n)}(c)}{n!}(x-c)^n$$

$$= f(c) + f'(c)(x-c) + \frac{f''(c)}{2!}(x-c)^2 + \cdots$$

where the power series converges on some interval of x values centered on $x = c$.

Proof Suppose that $f(x) = \sum_0^\infty a_n(x-c)^n$, convergent for all x with $|x-c| < R$, with $R > 0$. We can find the value of a_0 merely by setting $x = c$, for we then have

$$f(c) = a_0 + a_1(c-c) + a_2(c-c)^2 + \cdots$$
$$= a_0$$

If we do the same with the series obtained by differentiating $f(x)$, we have first (as in Theorem 3.15)

$$f'(x) = a_1 + 2a_2(x-c) + 3a_3(x-c)^2 + \cdots$$

and then

$$f'(c) = a_1 + 2a_2(c-c) + 3a_3(c-c)^2 + \cdots$$
$$= a_1$$

At the next stage, we find $f''(c) = 2a_2$ then $f^{(3)}(c) = 6a_3$ and in general, $f^{(n)}(c) = (n!)\,a_n$, which gives the general formula for a_n.

In particular, an analytic function can have only one power-series development around a given point $x = c$, since the coefficients can be calculated uniquely from the function itself. This brings us to an important fact about equality of power series.

COROLLARY *If* $\sum_0^\infty a_n(x-c)^n = \sum_0^\infty b_n(x-c)^n$, *with both series converging for x in some neighborhood of* $x = c$, *then* $a_n = b_n$ *for all values of n.*

This simple result turns out to be the key to the use of power series in the solution of differential equations.

Many of the standard functions that are met in practice are analytic at all values of x except those where the formula defining the function clearly breaks down. For example, e^x is analytic for every value of x, and $e^x/(1-x)$ is analytic at all values of x except $x = 1$. The function $f(x) = e^{\sec x}$ is analytic at all values of x except those where $\cos x = 0$, namely, $x = \pm\frac{1}{2}\pi, \pm\frac{3}{2}\pi, \pm\frac{5}{2}\pi, \ldots$.

Theorem 3.16 can be used directly to find power-series representations for analytic functions, if it is not too difficult to calculate the higher derivatives of the function. If $f(x) = e^x$, then $f^{(n)}(x) = e^x$ for all n, and $f^{(n)}(0) = 1$. Hence,

$$\exp(x) = e^x = 1 + x + \frac{x^2}{2!} + \frac{x^3}{3!} + \cdots + \frac{x^n}{n!} + \cdots \tag{3.11}$$

Actually, in more advanced courses where the emphasis is upon a consistent rigorous development of analysis, this series is taken as the basic definition of the exponential function, and all of its properties can be deduced from this series formula.

If we assume that we know that the sine function is analytic everywhere, then we can find its power-series expansion. Putting $g(x) = \sin x$, we have $g(0) = 0$, $g'(0) = 1$, $g''(0) = 0$, $g^{(3)}(0) = -1$, etc., leading to

$$\sin x = x - \frac{x^3}{3!} + \frac{x^5}{5!} - \frac{x^7}{7!} + \cdots \tag{3.12}$$

(This too can be taken as the basic definition of sine, and all the remaining properties can be obtained from it.)

In many ways, power series can be treated as though they were merely generalized polynomials. If we are careful to use only those values of x for which the series are simultaneously convergent, then power series can be added, subtracted, and multiplied just as one would add, subtract, and multiply polynomials, collecting like powers of x together. (Division is more complicated.) We can also make simple substitutions in power series if again we are careful to keep all the variables inside the intervals where the power series converge. For example, from the series (3.11) for e^x, we can find a series for e^{-x} by replacing x by $-x$,

$$\begin{aligned} e^{-x} &= 1 + (-x) + \frac{(-x)^2}{2!} + \frac{(-x)^3}{3!} + \cdots \\ &= 1 - x + \frac{x^2}{2!} - \frac{x^3}{3!} + \frac{x^4}{4!} - \cdots \end{aligned}$$

Hence, adding the series for e^x and e^{-x}, and multiplying by $\frac{1}{2}$, we arrive at

$$\begin{aligned} \cosh x &= \frac{e^x + e^{-x}}{2} = 1 + \frac{x^2}{2!} + \frac{x^4}{4!} + \frac{x^6}{6!} + \cdots \\ &= \sum_0^\infty \frac{x^{2n}}{(2n)!} \end{aligned} \tag{3.13}$$

Again we can find a power series for $(1+x)/(1-x)$:

$$\begin{aligned} \frac{1+x}{1-x} &= (1+x)\frac{1}{1-x} = (1+x)(1+x+x^2+x^3+x^4+\cdots) \\ &= (1+x+x^2+x^3+\cdots) + (x+x^2+x^3+x^4+\cdots) \\ &= 1 + 2x + 2x^2 + 2x^3 + 2x^4 + 2x^5 + \cdots \end{aligned}$$

As a further example, let us find a power series for the function $e^x/(1-x)$ about $x = 0$ by combining a series expansion for e^x and one for $1/(1-x)$.

We use (3.11) and (3.10).

$$e^x = 1 + x + \tfrac{1}{2}x^2 + \tfrac{1}{6}x^3 + \tfrac{1}{24}x^4 + \tfrac{1}{120}x^5 + \cdots$$

$$\frac{1}{1-x} = 1 + x + x^2 + x^3 + x^4 + x^5 + x^6 + \cdots$$

We multiply these together, collecting terms that involve the same powers of x.

$$\frac{e^x}{1-x} = 1 + (1+1)x + (1+1+\tfrac{1}{2})x^2 + (1+1+\tfrac{1}{2}+\tfrac{1}{6})x^3$$

$$+ (1+1+\tfrac{1}{2}+\tfrac{1}{6}+\tfrac{1}{24})x^4 + \cdots$$

$$= 1 + 2x + \tfrac{5}{2}x^2 + \tfrac{8}{3}x^3 + \tfrac{65}{24}x^4 + \tfrac{321}{120}x^5 + \cdots$$

The same series could have been found by obtaining the successive derivatives of the function $e^x/(1-x)$, but this would have been much more work.

Here are several more illustrations to show how one can manipulate power series. Suppose we want a series for the function $1/(5-3x)$ that converges near $x = 0$.

$$\frac{1}{5-3x} = \frac{1}{5}\,\frac{1}{1-\frac{3}{5}x} = \frac{1}{5}\,\frac{1}{1-t}$$

$$= \tfrac{1}{5}(1+t+t^2+t^3+t^4+\cdots)$$

$$= \tfrac{1}{5}[1+\tfrac{3}{5}x+(\tfrac{3}{5}x)^2+(\tfrac{3}{5}x)^3+\cdots]$$

$$= \tfrac{1}{5} + \tfrac{3}{25}x + \tfrac{9}{125}x^2 + \tfrac{27}{625}x^3 + \cdots$$

What about convergence of this series? It is clear that the series in powers of t converged for $|t| < 1$, and since we have $t = \frac{3}{5}x$, the final series converges for all x with $|\frac{3}{5}x| < 1$, which is equivalent first to $|3x| < 5$ and then to $|x| < \frac{5}{3}$. Thus, the series we obtained converges for all x with $-\frac{5}{3} < x < \frac{5}{3}$. This can also be shown by applying the standard ratio test to the series itself.

Again, suppose we wanted to find a series for the same function to express it as a power series in $x-1$, convergent on some neighborhood of $x = 1$. Put $t = x-1$ so that $x = t+1$. Then,

$$\frac{1}{5-3x} = \frac{1}{5-3(t+1)} = \frac{1}{2-3t}$$

Using the same technique as that explained for the previous example, we have

$$\frac{1}{2-3t} = \tfrac{1}{2}[1+\tfrac{3}{2}t+(\tfrac{3}{2}t)^2+\cdots]$$

$$= \tfrac{1}{2} + \tfrac{3}{4}t + \tfrac{9}{8}t^2 + \cdots$$

Replacing t by $x-1$, we obtain the desired series

$$\frac{1}{5-3x} = \tfrac{1}{2} + \tfrac{3}{4}(x-1) + \tfrac{9}{8}(x-1)^2 + \tfrac{27}{16}(x-1)^3 + \cdots$$

This series converges for all x with $\frac{1}{3} < x < \frac{5}{3}$. Notice that you cannot expect it to converge on an interval that contains the point $x = \frac{5}{3}$, for the function fails to be defined (indeed, it becomes infinite or "blows up") at this value.

It is also permissible to integrate power series termwise, and thus obtain new identities, provided that you stay within the interval of convergence and use definite rather than indefinite integrals. The following illustrates how this can be useful.

Start from the series (3.10) for $1/(1-t)$ and then replace t by $-t^2$, obtaining

$$\frac{1}{1+t^2} = 1 - t^2 + t^4 - t^6 + t^8 - t^{10} + \cdots$$

Then, integrate this between zero and x (where we remember to keep x between -1 and 1 to stay within the interval of convergence).

$$\int_0^x \frac{dt}{1+t^2} = \int_0^x (1-t^2+t^4-t^6+t^8-\cdots)\,dt$$

$$\arctan x = x - \tfrac{1}{3}x^3 + \tfrac{1}{5}x^5 - \tfrac{1}{7}x^7 + \tfrac{1}{9}x^9 - \cdots$$

The resulting series in fact converges for $|x| \leq 1$, and leads to the interesting identity

$$\tfrac{1}{2}\pi = 1 - \tfrac{1}{3} + \tfrac{1}{5} - \tfrac{1}{7} + \tfrac{1}{9} - \tfrac{1}{11} + \cdots$$

Lest the reader get the impression from these examples that in working with series *anything* that seems sensible is correct, we append two instructive illustrations to disprove this. Consider first the series

$$A(x) = 2x + x(1-x) + x(1-x)^2 + x(1-x)^3 + \cdots$$

and suppose that we want to find the limiting value of $A(x)$ as x approaches zero. It is clear that the general term $x(1-x)^n$ itself tends to zero as x approaches zero, so it would seem plausible that we have

$$\begin{aligned}\lim_{x\to 0} A(x) &= 0 + 0 + 0 + 0 + \cdots \\ &= 0\end{aligned}$$

However, this is quite incorrect, and in fact $\lim_{x\to 0} A(x) = 1$. The flaw lies in treating infinite sums as though they were finite sums. A correct treatment starts by finding the sum of the series, and thus calculating $A(x)$ directly.

$$\begin{aligned} A(x) &= x + x[1 + (1-x) + (1-x)^2 + (1-x)^3 + \cdots] \\ &= x + x\left[\frac{1}{1-(1-x)}\right] \\ &= x + 1 \end{aligned}$$

holding for all x with $0 < x < 2$. Letting $x \downarrow 0$ we have $\lim_{x\downarrow 0} A(x) = 1$.

Here is another example that illustrates what may seem to be even a greater paradox. Let us look at the following series.

$$\begin{aligned} B(x) &= x + x^2 + (2x-1)x^2 + (3x-2)x^3 + (4x-3)x^4 \\ &\quad + \cdots + ([n+1]x - n)x^{n+1} + \cdots \end{aligned}$$

What is the behavior of $B(x)$ as x approaches 1? Looking at the behavior of each term separately, it would seem plausible that we must have

$$\begin{aligned} \lim_{x\to 1} B(x) &= 1 + 1 + 1 + 1 + 1 + \cdots \\ &= \infty \end{aligned}$$

However, this tempting answer is again wrong. This time the correct answer is $\lim_{x\to 1} B(x) = 1$! To see this, we again calculate $B(x)$ exactly by summing the series when $|x| < 1$.

$$\begin{aligned} B(x) &= x + x^2 + (2x^3 - x^2) + (3x^4 - 2x^3) + (4x^5 - 3x^4) + \cdots \\ &= x + (x^2 - x^2) + (2x^3 - 2x^3) + (3x^4 - 3x^4) + \cdots \\ &= x \end{aligned}$$

[This is a series that is often called a **telescoping** series, since it can be summed by canceling terms and thus finding a simple formula for the partial sums of the series. See Exercise 3.1 for another example.] It should be observed that neither of the series used in these two illustrations was a power series.

Paradoxes can also arise that involve merely sums of numbers, if infinite series are involved. Recall first the standard "bookkeeper's check" in accounting. If you have a collection of sums such as the following

20.35	31.18	23.86
7.18	19.35	47.07
3.26	9.57	13.29
30.79	60.10	84.22

then a check on the accuracy of the addition is to add across the rows and then compare the total of the row sums with the total of the column sums; these final totals must agree, since each is the sum of all the entries in the table. In this example, we would arrive at the following:

				Row sums
20.35	31.18	23.86		75.39
7.18	19.35	47.07		73.60
3.26	9.57	13.29		26.12
30.79 +	60.10 +	84.22 =	175.11 =	175.11

However, this breaks down when we try it on infinite arrays. Table 3.1 shows an infinite array where the total of the row sums and the total of the column sums are different!

Table 3.1

							Row sums
5	−2	0	0	0	0	⋯	3
−3	5	−2	0	0	0	⋯	0
0	−3	5	−2	0	0	⋯	0
0	0	−3	5	−2	0	⋯	0
0	0	0	−3	5	−2	⋯	0
0	0	0	0	−3	5	⋯	0
⋮	⋮	⋮	⋮	⋮	⋮	⋮	⋮
2 +	0 +	0 +	0 +	0 +	0 +	⋯ = 2 ≠	3

Column sums

This section has been intended as a brief summary of certain basic techniques in the use of infinite series, and not as a comprehensive treatment of the theory, which is better left to other texts. In particular, many refinements and much of the more subtle theory have been omitted. These may be found in any text in advanced calculus or analysis, such as the following:

R. C. Buck and E. F. Buck, *Advanced Calculus*, 2nd ed. (McGraw-Hill, New York, 1965).

W. Kaplan, *Advanced Calculus*, 2nd ed. (Addison-Wesley, Reading, Mass., 1973).

Tom Apostol, *Mathematical Analysis*, 2nd ed. (Addison-Wesley, Reading, Mass., 1974).

Exercises

Discuss the convergence or divergence of each of the series in Exercises 1–9.

1. $\frac{3}{10} + \frac{4}{20} + \frac{5}{30} + \frac{6}{40} + \cdots$

2. (a) $\frac{3}{4} + \frac{5}{8} + \frac{7}{16} + \frac{9}{32} + \frac{11}{64} + \frac{13}{128} + \cdots$

   (b) $\frac{3}{4} + \frac{5}{9} + \frac{7}{16} + \frac{9}{25} + \frac{11}{36} + \frac{13}{49} + \frac{15}{64} + \cdots$

3. $\dfrac{1}{6} + \dfrac{\sqrt{2}}{12} + \dfrac{\sqrt{3}}{20} + \dfrac{\sqrt{4}}{30} + \dfrac{\sqrt{5}}{42} + \dfrac{\sqrt{6}}{56} + \cdots$

4. $\displaystyle\sum_{1}^{\infty} \frac{6n}{1+n^3}$

5. $\displaystyle\sum_{1}^{\infty} \frac{7+5n+n^2}{1+n^2+n^4}$

6. $\displaystyle\sum_{1}^{\infty} \frac{2+5n}{3+4n+5n^2}$

7. $\displaystyle\sum_{1}^{\infty} \frac{(-1)^n}{1+3n}$

8. $\frac{1}{2} - \frac{3}{4} + \frac{1}{8} - \frac{3}{16} + \frac{1}{32} - \frac{3}{64} + \cdots$

9. $\frac{1}{3} - \frac{2}{5} + \frac{3}{7} - \frac{4}{9} + \frac{5}{11} - \frac{6}{13} + \cdots$

For which values of x do the series in Exercises 10–19 converge?

10. (a) $\displaystyle\sum_{1}^{\infty} \frac{x^n}{n^2}$

    (b) $\displaystyle\sum_{1}^{\infty} \frac{x^{n-1}}{n}$

    (c) $\dfrac{1}{2} + \dfrac{2}{3}x + \dfrac{3}{4}x^2 + \dfrac{4}{5}x^3 + \cdots + \dfrac{n-1}{n}x^{n-2} + \cdots$

11. $\displaystyle\sum_{1}^{\infty} \frac{x^n}{n+n^2}$

12. $2 + 3x + 5x^2 + 9x^3 + 17x^4 + 33x^5 + 65x^6 + 129x^7 + \cdots$

13. $\displaystyle\sum_{1}^{\infty} n(x+5)^n$

14. $\displaystyle\sum_{1}^{\infty} \frac{2^n(x-1)^n}{n^2}$

15. $\displaystyle\sum_{0}^{\infty} \frac{(3x+1)^n}{1+2^n}$

16. $1 + x + 2x^2 + 6x^3 + 24x^4 + \cdots + (n!)\,x^n + \cdots$

17. $\displaystyle\sum_{1}^{\infty} \frac{(x+3)^n}{(2n)!}$

18. $1 - 4x + 9x^2 - 16x^3 + 25x^4 - 36x^5 + 49x^6 - \cdots$

*19. $\frac{1}{2}+\left(\frac{1\cdot 3}{2\cdot 4}\right)x+\left(\frac{1\cdot 3\cdot 5}{2\cdot 4\cdot 6}\right)x^2+\left(\frac{1\cdot 3\cdot 5\cdot 7}{2\cdot 4\cdot 6\cdot 8}\right)x^3+\cdots$

20. Show that the series for e^x is correct by proving that the function defined by $F(x)=\sum_0^\infty x^n/n!$ obeys the differential equation $F'(x)=F(x)$, and has $F(0)=1$.

21. (a) Find the Taylor series for $\cos x$, with $c=0$.
(b) Do the same as in part (a), but with $c=\frac{1}{2}\pi$.

22. By any method, find a power series for $1/(2x+1)$ of the form $\sum_0^\infty a_n x^n$, and test it for convergence.

23. By any method, find a series for $(x+x^2)/(x-2)$ of the form $\sum a_n x^n$.

24. Find a series for $1/(1-x^2)$ by multiplying the series for $1/(1+x)$ and $1/(1-x)$.

25. Check that the square of e^x is e^{2x} by squaring the series for e^x.

26. Show that the initial terms in the power series about $x=0$ for the function $e^{-x}/(1-x^2)$ are

$$1-x+\tfrac{3}{2}x^2-\tfrac{7}{6}x^3+\tfrac{37}{24}x^4-\tfrac{141}{120}x^5+\cdots$$

27. Find a formula for the sum of the power series

$$x+x^4+x^7+x^{10}+x^{13}+\cdots$$

28. Use a power series to find the numerical sum of the series

$$1+\tfrac{2}{3}+\tfrac{3}{9}+\tfrac{4}{27}+\tfrac{5}{81}+\tfrac{6}{243}+\cdots$$

29. Use the series expansion of $1/(1+x)$ to obtain the formula

$$\log(1+x)=x-\tfrac{1}{2}x^2+\tfrac{1}{3}x^3-\tfrac{1}{4}x^4+\tfrac{1}{5}x^5-\cdots$$

30. Use the fact that $\log(A/B)=\log A-\log B$ to find a series expansion for $\log[(1-x)/(1+x)]$ of the form $\sum_0^\infty a_n x^n$. Test your series for convergence.

31. (a) Find the sum of the special series

$$\frac{1}{2}+\frac{1}{6}+\frac{1}{12}+\frac{1}{20}+\frac{1}{30}+\cdots+\frac{1}{n(n+1)}+\cdots$$

by writing $\frac{1}{2}$ as $1-\frac{1}{2}$, $\frac{1}{6}$ as $\frac{1}{2}-\frac{1}{3}$, $\frac{1}{12}$ as $\frac{1}{3}-\frac{1}{4}$, etc.
(b) Apply the same idea as in part (a) to find the sum of the special (telescoping) series

$$\tfrac{1}{3}+\tfrac{1}{15}+\tfrac{1}{35}+\tfrac{1}{63}+\tfrac{1}{99}+\cdots$$

32. (a) Formulate a definition for an infinite product

$$(a_1)(a_2)(a_3)(a_4)(a_5)\cdots$$

(b) Show that

$$\left(\frac{1\cdot 3}{4}\right)\left(\frac{2\cdot 4}{9}\right)\left(\frac{3\cdot 5}{16}\right)\left(\frac{4\cdot 6}{25}\right)\cdots=\frac{1}{2}$$

3.3 Complex-Valued Series and Functions

Many of the most useful techniques involved in the solution of differential equations depend upon being able to use both real numbers and complex numbers for the values of various constants that enter into the solutions. This section is a brief review of the needed information.

A student's first introduction to complex numbers has usually been given by a teacher who said something like this: We know that x^2 can never be negative—the square of a negative number is positive and the square of a positive is positive. So, the equation $x^2 = -1$ doesn't have any solutions. Let us invent a number i with the property that $i^2 = -1$. Then, numbers like $a+bi$, where a and b are ordinary numbers, will be called *complex numbers*. The laws for working with these new numbers are the same ones that we already know from algebra; for example, $(2-3i)(4+i) = 8-12i+2i-3i^2 = 11-10i$. The **absolute value** of a complex number is defined to be $|a+bi| = \sqrt{a^2+b^2}$.

This approach, while aping the historical process, leaves unanswered many reasonable questions. What is i, really? What justification is there for creating something like this? How do we know it obeys all the rules of arithmetic?

The more modern approach, by now quite common, was introduced by Hamilton in about 1840. The field of complex numbers can be taken to be the ordinary two-dimensional plane, all points of which are given as pairs (a, b) with a special definition of the operations of addition and multiplication. The sum of two points P and Q is defined in the usual vector way: If $P = (a, b)$ and $Q = (c, d)$, then $P+Q = (a+c, b+d)$. The product of P and Q is the point whose coordinates are $(ac-bd, ad+bc)$. In this context we observe that $|a+bi| = \sqrt{a^2+b^2}$ is the distance from P to the origin. Using these definitions, it is merely a tedious calculation to verify that all the rules of arithmetic that describe a field are satisfied. A crucial step is to verify that if (a, b) is not the origin $(0, 0)$, then there is a unique point (x, y) such that $(a, b)(x, y) = (c, d)$ for any choice of (c, d).

Using the definition of multiplication, one must solve the equations

$$ax - by = c$$
$$bx + ay = d$$

obtaining $x = (ac+bd)/(a^2+b^2)$, $y = (ad-bc)/(a^2+b^2)$, which are defined, since $a^2+b^2 \neq 0$.

With this approach, a correct answer to the question "What is i?" is given by "$i = (0, 1)$." In passing it should be mentioned that other models for the complex numbers can also be given. Even though they look entirely different, all the models are isomorphic—meaning that they can be put in a one-to-one structure-preserving correspondence with the points of the plane. One such additional model consists of all the 2-by-2 matrices of the form

$$\begin{bmatrix} a & -b \\ b & a \end{bmatrix}$$

where a and b are ordinary (real) numbers.

From a naive viewpoint, the "number" i was created to enable us to solve the equation $x^2+1 = 0$. The complex numbers which resulted then allowed us to solve any quadratic equation of the form $ax^2+bx+c = 0$ where the coefficients a, b, c are real numbers. One might have expected that a new collection of numbers would have to be invented to solve an equation like $x^2 = i$; however, this is solved by

$$x = \pm[(1/\sqrt{2})+(1/\sqrt{2})i]$$

It was a pleasant surprise to learn that indeed one did not have to go outside the complex numbers to solve *any* quadratic equation with arbitrary complex coefficients, such as $x^2 + (3-7i)x + (4+5i) = 0$. But it was astounding to discover that this extension of the number system made it possible to find solutions of every polynomial equation of any degree. This result, called by tradition the "fundamental theorem of algebra"—which it certainly is not—does not yet have a proof that is sufficiently easy to present in an elementary class.

Another simple algebraic aspect of the field of complex numbers is the fact that it has a self-isomorphism, the mapping that sends the complex number $a+bi$ into its **conjugate**, the complex number $a-bi$. Thus, the conjugate of $-2+4i$ is $-2-4i$ and the conjugate of $3-7i$ is $3+7i$; of course, the conjugate of the real number 5 is 5, since $5 = 5+0i$. If z is any complex number, its conjugate is usually denoted by $\bar{z}$.

The reason for the importance of this concept is contained in the word "isomorphism," which means that the mapping preserves the algebraic structure of the field of complex numbers. If we write $\bar{z} = \mathbf{h}(z)$, then it can be checked easily that $\mathbf{h}$ satisfies the identities

$$\mathbf{h}(\alpha-\beta) = \mathbf{h}(\alpha) - \mathbf{h}(\beta)$$

$$\mathbf{h}(\alpha\beta) = \mathbf{h}(\alpha)\mathbf{h}(\beta)$$

$$\mathbf{h}\left(\frac{1}{\gamma}\right) = \frac{1}{\mathbf{h}(\gamma)}$$

for any complex numbers α, β, γ with $\gamma \neq 0$.

There is one immediate consequence of this property that is very important in the solution of polynomial equations. If $P(x) = a_0 x^n + a_1 x^{n-1} + \cdots + a_n$ is a polynomial with *all of its coefficients real*, and if γ is any root of $P(x) = 0$, then so is the complex conjugate $\bar{\gamma} = \mathbf{h}(\gamma)$. The key is the fact that if c is real, then $\mathbf{h}(c) = \bar{c} = c$. Hence, if $P(\gamma) = 0$, then $0 = \mathbf{h}(P(\gamma)) = P(\mathbf{h}(\gamma)) = P(\bar{\gamma}) = 0$, so $\bar{\gamma}$ is also a root. Note also that this property fails for polynomials with complex coefficients.

It is customary to concentrate on real-valued functions in developing the elements of calculus. However, it takes very little work to extend everything needed to make it possible to deal with functions that yield complex values instead. For example, the function defined by

$$f(x) = (2-3i)x^2 + 5ix + (7-i) \tag{3.14}$$

is a well-defined function, defined for every (real) value of x, and the values it yields are complex numbers. We note that we could rewrite this formula as

$$f(x) = 2x^2 + 7 + i(-3x^2+5x-1) \tag{3.15}$$

In the same way, any complex-valued function of one variable can be written in the form

$$f(x) = g_1(x) + ig_2(x)$$

where g_1 and g_2 are ordinary real-valued functions of x.

Differentiation and integration can be carried out as usual. For example, we have either

$$f'(x) = 2(2-3i)x + 5i$$

using the equation (3.14), or

$$f'(x) = g_1'(x) + ig_2'(x) = 4x + 0 + i(-6x+5)$$

using formula (3.15).

Integration can be carried out the same way, either leaving the complex-valued function as it is, or separating it into real and imaginary parts. Thus, we have

$$\int_{-1}^{2} [(2-3i)x^2 + 5ix + (7-i)]\,dx$$

$$= \left[(2-3i)\frac{x^3}{3} + 5i\frac{x^2}{2} + (7-i)x\right]\Bigg|_{-1}^{2}$$

$$= (2-3i)\left(\frac{8}{3} - \frac{-1}{3}\right) + 5i\left(\frac{4}{2} - \frac{1}{2}\right) + (7-i)[2-(-1)]$$

$$= 6 - 9i + \frac{15}{2}i + 21 - 3i$$

$$= 27 - \frac{9}{2}i$$

It is also useful to develop a theory to handle infinite series whose terms are complex numbers. Very little of the theory of series of real numbers need be modified. If α_n is the complex number $a_n + ib_n$, then the series

$$\sum_1^\infty \alpha_n = \sum_1^\infty (a_n + ib_n) = (a_1 + ib_1) + (a_2 + ib_2) + \cdots$$

is convergent if and only if the two series $\sum_1^\infty a_n$ and $\sum_1^\infty b_n$ are convergent, and then one has

$$\sum_1^\infty \alpha_n = \sum_1^\infty a_n + i\sum_1^\infty b_n$$

The definition of the general exponential function is a very important special application of the above. If z is any complex number, then

$$\exp(z) = 1 + z + \frac{z^2}{2!} + \frac{z^3}{3!} + \frac{z^4}{4!} + \cdots \tag{3.16}$$

A special identity, originally discovered by De Moivre and Cotes about 1710 but usually ascribed to Euler, can be obtained from (3.16) by setting $z = i\theta$. Observe that $(i\theta)^n$ is always either θ^n, $i\theta^n$, $-\theta^n$, or $-i\theta^n$, depending upon the remainder after dividing n by 4; then

$$\exp(i\theta) = 1 + i\theta + \frac{(i\theta)^2}{2!} + \frac{(i\theta)^3}{3!} + \frac{(i\theta)^4}{4!} + \frac{(i\theta)^5}{5!} + \cdots$$

$$\begin{aligned} &= 1 + i\theta - \frac{\theta^2}{2!} - i\frac{\theta^3}{3!} + \frac{\theta^4}{4!} + i\frac{\theta^5}{5!} - \frac{\theta^6}{6!} - \cdots \\ &= \left(1 - \frac{\theta^2}{2!} + \frac{\theta^4}{4!} - \frac{\theta^6}{6!} + \cdots\right) + i\left(\theta - \frac{\theta^3}{3!} + \frac{\theta^5}{5!} - \frac{\theta^7}{7!} + \cdots\right) \end{aligned} \tag{3.17}$$

$$= \cos\theta + i\sin\theta$$

where we have replaced the well-known power series (3.11), (3.12), and the solution series to Exercise 21 of Section 3.2 by their sums. If we now replace θ by $-\theta$, we obtain

$$\begin{aligned} \exp(-i\theta) &= \cos(-\theta) + i\sin(-\theta) \\ &= \cos\theta - i\sin\theta \end{aligned} \tag{3.18}$$

One at once obtains the Euler identities

$$\begin{aligned} \cos\theta &= \frac{\exp(i\theta)+\exp(-i\theta)}{2} \\ \sin\theta &= \frac{\exp(i\theta)-\exp(-i\theta)}{2i} \end{aligned} \tag{3.19}$$

Another interesting identity can be obtained from the series (3.16) for the exponential function. Suppose we calculate the product $\exp(u)\exp(v)$, using the series for each. Thus,

$$\exp(u) = 1 + u + \frac{u^2}{2!} + \frac{u^3}{3!} + \frac{u^4}{4!} + \frac{u^5}{5!} + \cdots$$

$$\exp(v) = 1 + v + \frac{v^2}{2!} + \frac{v^3}{3!} + \frac{v^4}{4!} + \frac{v^5}{5!} + \cdots$$

$$\begin{aligned} \exp(u)\exp(v) &= 1 + (u+v) + \left(\frac{u^2}{2!} + uv + \frac{v^2}{2!}\right) + \left(\frac{u^3}{3!} + \frac{u^2v}{2!} + \frac{uv^2}{2!} + \frac{v^3}{3!}\right) \\ &\quad + \left(\frac{u^4}{4!} + \frac{u^3v}{3!\,1!} + \frac{u^2v^2}{2!\,2!} + \frac{uv^3}{1!\,3!} + \frac{v^4}{4!}\right) + \cdots \\ &= 1 + (u+v) + \frac{1}{2!}(u^2+2uv+v^2) \\ &\quad + \frac{1}{3!}(u^3+3u^2v+3uv^2+v^3) \\ &\quad + \frac{1}{4!}(u^4+4u^3v+6u^2v^2+4uv^3+v^4) + \cdots \\ &= 1 + (u+v) + \frac{(u+v)^2}{2!} + \frac{(u+v)^3}{3!} + \frac{(u+v)^4}{4!} + \cdots \\ &= \exp(u+v) \end{aligned}$$

It is this formula, $\exp(u)\exp(v) = \exp(u+v)$, which corresponds to the law of exponents and which justifies our adoption of the notation e^z for $\exp(z)$.

The identity just proved thus becomes

$$e^u e^v = e^{u+v} \tag{3.20}$$

holding now for any real or complex numbers u and v.

An interesting special case casts light on the mathematical origins of the trigonometric addition formulas. We have $e^{i\alpha}e^{i\beta} = e^{i(\alpha+\beta)}$. Rewritten in the De Moivre form, it becomes

$$(\cos\alpha + i\sin\alpha)(\cos\beta + i\sin\beta) = \cos(\alpha+\beta) + i\sin(\alpha+\beta)$$

Calculating the product on the left, we have

$$\cos\alpha\cos\beta - \sin\alpha\sin\beta + i(\sin\alpha\cos\beta + \cos\alpha\sin\beta)$$
$$= \cos(\alpha+\beta) + i\sin(\alpha+\beta)$$

from which we deduce at once that

$$\begin{aligned} \cos(\alpha+\beta) &= \cos\alpha\cos\beta - \sin\alpha\sin\beta \\ \sin(\alpha+\beta) &= \sin\alpha\cos\beta + \cos\alpha\sin\beta \end{aligned} \tag{3.21}$$

Exercises

1. Show that one value for $\sqrt{3+4i}$ is $2+i$.
2. Show that one value of $\sqrt{9-12i}$ is $\sqrt{12}-\sqrt{3}\,i$.
3. Find a value for $\sqrt{1+3i}$.
4. Solve the equation $x^2+(3-i)x-(5i+10) = 0$.
5. Solve $x^2+3x-10i = 0$.
6. Let $f(x) = (3x+7i)(x^2-5ix+1)$. Find $f'(x)$ and $f''(x)$.
7. Evaluate $\int_0^1 (2x-i)(x+i)\,dx$.
8. Evaluate $\int_0^1 (t-i)(t+i)\,dt$.
9. Evaluate $\int_0^{\pi/2} [(\theta-i)\sin\theta + 2i\cos\theta]\,d\theta$.
10. Verify that $|e^{i\theta}| = 1$ for all real θ.
11. Show that $|e^{iy+x}| = e^x$.

12. Let $\omega = -\frac{1}{2}+\frac{1}{2}\sqrt{3}\, i$. Verify that $\omega^3 = 1$ and that $1+\omega+\omega^2 = 0$.

13. With ω as defined in Exercise 12, show that

$$\frac{e^x+e^{\omega x}+e^{\omega^2 x}}{3} = 1+\frac{x^3}{3!}+\frac{x^6}{6!}+\frac{x^9}{9!}+\frac{x^{12}}{12!}+\cdots$$

14. With ω as defined in Exercise 12, show that

$$\frac{1}{3}\left(\frac{1}{1-x}+\frac{1}{1-\omega x}+\frac{1}{1-\omega^2 x}\right) = 1+x^3+x^6+x^9+\cdots$$

15. Show that $\sin(x+iy) = \sin x \cosh y + i \cos x \sinh y$.

16. Use (3.19) to verify that $(\sin\theta)^2+(\cos\theta)^2 = 1$.

*17. Can you find an argument to justify the equation

$$\arctan x = \frac{1}{2i}\log\left(\frac{1+ix}{1-ix}\right)$$

18. Show that any function of the form $A\sin\theta + B\cos\theta$ can be written as $C\sin(\theta+\alpha)$ for appropriate choices of C and α.

3.4 Polynomial Approximation

The idea behind the method of polynomial approximation is quite simple. Suppose we are interested in solving the equation $dy/dt = f(t, y)$ with initial condition $y(t_0) = y_0$. We already know something about the solution from the equation itself, namely, that $y'(t) = f(t, y(t))$, and in particular that $y'(t_0) = f(t_0, y_0)$. Letting $m = f(t_0, y_0)$, this tells us that the solution curve must go through the point (t_0, y_0) with slope m, and therefore that the straight line

$$y = y_0 + m(t-t_0)$$

is tangent to the solution there. (See Figure 3.3.)

An obvious generalization of this is to find the best-fitting parabola or cubic polynomial, defined as the one which best matches the true curve $y = y(t)$ at the point (t_0, y_0). Such polynomials are easily found if we know the higher derivatives of $y(t)$, since they are merely the partial sums of the Taylor Series for $y(t)$, namely,

$$P_n(t) = y(t_0) + y'(t_0)(t-t_0) + y''(t_0)(t-t_0)^2/2! + \cdots + y^{(n)}(t_0)(t-t_0)^n/n! \tag{3.22}$$

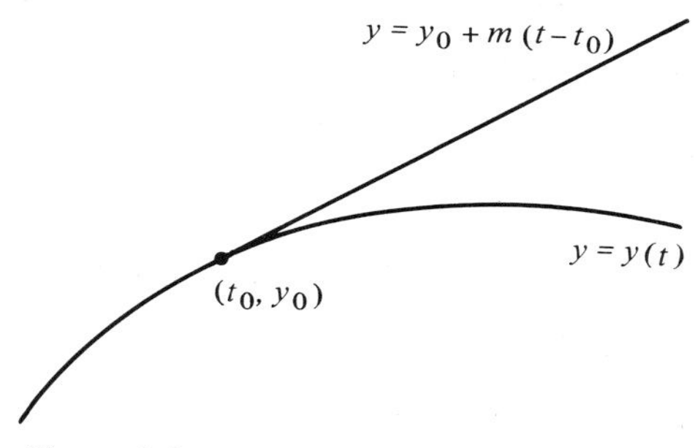

Figure 3.3

The hope is that such a polynomial will approximate the value of $y(t)$ for some range of values of t extending away from $t = t_0$ in a much more accurate way than the mere tangent line did.

To see how this works in practice, let us start by trying it on an equation whose solution we can find by other methods, so that we can check the accuracy. Consider the following equation

$$y' = -2ty, \qquad y(0) = 1 \tag{3.23}$$

From the equation, we immediately have $y'(0) = -2(0)(1) = 0$. To find y'', we differentiate both sides of (3.23)

$$y'' = -2y - 2ty' \tag{3.24}$$

and then substituting the values $t = 0$, $y = 1$, $y' = 0$, we arrive at $y''(0) = -2$. Differentiate (3.24) again and have

$$\begin{aligned} y^{(3)} &= -2y' - 2y' - 2ty'' \\ &= -4y' - 2ty'' \end{aligned} \tag{3.25}$$

Substituting $t = 0$, $y' = 0$, $y'' = -2$, we have $y^{(3)}(0) = 0$. The next several derivatives of (3.25) are as follows.

$$\begin{aligned} y^{(4)} &= -6y'' - 2ty^{(3)} \\ y^{(5)} &= -8y^{(3)} - 2ty^{(4)} \\ y^{(6)} &= -10y^{(4)} - 2ty^{(5)} \end{aligned} \tag{3.26}$$

and the corresponding values of $y^{(k)}(0)$ are

$$y^{(4)}(0) = 12 \qquad y^{(5)}(0) = 0 \qquad y^{(6)}(0) = 120$$

The corresponding polynomials $P_n(t)$ are, according to (3.22),

$$
\begin{aligned}
P_1(t) &= 1 + (0)t = 1 \\
P_2(t) &= 1 + (0)t + (-2)t^2/2! \\
&= 1 - t^2 \\
P_3(t) &= 1 + (0)t + (-2)t^2/2! + (0)t^3/3! \\
&= 1 - t^2 \\
P_4(t) &= 1 - t^2 + (12)t^4/4! \\
&= 1 - t^2 + \tfrac{1}{2}t^4 \\
P_5(t) &= 1 - t^2 + \tfrac{1}{2}t^4 \\
P_6(t) &= 1 - t^2 + \tfrac{1}{2}t^4 - \tfrac{1}{6}t^6
\end{aligned}
$$

The exact solution of the differential equation is easily seen to be $y = e^{-t^2}$. In Figure 3.4 we have shown a graph of this and contrasted it with the graphs of P_1, P_2, and P_4. As you see, there seems to be an improvement in the approximation near the initial point (0, 1) as the degree of the Taylor polynomial is increased, but none of these polynomials is a very good approximation for large values of t.

In this particularly simple differential equation, we can in fact find a formula for all the higher derivatives of $y(t)$ from the equation. We are able to observe from (3.26) the general recursion, which has the form

$$y^{(n+1)} = -2ny^{(n-1)} - 2ty^{(n)}$$

(Check this by differentiating it to see that the relationship persists.) Set

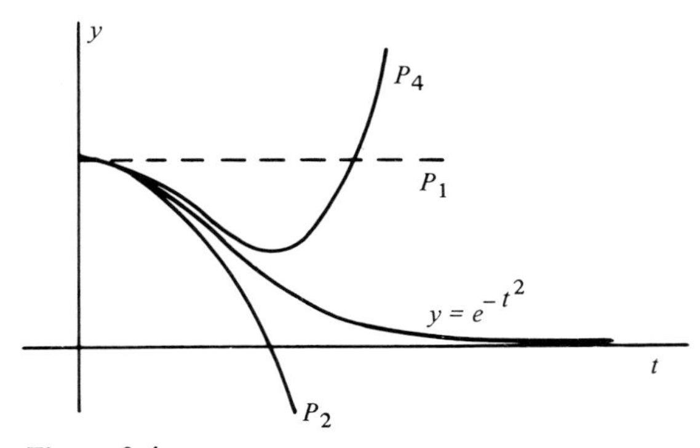

Figure 3.4

$t = 0$ and $Y_n = y^{(n)}(0)$, and we have the recursion

$$Y_{n+1} = -2nY_{n-1}, \qquad Y_0 = 1, \quad Y_1 = 0$$

The solution of this turns out to be

$$Y_{2k+1} = 0 \qquad \text{for } k = 0, 1, 2, \ldots$$
$$Y_{2k} = (-1)2^k[1 \cdot 3 \cdot 5 \cdot 7 \cdots (2k-1)]$$

If we use this to find the general polynomial $P_n(t)$, we can use the fact that $Y_{2k}/(2k)! = (-1)^k/k!$, and obtain the simple formula

$$P_{2n}(t) = 1 - t^2 + t^4/2! - t^6/3! + t^8/4! + \cdots + (-1)^n t^{2n}/n!$$

This is not at all surprising, since the Taylor series for $\exp(-t^2)$ around $t = 0$ is $\sum_0^\infty (-1)^n t^{2n}/n!$. Since this series converges for all t, we know that the polynomials $P_{2n}(t)$ which are its partial sums must converge to the solution $\exp(-t^2)$ at any chosen point t, as n increases. Thus, if we were to take n very large, we know that the approximating polynomial would in fact follow the true solution very closely for some distance before departing from it. No polynomial, of course, could stay close to $\exp(-t^2)$ all the way, since any polynomial that is not a constant must ultimately become large as t increases, while $\exp(-t^2)$ approaches zero as t increases.

The next example will show some of the drawbacks in this method. Let us take an equation which does not have a simple solution and see how the method applies. Consider the equation

$$y' = t^2 + y^2, \qquad t_0 = 0, \quad y(0) = 1 \tag{3.27}$$

If we calculate higher derivatives as before, we arrive at

$$\begin{aligned} y'' &= 2t + 2yy' \\ y^{(3)} &= 2 + 2(y')^2 + 2yy'' \\ y^{(4)} &= 6y'y'' + 2yy^{(3)} \\ y^{(5)} &= 6(y'')^2 + 8y'y^{(3)} + 2yy^{(4)} \\ y^{(6)} &= 20y''y^{(3)} + 10y'y^{(4)} + 2yy^{(5)} \\ y^{(7)} &= 20(y^{(3)})^2 + 30y''y^{(4)} + 12y'y^{(5)} + 2yy^{(6)} \end{aligned} \tag{3.28}$$

It does not seem feasible to find a simple formula for $y^{(n)}$ as in the previous case. However, we can substitute $t = 0$, $y(0) = 1$, and recursively find the numerical values of these derivatives.

n	0	1	2	3	4	5	6	7
$y^{(n)}(0)$	1	1	2	8	28	144	888	6464

Thus, we obtain an approximating polynomial

$$P_7(t) = 1 + t + t^2 + \tfrac{4}{3}t^3 + \tfrac{7}{6}t^4 + \tfrac{6}{5}t^5 + \tfrac{37}{30}t^6 + \tfrac{404}{315}t^7$$

We can reasonably hope that this polynomial is a good approximation to the true solution $y = y(t)$ of (3.27), at least for small values of t. But we do not know how large we can take t and still have the approximation fairly good, nor do we know how big the error is. Moreover, this polynomial $P_7(t)$ is presumably the first few terms of the Taylor series for the solution, in the form $\sum_0^\infty a_n t^n$. However, we do not have a formula for the $\{a_n\}$ and so have no way to estimate the radius of convergence of this power series.

It is possible to call upon some mathematical theory to help us at this point. The following useful result is usually called Taylor's Theorem with Remainder, to distinguish it from Theorem 3.16, which dealt with infinite series. Since it is a standard result, generalizing the mean-value theorem, we do not give a proof but refer to any book on advanced calculus.

THEOREM 3.17 *Suppose that f is a function having continuous derivatives up to order $N+1$. Then, for any t there is a number τ between t_0 and t such that*

$$f(t) = f(t_0) + f'(t_0)(t-t_0) + f''(t_0)(t-t_0)^2/2! + \cdots + f^{(N-1)}(t_0)(t-t_0)^{N-1}/(N-1)! + R_N \tag{3.29}$$

where

$$R_N = \frac{f^{(N)}(\tau)}{N!}(t-t_0)^N$$

Observe that the first part of this expression is merely the Taylor polynomial for f at t_0. The importance of this result is that it gives a formula for R_N, which is the difference between a function and its Taylor polynomial.

If we use Theorem 3.17, with $N = 4$, for the example (3.27) we have been considering, we obtain an exact equation for the solution $y(t)$ involving the polynomial $P_3(t)$.

$$y(t) = 1 + t + t^2 + \frac{4}{3}t^3 + \frac{y^{(4)}(\tau)}{4!}t^4 \tag{3.30}$$

Unfortunately, all that we know about τ is that $0 < \tau < t$. We do know that $y^{(4)}(0) = 28$, but this tells us little about the possible value of $y^{(4)}(\tau)$. The form of the remainder term—a multiple of t^4—suggests strongly that the approximation made by using $P_3(t)$ to represent $y(t)$ will be quite good if t is sufficiently small, but it is not obvious that we can make this precise or quantitative.

The following special argument works for the equation we are dealing with; it is an example of the mixture of theoretical and applied reasoning that is typical in numerical analysis and in obtaining usable error bounds in the solution of differential equations.

We start by observing that the right-hand side of the differential equation (3.27) is always positive. This being $y'(t)$, we know that $y(t)$ is an increasing function. In particular, since $y(0) = 1$, it must be true that $y(t) \geq 1$ for all $t > 0$. Returning to the differential equation, we observe that

$$y'(t) = t^2 + y^2 \geq t^2 + 1$$

for all $t > 0$. If we integrate this inequality between 0 and t, we have

$$\int_0^t y'(t)\,dt = y(t) - y(0) \geq \int_0^t (t^2+1)\,dt = \tfrac{1}{3}t^3 + t$$

or

$$y(t) > y(0) + t + \tfrac{1}{3}t^3$$

This shows clearly that $y(t)$ becomes arbitrarily large as t increases.

There must then exist a point $t_1 > 0$ such that $y(t_1) = 2$. Because $y(t)$ is increasing, we then have $y(t) \leq 2$ for all t, $0 \leq t \leq t_1$. Our first step is to obtain a numerical estimate for t_1.

We have

$$y'(t) = t^2 + y^2 \leq t^2 + 4$$

for all t in the interval $[0, t_1]$. Integrate this from zero to t, as before, recalling that $y(0) = 1$.

$$\int_0^t y'(t)\,dt = y(t) - 1 \leq \int_0^t (t^2+4)\,dt = \tfrac{1}{3}t^3 + 4t$$

Then set $t = t_1$, and use the fact that $y(t_1) = 2$ to obtain

$$2 - 1 \leq \tfrac{1}{3}(t_1)^3 + 4t_1$$

If we try $t_1 = 0.2$, the right-hand side becomes 0.8026, which is not big enough. Thus, t_1 must be larger than 0.2. (More accurate arithmetic shows that t_1 must be larger than 0.248.)

We can now return to the formulas (3.27) and (3.28) for the derivatives of $y(t)$, and arrive at the following rather crude estimates. For every t obeying $0 \leq t \leq 0.20$,

$$
\begin{aligned}
y(t) &\leq 2 \\
y'(t) &\leq (0.2)^2 + (2)^2 &&= 4.04 < 5 \\
y''(t) &\leq 2(0.2) + 2(2)(4.04) &&< 17 \\
y^{(3)}(t) &\leq 2 + 2(5)^2 + 2(2)(17) &&= 120 \\
y^{(4)}(t) &\leq 6(5)(17) + 2(2)(120) &&= 990
\end{aligned}
\tag{3.31}
$$

This gives an estimate for the size of the remainder term in (3.30), namely,

$$\left| \frac{y^{(4)}(\tau)}{4!} t^4 \right| < \frac{990}{24} t^4 < 42t^4$$

This enables us to state that on the interval $0 \leq t \leq 0.2$,

$$y(t) \approx 1 + t + t^2 + \tfrac{4}{3}t^3$$

with an error of at most $42t^4$, which for this range of t never exceeds 0.07. The calculations tell us nothing about the accuracy of this approximation outside this small interval $[0, 0.2]$, but they provide explicit bounds on the error involved in the approximation.

Estimates like those we obtained in (3.31) are often called *a priori estimates*, since they are obtained before one knows much about the actual solution function $y(t)$. Once the approximate solution, with some estimate for the error, has been obtained, it is sometimes possible to sharpen the estimates considerably and thus obtain more accurate estimates for the error of the approximation.

The key fact which made this method possible, and which we have not mentioned explicitly, is that the solution functions of these differential equations were analytic at the point t_0, and therefore could be expressed near there by a power series, namely, their Taylor Series. It is not easy to tell if this is the case by looking at the form of a differential equation. To illustrate this, consider the following apparently simple equation.

$$y' = 2\sqrt{y}, \qquad y(0) = 0 \tag{3.32}$$

If we try to apply the method developed above, we differentiate the equation, obtaining

$$y'' = \frac{1}{\sqrt{y}}$$

However, if we now attempt to compute the values of $y'(0)$ and $y''(0)$, we are successful with the first but in trouble with the second, since we obtain $y''(0) = 1/0$. This suggests that the true solution $y(t)$ is not a function that is analytic at $t = 0$, since it would then have to have derivatives of all orders as is the case with any convergent power series. However, as can be checked easily (or obtained by the methods of the preceding chapter), a solution of (3.32) for $t \geq 0$ is $y = t^2$, which certainly seems like a well-behaved function!

The resolution of this apparent paradox lies partly in the fact that *the equation* (3.32) *does not have a unique solution passing through the origin.* In fact, it fails to satisfy the hypotheses needed for the basic existence and uniqueness theorem (Theorem 1.1), which required that the function f in the general equation $y' = f(t, y)$ have partial derivatives with respect to both t and y that are continuous at the point (t_0, y_0). In our case, $f_y = 1/\sqrt{y}$, which is certainly not continuous for $y = 0$.

To see that the uniqueness breaks down, we need only observe that $y \equiv 0$ is also a solution of (3.32). Indeed, the following functions are also solutions obeying the given initial conditions.

$$y(t) = \begin{cases} 0 & 0 \leq t \leq C \\ (t-C)^2 & C < t < \infty \end{cases}$$

These functions are continuous and have a continuous first derivative $y'(t)$, but they do not have a continuous second derivative, since

$$y''(t) = \begin{cases} 0 & 0 \leq t \leq C \\ 2 & C < t < \infty \end{cases}$$

The solution $y(t) \equiv 0$ is in fact analytic, and has derivatives of all orders.

What about the function $y = t^2$, which certainly has derivatives of all orders? If we return to the equation (3.32), we will see that this function is a solution only for $t \geq 0$. With $y = t^2$, we have $\sqrt{y} = |t|$, and $y' = 2t = 2|t|$ holds only when $t \geq 0$.

Summarizing this discussion, the equation (3.32) has only one solution that is analytic at $t = 0$, namely, $y(t) \equiv 0$. However, this solution cannot be obtained by the method of the present section, since we cannot carry out the steps required to find $y^{(n)}(0)$.

As a second illustration, consider the equation

$$y' = \sqrt{t} + y, \qquad y(0) = 1 \tag{3.33}$$

This is a linear equation of the sort discussed in Section 2.2, and has a solution

$$y(t) = e^t + e^t \int_0^t \sqrt{s}\, e^{-s}\, ds \tag{3.34}$$

However, if we try to apply the method of the present section to find this solution, we are immediately in trouble again. From (3.33), we find $y'(0) = 0 + 1 = 1$. However, when we differentiate (3.33), we have

$$y'' = \frac{1}{2\sqrt{t}} + y'$$

and we cannot calculate a value for $y''(0)$ because of the first term.

What is happening here is that although the solution given in (3.34) is defined for every choice of $t \geq 0$, it too is not analytic at $t = 0$ and therefore cannot be written as a power series around that point. To see this intuitively, consider trying to find the value of $y(t)$ for $t = -0.01$; the first term in (3.34) is satisfactory, but the integral cannot be evaluated since s must run through negative values. A change of variables may make this even more evident. Set $s = u^2$ in the integral, and we have

$$y(t) = e^t + e^t \int_0^{\sqrt{t}} 2u^2 e^{-u^2}\, du$$

A function that is analytic at a point t_0 must be defined on an interval extending to both sides of t_0, since it is represented by a power series that converges on such an interval. In addition, it must have continuous derivatives at t_0 of all orders.

The method of the present section can be expected to be most useful when the solution function is smooth enough on a neighborhood of t_0 to be infinitely differentiable there; it is even better if the solution is analytic.

The method can also be used for higher-order equations. Suppose we wish to solve the equation

$$y'' = ty + 2y', \qquad y(0) = 1, \quad y'(0) = 2 \tag{3.35}$$

and from (3.35), $y''(0) = 4$. As before, we differentiate (3.35) and have

$$y^{(3)} = y + ty' + 2y''$$
$$y^{(4)} = 2y' + ty'' + 2y^{(3)}$$
$$y^{(5)} = 3y'' + ty^{(3)} + 2y^{(4)}$$

from which we find the numerical values

$$y^{(3)}(0) = 9 \qquad y^{(4)}(0) = 22 \qquad y^{(5)}(0) = 56$$

and can construct the approximating Taylor polynomial

$$y(t) \approx 1 + 2t + \frac{4}{2!}t^2 + \frac{9}{3!}t^3 + \frac{22}{4!}t^4 + \frac{56}{5!}t^5$$

It is possible to obtain explicit error estimates again, but the process is more complicated since we must first find an interval $0 \le t \le t_1$ on which we have *a priori* estimates for the size of $y(t)$ and $y'(t)$.

Sometimes it is a simpler process to rewrite the differential equation before going through the repeated differentiations. A sample of this is Exercise 8. Another device is to substitute the expression for y' into successive differentiations as you go along. Exercises 3, 6, 7, and also 8 are examples of equations that can be solved by this procedure.

Exercises

Use the method of repeated differentiation to obtain the indicated polynomial approximations for the solution to each of the following differential equations.

1. $y' = e^t + 2y, \quad t_0 = 0, \ y(0) = 2$

$$y \approx 2 + 5t + \frac{11}{2}t^2 + \frac{23}{6}t^3 + \frac{47}{24}t^4 + \frac{95}{120}t^5 + \cdots + \frac{3(2^n) - 1}{n!}t^n$$

2. $y'' = t + y, \quad t_0 = 0, \ y(0) = 2, \ y'(0) = 1$

$$y \approx 2 + t + t^2 + \frac{2}{3!}t^3 + \frac{2}{4!}t^4 + \cdots + \frac{2}{n!}t^n$$

3. $y' = y^3, \quad t_0 = 1, \ y(1) = 1$

$$y \approx t + \frac{1 \cdot 3}{2!}(t-1)^2 + \frac{1 \cdot 3 \cdot 5}{3!}(t-1)^3 + \cdots + \frac{1 \cdot 3 \cdot 5 \cdots (2n-1)}{n!}(t-1)^n$$

4. $y' = ty^2, \quad t_0 = 0, \ y(0) = 1$

$$y \approx 1 + \tfrac{1}{2}t^2 + \tfrac{1}{4}t^4 + \tfrac{1}{8}t^6$$

For each of Exercises 5–12, obtain a polynomial approximation to the solution of degree at least 6.

5. $y' = y + y^2, \quad t_0 = 0, \ y(0) = 1$

6. $y' = t^2\sqrt{y}, \quad t_0 = 0, \ y(0) = 4$

7. $y' = y/t, \quad t_0 = 1, \ y(1) = 3$

8. $y' = t/y, \quad t_0 = 1, \ y(1) = 2$ [*Hint:* Use form $y'y = t$.]

9. $y'' = ty + t^2, \quad t_0 = 0, \ y(0) = 2, \ y'(0) = 1$

10. $y'' = ty^2, \quad t_0 = 0, \; y(0) = 1, \; y'(0) = 2$

11. $y' = ty + t^2, \quad t_0 = 1, \; y(1) = 1$

12. $y'' = y^2, \quad t_0 = 0, \; y(0) = -1, \; y'(0) = 2$

13. Find a cubic polynomial approximating the solution of $y' = y^2$ through $(0, 1)$. Find the exact solution and sketch both on the interval $0 \leq t \leq 2$.

14. (a) Using a method from Chapter 2, solve the differential equation $y'' = \sqrt{y}$ with $t_0 = 0$, $y(0) = 0$, $y'(0) = 0$.
    (b) Explain why the method of this section fails.

15. Using only a cubic as estimate for the solution in Exercise 5, show that the error term $R_4 < 0.07$ when $0 \leq t < \frac{1}{6}$.

16. (a) For the differential equation $y' = \frac{1}{2}y^2 + 2t + 1$ where $y(0) = 0$, show that $y \leq 1$ for $0 \leq t \leq \frac{1}{2}$.
    (b) Find the cubic Taylor approximation for the solution.
    (c) Show that the error made by using the approximation in part (b) is less than $\frac{1}{8}$ when $0 \leq t \leq \frac{1}{2}$.

17–19. Solve directly the equations in Exercises 1, 3, and 4, and expand the solutions by means of known series. Check with the polynomials found earlier in Exercises 1, 3, 4.

20. Show that if $|x| < A/10$, then $\sqrt{A + x}$ can be approximated by

$$\sqrt{A} + \frac{x}{2\sqrt{A}} - \frac{x^2}{8A\sqrt{A}}$$

with an error less than $\sqrt{A}/10^4$.

3.5 Solution by Power Series

The approximation method described in the previous section produced a polynomial that was an initial segment of the Taylor series for the solution function $y(t)$. In some cases, it was possible to obtain the general term in the series. In others, the effort required to compute higher derivatives limited the practical efficiency of the method and made it difficult or impossible to obtain a formula for the general term in the Taylor series.

There is another procedure for achieving the same result which is simpler in many cases. It is called the method of **undetermined coefficients**. We make the assumption that $y(t)$ is analytic at $t = t_0$, and thus that there is a power-series expansion

$$y(t) = \sum_0^\infty a_n(t-t_0)^n$$
$$= a_0 + a_1(t-t_0) + a_2(t-t_0)^2 + \cdots$$

which converges on some interval centered on t_0. The basic procedure is to substitute the series for y in the differential equation, and then solve for the unknown coefficients $a_0, a_1, a_2, \ldots$. If we have an initial-value problem, and are given the value $y_0 = y(t_0)$, then we can start with a knowledge of a_0, and use this to determine the remaining coefficients $\{a_n\}$. An example may make this description clearer; we take an equation which can be solved easily in other ways so that we can check the answer.

Suppose we wish to solve the equation

$$y' = -2ty, \qquad t_0 = 0, \ y(0) = 3 \tag{3.36}$$

We assume that $y(t)$ has the form

$$y = a_0 + a_1 t + a_2 t^2 + a_3 t^3 + \cdots \tag{3.37}$$

Since $y(0) = 3$, we already know that $a_0 = 3$. The next step is to write a series for the right-hand side of (3.36).

$$-2ty = -2a_0 t - 2a_1 t^2 - 2a_2 t^3 - 2a_3 t^4 - \cdots$$

We then obtain a series for the left-hand side of (3.36) by differentiating (3.37).

$$y' = a_1 + 2a_2 t + 3a_3 t^2 + 4a_4 t^3 + 5a_5 t^4 + \cdots$$

By the corollary to Theorem 3.16, two convergent power series which represent functions that are identically equal, such as y' and $-2yt$, must have exactly the same coefficients. Comparing the coefficients of like powers of t, we generate the following system of equations:

$$a_1 = 0$$
$$2a_2 = -2a_0$$
$$3a_3 = -2a_1$$
$$4a_4 = -2a_2$$
$$5a_5 = -2a_3$$
etc.

If we add the fact that $a_0 = 3$, we can summarize this infinite system of equations by writing

$$\begin{cases} a_0 = 3 \\ a_1 = 0 \\ a_{n+1} = -\dfrac{2}{n+1} a_{n-1} \end{cases}$$

As in Section 3.1, we can use this recursion to calculate the successive values of a_n.

$$\begin{array}{llll} a_0 = 3 & a_2 = -3 & a_4 = \frac{3}{2} & a_6 = -\frac{1}{2} \\ a_1 = 0 & a_3 = 0 & a_5 = 0 & a_7 = 0 \text{ etc.} \end{array}$$

Indeed, we can find a formula for the general terms.

$$a_{2k+1} = 0 \qquad a_{2k} = (-1)^k \frac{3}{k!}$$

To check this by induction, we make the following calculation:

$$-\frac{2}{2k+2} a_{2k} = -\frac{2}{2k+2}(-1)^k \frac{3}{k!} = (-1)^{k+1} \frac{3}{(k+1)!} = a_{2k+2}$$

The desired solution of (3.36) is

$$\begin{aligned} y(t) &= 3 + (0)t + (-3)t^2 + (0)t^3 + (\tfrac{3}{2})t^4 + (0)t^5 + \cdots \\ &= 3(1 - t^2/1! + t^4/2! - t^6/3! + t^8/4! - \cdots) \end{aligned}$$

This can be recognized as a familiar series, so that $y(t) = 3e^{-t^2}$. [See formula (3.11).]

Now, let us try this method with a slightly more difficult equation. Suppose we want to solve

(3.38) $$y' = 5t^2 - 2ty, \qquad t_0 = 0, \; y(0) = C$$

Since this is a linear equation, the method of Section 2.2 gives us a solution:

(3.39) $$y(t) = Ce^{-t^2} + e^{-t^2} \int_0^t 5s^2 e^{s^2} \, ds$$

To find the solution by the method of power series, we assume as before that we have

$$y = a_0 + a_1 t + a_2 t^2 + \cdots$$

Since $y(0) = C$, we have $a_0 = C$. Calculating the power series for the right-hand side of (3.38) we have

$$5t^2 - 2ty = 5t^2 - 2a_0 t - 2a_1 t^2 - 2a_2 t^3 - 2a_3 t^4 - \cdots$$
$$= -2a_0 t + (5-2a_1)t^2 - 2a_2 t^3 - 2a_3 t^4 - \cdots$$

For the left-hand side, we again have

$$y' = a_1 + 2a_2 t + 3a_3 t^2 + 4a_4 t^3 + 5a_5 t^4 + \cdots$$

Equating coefficients of like powers, we arrive at the system of equations

$$\begin{aligned} a_1 &= 0 \\ 2a_2 &= -2a_0 \\ 3a_3 &= 5 - 2a_1 \\ 4a_4 &= -2a_2 \\ 5a_5 &= -2a_3 \\ 6a_6 &= -2a_4 \quad \text{etc.} \end{aligned} \tag{3.40}$$

Recalling that $a_0 = C$, we calculate the first few a_n to be

$$\begin{aligned} a_0 &= C \\ a_1 &= 0 \\ a_2 &= -a_0 = -C \\ a_3 &= \frac{5-0}{3} = \tfrac{5}{3} \\ a_4 &= -\tfrac{2}{4}a_2 = \tfrac{1}{2}C \\ a_5 &= -\tfrac{2}{5}a_3 = -\tfrac{2}{3} \\ a_6 &= -\tfrac{2}{6}a_4 = -\tfrac{1}{6}C \\ a_7 &= -\tfrac{2}{7}a_5 = \tfrac{4}{21} \\ a_8 &= -\tfrac{2}{8}a_6 = \tfrac{1}{24}C \\ a_9 &= -\tfrac{2}{9}a_7 = -\tfrac{8}{189} \end{aligned}$$

Thus, we have

$$y \approx C + (0)t - Ct^2 + \tfrac{5}{3}t^3 + \tfrac{1}{2}Ct^4 - \tfrac{2}{3}t^5 - \tfrac{1}{6}Ct^6 + \tfrac{4}{21}t^7 + \tfrac{1}{24}Ct^8 - \tfrac{8}{189}t^9$$

If we rewrite this, we can compare it with the exact solution (3.39) given above. Factor out C, and have

$$y \approx C(1 - t^2 + \tfrac{1}{2}t^4 - \tfrac{1}{6}t^6 + \tfrac{1}{24}t^8) + (\tfrac{5}{3}t^3 - \tfrac{2}{3}t^5 + \tfrac{4}{21}t^7 - \tfrac{8}{189}t^9)$$

Looking back at (3.39), we recognize that the polynomial multiplying C is a partial sum for the Taylor Series of e^{-t^2}. The remaining polynomial should then be a power series for the second term in (3.39), which is there defined by a complicated integral.

It is natural to ask if we can solve the system of equations for the $\{a_n\}$ to find a general formula for the terms. Examining (3.40), we see that we can write two simple recursions, one for the even-index terms and one for the odd. For the even case, we have

$$a_{2k+2} = -\frac{2}{2k+2}a_{2k} \qquad a_0 = C$$

and it is easy to see that this has the solution $a_{2k} = (-1)^k C/k!$. This of course yields the complete series for the function $C\exp(-t^2)$. The recursion for the odd-index terms is more complicated to solve. From (3.40) we have

$$\begin{cases} a_3 = \frac{5}{3} \\ a_{2k+1} = -\dfrac{2}{2k+1}a_{2k-1} \end{cases} \tag{3.41}$$

If we write a few of the values of successive a_n without canceling numerical factors, we can arrive at a general formula. We have

$$a_3 = \frac{5}{3}$$

$$a_5 = -\frac{(2)(5)}{(5)(3)}$$

$$a_7 = \frac{(2)(2)(5)}{(7)(5)(3)}$$

$$a_9 = -\frac{(2)(2)(2)(5)}{(9)(7)(5)(3)}$$

This leads us to the formula

$$a_{2k+1} = (-1)^{k+1}\frac{(5)(2)^{k-1}}{(3)(5)(7)\cdots(2k-1)(2k+1)} \tag{3.42}$$

which can be shown to satisfy (3.41) for $k = 1, 2, 3, \ldots$. It is possible to obtain a more compressed formula by using factorials. The denominator in (3.42) suggests this, since it differs from $(2k+1)!$ only in not having the factors $2, 4, 6, \ldots, 2k$. Supply these by multiplying both on top and bottom.

$$a_{2k+1} = (-1)^{k+1}(5)(2)^{k-1}\frac{(2)(4)(6)\cdots(2k)}{(2)(3)(4)\cdots(2k)(2k+1)}$$

Then, factor a 2 out of each term on top, and reduce this formula to a slightly better expression.

$$a_{2k+1} = (-1)^{k+1}(5)(2)^{k-1}(2)^k \frac{k!}{(2k+1)!}$$
$$= \frac{5}{2}(-1)^{k+1}4^k \frac{k!}{(2k+1)!}$$

The final solution of (3.38) can thus be given as

$$y(t) = Ce^{-t^2} + \frac{5}{2}\sum_{k=1}^{\infty}(-1)^{k+1}4^k \frac{k!}{(2k+1)!}t^{2k+1}$$

Let us next apply this method to Exercise 1 in Section 3.4. The equation to be solved was

$$y' = e^t + 2y, \qquad t_0 = 0, \ y(0) = 2 \tag{3.43}$$

The only difference here is that we must use a power series for e^t. With $y = \sum a_n t^n$, we have

$$e^t = 1 + t + \tfrac{1}{2}t^2 + \tfrac{1}{6}t^3 + \tfrac{1}{24}t^4 + \cdots$$
$$2y = 2a_0 + 2a_1 t + 2a_2 t^2 + 2a_3 t^3 + 2a_4 t^4 + \cdots$$

so that the right-hand side of (3.43) is

$$e^t + 2y = (1+2a_0) + (1+2a_1)t + (\tfrac{1}{2}+2a_2)t^2 + (\tfrac{1}{6}+2a_3)t^3 + \cdots$$

As before,

$$y' = a_1 + 2a_2 t + 3a_3 t^2 + 4a_4 t^3 + \cdots$$

Equating coefficients, we have

$$a_1 = 1 + 2a_0$$
$$2a_2 = 1 + 2a_1$$
$$3a_3 = \tfrac{1}{2} + 2a_2$$
$$4a_4 = \tfrac{1}{6} + 2a_3$$
$$5a_5 = \tfrac{1}{24} + 2a_4$$
$$6a_6 = \tfrac{1}{120} + 2a_5$$

Since we have the initial condition $y(0) = 2$, we know that $a_0 = 2$. Proceeding from here, we can generate numerical values for the first few a_n:

$$a_0 = 2 \qquad a_1 = 5 \qquad a_2 = \tfrac{11}{2} \qquad a_3 = \tfrac{23}{6} \qquad a_4 = \tfrac{47}{24}$$

Examining the set of equations, we see that the general recursion is

$$\begin{cases} a_{n+1} = \dfrac{1}{(n+1)!} + \dfrac{2}{n+1} a_n \\ a_0 = 2 \end{cases}$$

and it can be verified that a general formula for a_n is

$$a_n = \frac{3(2)^n - 1}{n!}$$

The solution of (3.43) is therefore

$$y(t) = 2 + 5t + \tfrac{11}{2}t^2 + \tfrac{23}{6}t^3 + \tfrac{47}{24}t^4 + \tfrac{95}{120}t^5 + \tfrac{191}{720}t^6 + \cdots$$

$$= \sum_0^\infty \frac{3(2)^n - 1}{n!} t^n$$

If we rewrite the last expression for y as

$$y(t) = 3 \sum_0^\infty \frac{(2t)^n}{n!} - \sum_0^\infty \frac{t^n}{n!}$$

we recognize that we simply have $y = 3e^{2t} - e^t$, which can be checked immediately to be the correct solution.

All the examples used so far have been linear equations. Although the method of power series can be used if the right-hand side of the equation is not of the simple form $a(t) + b(t)y$, the difficulties are at times massive, and it becomes extremely difficult or impossible to obtain a general formula for the coefficients of the power series for y. As in the case of the method of repeated differentiation, explained in the preceding section, there are times when it is useful to obtain a polynomial approximation that is the first few terms of the Taylor expansion of $y(t)$; in such cases, the method of power series offers an alternative to the method of differentiation.

To illustrate both aspects of this, consider again the nonlinear equation used as an example in the last section:

(3.44) $$y' = t^2 + y^2, \qquad t_0 = 0,\ y(0) = 1$$

Using the initial condition to give us $a_0 = 1$, we write

$$y = 1 + a_1 t + a_2 t^2 + a_3 t^3 + a_4 t^4 + a_5 t^5 + \cdots$$

To find a series for y^2, we must multiply this power series by itself, obtaining

$$y^2 = 1 + 2a_1 t + (2a_2 + a_1{}^2) t^2 + (2a_3 + 2a_1 a_2) t^3$$
$$+ (2a_4 + 2a_1 a_3 + a_2{}^2) t^4 + (2a_5 + 2a_1 a_4 + 2a_2 a_3) t^5 + \cdots$$

Then, the power-series expansion of the right-hand side is

$$t^2 + y^2 = 1 + 2a_1 t + (1+2a_2+a_1{}^2)t^2 + (2a_3+2a_1 a_2)t^3 + \cdots$$

Comparing this with the usual series for $y'(t)$, we arrive at a much more complicated set of equations for the unknown coefficients a_n. [The fact that they are not linear equations reflects the nonlinearity of the differential equation (3.44).]

$$\begin{aligned} a_1 &= 1 \\ 2a_2 &= 2a_1 \\ 3a_3 &= 1 + 2a_2 + a_1{}^2 \\ 4a_4 &= 2a_3 + 2a_1 a_2 \\ 5a_5 &= 2a_4 + 2a_1 a_3 + a_2{}^2 \\ 6a_6 &= 2a_5 + 2a_1 a_4 + 2a_2 a_3 \\ 7a_7 &= 2a_6 + 2a_1 a_5 + 2a_2 a_4 + a_3{}^2 \\ &\text{etc.} \end{aligned}$$

However, they can still be solved recursively, using the values of a_n for earlier n to obtain the values for later n. The values obtained yield the following approximation for $y(t)$:

$$y(t) \approx 1 + t + t^2 + \tfrac{4}{3}t^3 + \tfrac{7}{6}t^4 + \tfrac{6}{5}t^5 + \tfrac{37}{30}t^6 + \tfrac{404}{315}t^7$$

Since we do not have a general form for the coefficients, we have no way of testing the series for convergence. Nor do we this time have a method for finding *a priori* bounds which would enable us to use Taylor's Theorem with Remainder to estimate the accuracy of this polynomial approximation.

With more complicated nonlinear equations, the method of power series encounters formidable difficulties. Consider, for instance, the pair of differential equations

$$y' = \frac{t-y}{t+y} \qquad y(0) = 1$$

$$y' = t^3 + \sqrt{y}, \qquad y(0) = 1$$

While we can start as before with an assumed series for y of the form $y = 1+a_1 t+a_2 t^2 + \cdots$, it is a challenging task to substitute this into the right-hand side of either equation and then find the corresponding power series, at least for more than one or two terms. (It should be noted that the first equation in this pair is homogeneous so that it is one that in theory can be solved in closed form by the special method of Section 2.1.)

The method of power series is not limited to first-order equations. For the present time, one illustration should show how to proceed. Consider the equation

$$y'' = e^{-t} - y + 2y', \qquad y(0) = 1,\ y'(0) = 2 \tag{3.45}$$

We make one slight change to make the algebraic work easier by assuming that the solution function $y(t)$ has a series of the form

$$y = a_0 + a_1 t + a_2 t^2/2! + a_3 t^3/3! + \cdots + a_n t^n/n! + \cdots$$

Then, with this special form—patterned after the format of Taylor Series—it is easy to write down series for y' and y''.

$$y' = a_1 + a_2 t + a_3 t^2/2! + a_4 t^3/3! + a_5 t^4/4! + \cdots$$

$$y'' = a_2 + a_3 t + a_4 t^2/2! + a_5 t^3/3! + a_6 t^4/4! + \cdots$$

For e^{-t}, as usual we have

$$e^{-t} = 1 - t + t^2/2! - t^3/3! + t^4/4! - \cdots$$

The right-hand side of (3.45) has the series expansion

$$(1-a_0+2a_1) + (-1-a_1+2a_2)t + (1-a_2+2a_3)t^2/2! + (-1-a_3+2a_4)t^3/3! + (1-a_4+2a_5)t^4/4! + \cdots$$

Comparing this with the series for y'', we obtain the following system of algebraic equations:

$$\begin{aligned} a_2 &= 1 - a_0 + 2a_1 \\ a_3 &= -1 - a_1 + 2a_2 \\ a_4 &= 1 - a_2 + 2a_3 \\ a_5 &= -1 - a_3 + 2a_4 \\ a_6 &= 1 - a_4 + 2a_5 \\ a_7 &= -1 - a_5 + 2a_6 \\ a_8 &= 1 - a_6 + 2a_7 \end{aligned}$$

etc.

To get a start on the recursive solution of these, we use the initial conditions for the differential equation; since $y(0) = 1$ and $y'(0) = 2$, we have $a_0 = 1$,

$a_1 = 2$. In the usual fashion, we then obtain in turn

$$a_2 = 4 \qquad a_3 = 5$$
$$a_4 = 7 \qquad a_5 = 8$$
$$a_6 = 10 \qquad a_7 = 11$$

There is pattern to these numbers, made evident by, for instance, calculating the first and second differences of the sequence. If one looks first at the terms with even index and then at those with odd index, the following formula seems to fit:

$$\begin{aligned} a_{2k} &= 3k + 1 \\ a_{2k+1} &= 3k + 2 \end{aligned} \qquad k = 0, 1, 2, \ldots \tag{3.46}$$

These can now be checked in the preceding equations to see that they are satisfied. To see how this is done, we first observe that the system of equations to be satisfied can be given a general form

$$a_n = (-1)^n - a_{n-2} + 2a_{n-1} \tag{3.47}$$

Suppose we take the case of n even, and set $n = 2k+2$. Then, the relation to be verified in (3.47) becomes

$$a_{2k+2} \stackrel{?}{=} 1 - a_{2k} + 2a_{2k+1}$$

We use (3.46) to substitute for each term in this, and arrive at

$$3(k+1) + 1 \stackrel{?}{=} 1 - (3k+1) + 2(3k+2)$$

Since both sides are seen to be $3k+4$, we have checked (3.46) for n even. The case of n odd is similar; take $n = 2k+1$, so that (3.47) becomes

$$a_{2k+1} \stackrel{?}{=} -1 - a_{2k-1} + 2a_{2k}$$

If we substitute from (3.46), both sides turn out to be $3k+2$.

We can now write a Taylor polynomial approximating the solution $y(t)$ of arbitrarily high degree, using (3.46) to calculate the coefficients.

$$y(t) = 1 + 2t + 4\frac{t^2}{2!} + 5\frac{t^3}{3!} + 7\frac{t^4}{4!} + 8\frac{t^5}{5!} + \cdots$$

In this particular case, we can also express the sum of this convergent series in terms of familiar functions, and thus check directly that we have found the solution of (3.45). We observe first that we can rewrite (3.46) as

$$a_n = \begin{cases} \frac{3}{2}n + 1 & \text{when } n \text{ is even} \\ \frac{3}{2}n + \frac{1}{2} & \text{when } n \text{ is odd} \end{cases} \tag{3.48}$$

It is convenient to introduce a special sequence $\{\sigma_n\}$, defined by

$$\sigma_n = \begin{cases} 1 & \text{when } n \text{ is even} \\ 0 & \text{when } n \text{ is odd} \end{cases}$$

[A formula for σ_n is $\frac{1}{2}[1+(-1)^n]$, but it is not necessary to use this.] Using (3.48), we can write a single definition for $\{a_n\}$, namely, $a_n = \frac{3}{2}n + \frac{1}{2} + \frac{1}{2}\sigma_n$. Then

$$y(t) = \sum_0^\infty a_n \frac{t^n}{n!} = \frac{3}{2}\sum_0^\infty n\frac{t^n}{n!} + \frac{1}{2}\sum_0^\infty \frac{t^n}{n!} + \frac{1}{2}\sum_0^\infty \sigma_n \frac{t^n}{n!}$$

$$= \frac{3}{2}\sum_1^\infty \frac{t^n}{(n-1)!} + \frac{1}{2}e^t + \frac{1}{2}\cosh t$$

Using the series for exponential, and also (3.13), this is

$$y(t) = \tfrac{3}{2}te^t + \tfrac{1}{2}e^t + \tfrac{1}{4}(e^t + e^{-t})$$

$$= \tfrac{3}{2}te^t + \tfrac{3}{4}e^t + \tfrac{1}{4}e^{-t}$$

Exercises

1. By the method of undetermined coefficients find the power-series solution of the equation $y' = y + t^2$ satisfying the initial conditions $y = 1$ when $t = 0$. Check your result by direct integration.

2. Find the power-series solution for the differential equation $y' = t^2y + 3t - 1$, $y = C$ when $t = 0$. Use the method of undetermined coefficients.

Using the method of undetermined coefficients, find the power-series terms through t^6 for solutions of the equations of Exercises 3–5.

3. $y' = y + 1/(1-t), \quad t = 0, \ y = C$

4. $y' = e^{-t}y + t, \quad t = 0, \ y = 1$

5. $y' = 1 + ty^2, \quad t = 0, \ y = 1$

Using the method of undetermined coefficients, find the power-series solutions for the equations of Exercises 6 and 7.

6. $y'' = t^2 - y, \quad t = 0, \ y = 0, \ y' = 1$

7. $y'' = ty, \quad t = 0, \ y = 1, \ y' = 0$

8. Carry out the algebraic solution of the coefficient equations for the differential equation (3.44) to obtain the terms of the power series shown in the text.

9. (a) If $\phi(t) = \sum C_n t^n$ and $C_0 \neq 0$, then the series expansion for $1/\phi(t)$ begins thus:

$$\frac{1}{C_0}\left[1 - C_1 \frac{t}{C_0} - (C_0 C_2 - C_1{}^2)\left(\frac{t}{C_0}\right)^2 - (C_3 C_0{}^2 + C_1{}^3 - 2C_0 C_1 C_2)\left(\frac{t}{C_0}\right)^3 - \cdots\right]$$

Verify this by multiplying the series.

*(b) Find the next term in the expansion of $1/\phi(t)$.

10. Apply Exercise 9 to show that $1/(1+e^t) = \frac{1}{2} - \frac{1}{4}t + \frac{1}{48}t^3 + \cdots$.

11. Use Exercise 9 to find the terms through t^3 of the power-series solution of $y' = 2 + t^2 + (1/y)$, with initial conditions $t = 0$, $y = 1$.

3.6 Picard Iteration

The final approximation method to be discussed in this chapter is one which is of great use in the theory of differential equations but of somewhat limited use as a practical way to obtain approximate solutions. The idea behind it is very simple. Suppose we are concerned with the initial-value problem

$$y' = f(t, y), \qquad y(t_0) = y_0 \tag{3.49}$$

The solution is a function ϕ defined on a neighborhood of t_0, which is such that $\phi(t_0) = y_0$, and for all t near t_0,

$$\phi'(t) = f(t, \phi(t))$$

holds identically.

The Picard method produces a sequence of successive approximations ϕ_n, each a continuous function that obeys the initial condition $\phi_n(t_0) = y_0$, and such that $\{\phi_n\}$ converges to the true solution ϕ. The functions ϕ_n are defined recursively. The first function ϕ_1 can be taken to be the constant function with value y_0, and all the rest are then obtained by the formula

$$\phi'_{n+1}(t) = f(t, \phi_n(t))$$

At each stage, the right-hand side is completely determined so that (in theory) we can obtain the function ϕ_{n+1} by a single ordinary integration, choosing the constant of integration so that we satisfy the condition $\phi_{n+1}(t_0) = y_0$. An alternative formula is therefore

$$\phi_{n+1}(t) = y_0 + \int_{t_0}^{t} f(s, \phi_n(s))\, ds$$

Perhaps an example will clarify this description of the Picard process. We take the simple equation

$$y' = 2t + y, \qquad y(0) = 1 \tag{3.50}$$

Then, taking $\phi_1 \equiv 1$, we have

$$\phi'_2 = 2t + \phi_1 = 2t + 1$$

so that

$$\phi_2 = t^2 + t + 1$$

Continuing, we have

$$\begin{aligned} \phi'_3 &= 2t + \phi_2 \\ &= t^2 + 3t + 1 \end{aligned}$$

and then

$$\begin{aligned} \phi_3 &= \tfrac{1}{3}t^3 + \tfrac{3}{2}t^2 + t + 1 \\ \phi'_4 &= 2t + \phi_3 \\ &= \tfrac{1}{3}t^3 + \tfrac{3}{2}t^2 + 3t + 1 \\ \phi_4 &= \tfrac{1}{12}t^4 + \tfrac{1}{2}t^3 + \tfrac{3}{2}t^2 + t + 1 \\ \phi'_5 &= 2t + \phi_4 \\ &= \tfrac{1}{12}t^4 + \tfrac{1}{2}t^3 + \tfrac{3}{2}t^2 + 3t + 1 \\ \phi_5 &= \tfrac{1}{60}t^5 + \tfrac{1}{8}t^4 + \tfrac{1}{2}t^3 + \tfrac{3}{2}t^2 + t + 1 \\ \phi'_6 &= \tfrac{1}{60}t^5 + \tfrac{1}{8}t^4 + \tfrac{1}{2}t^3 + \tfrac{3}{2}t^2 + 3t + 1 \\ \phi_6 &= \tfrac{1}{360}t^6 + \tfrac{1}{40}t^5 + \tfrac{1}{8}t^4 + \tfrac{1}{2}t^3 + \tfrac{3}{2}t^2 + t + 1 \end{aligned}$$

Note that each time we have been careful to add a constant to satisfy the condition $\phi_n(0) = 1$.

Now let us compare these polynomials ϕ_n with the exact solution of the equation (3.50). Since it is linear, the method of Section 2.2 applies, and we find

$$y(t) = 3e^t - 2 - 2t$$

If we expand this into its Taylor Series, we have

$$\begin{aligned} y(t) &= 3(1+t+t^2/2+t^3/6+\cdots) - 2 - 2t \\ &= 1 + t + \tfrac{3}{2}t^2 + \tfrac{1}{2}t^3 + \tfrac{1}{8}t^4 + \tfrac{1}{40}t^5 + \tfrac{1}{240}t^6 + \cdots \end{aligned}$$

If this is compared with the functions ϕ_n obtained above, we see that each seems successively closer to the true solution, with ϕ_6 agreeing up to terms in t^5. This indeed suggests that this is an effective way to obtain approximate solutions of a differential equation when t is quite small.

Before trying to understand why this scheme works, let us look at several other examples. Consider next the equation

$$(3.51) \qquad y' = e^t + 2y, \qquad y(0) = 1$$

This is again linear, and can be verified to have the solution

$$y = 2e^{2t} - e^t$$

We can again start with the trial solution $\phi_1 \equiv 1$, so that

$$\phi_2' = e^t + 2\phi_1 = e^t + 2$$

Integrating [and checking to satisfy $\phi_2(0) = 1$], we have

$$\phi_2 = e^t + 2t$$

Proceeding as before, we obtain in succession:

$$\begin{aligned} \phi_3' &= e^t + 2\phi_2 \\ &= 3e^t + 4t \\ \phi_3 &= 3e^t + 2t^2 - 2 \\ \phi_4' &= e^t + 2\phi_3 \\ &= 7e^t + 4t^2 - 4 \\ \phi_4 &= 7e^t + \tfrac{4}{3}t^3 - 4t - 6 \\ \phi_5' &= e^t + 2\phi_4 \\ &= 15e^t + \tfrac{8}{3}t^3 - 8t - 12 \\ \phi_5 &= 15e^t + \tfrac{2}{3}t^4 - 4t^2 - 12t - 14 \\ \phi_6' &= e^t + 2\phi_5 \\ &= 31e^t + \tfrac{4}{3}t^4 - 8t^2 - 24t - 28 \\ \phi_6 &= 31e^t + \tfrac{4}{15}t^5 - \tfrac{8}{3}t^3 - 12t^2 - 28t - 30 \end{aligned}$$

At first glance, it does not look as though this sequence is converging to anything; it certainly does not look like the true solution of (3.51), which is $2e^{2t}-e^t$. However, let us again compare the Taylor series expansion of both. For ϕ_6, we have

$$\begin{aligned}
&31(1+t+t^2/2!+t^3/3!+\cdots)+\tfrac{4}{15}t^5-\tfrac{8}{3}t^3-12t^2-28t-30\\
&\quad=(31-30)+(31-28)t+(31-24)t^2/2!+(31-16)t^3/3!\\
&\qquad+31t^4/4!+(31+32)t^5/5!+31t^6/6!+\cdots\\
&\quad=1+3t+7t^2/2!+15t^3/3!+31t^4/4!+63t^5/5!+31t^6/6!+\cdots
\end{aligned}$$

In comparison, we can calculate the series for the solution

$$\begin{aligned}
y(t)=2e^{2t}-e^t&=2\sum_0^\infty\frac{(2t)^n}{n!}-\sum_0^\infty\frac{t^n}{n!}\\
&=\sum_0^\infty(2^{n+1}-1)\frac{t^n}{n!}
\end{aligned}$$

and we see that in fact the series agree out to terms in t^5. This again seems to confirm that the functions $\{\phi_n\}$ produced by this iterative process do converge to a solution of the differential equation, obeying the initial condition.

We have illustrated the method with two rather simple equations, both linear. It should be clear, however, that there would be great practical difficulties in applying this method to either of the following equations.

$$y'=e^{t^2}+y,\qquad y(0)=1$$

$$y'=1+\sqrt{1+y},\qquad y(0)=1$$

In the former, we are stymied at the second step. If we start with $\phi_1\equiv1$, then we have

$$\phi_2'=e^{t^2}+1$$

and we cannot write a formula for ϕ_2 because we cannot integrate the right-hand side in elementary functions. In the second example, we can carry out the first several steps, but we quickly get hung up on another integration which is nonelementary. With ϕ_1 again 1, we have

$$\phi_2'=1+\sqrt{2}$$

so that

$$\phi_2=(1+\sqrt{2})t+1$$

Then

$$\phi_3' = 1 + \sqrt{2 + (1+\sqrt{2})t}$$

so that

$$\phi_3 = t + \frac{2}{3(1+\sqrt{2})}[2 + (1+\sqrt{2})t]^{3/2}$$

but it seems hopeless to seek a simple expression for ϕ_4.

However, since any continuous function possesses an indefinite integral, we could in theory continue this process, no matter how complicated the formulas became. This explains why this iterative process is suited better to proving results dealing with the *theory* of differential equations than to finding useful approximate solutions.

It still remains a useful procedure for discovering something about the nature of a solution in the immediate neighborhood of the initial point t_0, as the following example shows. Consider the equation

$$y' = \sqrt{t} + \sqrt{y}, \qquad y(0) = 0 \tag{3.52}$$

Applying the Picard process, and starting with $\phi_1 \equiv 0$,

$$\phi_2' = \sqrt{t} + 0 = t^{1/2}$$

so

$$\phi_2 = \tfrac{2}{3}t^{3/2}$$

We next have

$$\phi_3' = t^{1/2} + \sqrt{\tfrac{2}{3}}\,t^{3/4}$$

so that

$$\begin{aligned}\phi_3 &= \tfrac{2}{3}t^{3/2} + \tfrac{4}{7}\sqrt{\tfrac{2}{3}}\,t^{7/4} \\ &= t^{3/2}\left(\tfrac{2}{3} + \tfrac{4}{7}\sqrt{\tfrac{2}{3}}\,t^{1/4}\right)\end{aligned}$$

At the next stage, we should take

$$\phi_4' = t^{1/2} + \sqrt{\phi_3}$$

In order to work with the square root, we use the convenient approximation (see Exercise 20, Section 3.4) which holds if $a \ll A$:

$$\sqrt{A+a} = \sqrt{A} + \frac{a}{2\sqrt{A}} - \frac{a^2}{8A\sqrt{A}} + \cdots$$

This gives us

$$\phi_4' = t^{1/2} + t^{3/4}\sqrt{\tfrac{2}{3}+\tfrac{4}{7}\sqrt{\tfrac{2}{3}}\,t^{1/4}}$$

$$= t^{1/2} + t^{3/4}\left(\sqrt{\frac{2}{3}} + \frac{(4/7)\sqrt{2/3}\,t^{1/4}}{2\sqrt{2/3}} - \cdots\right)$$

$$= t^{1/2} + \sqrt{\tfrac{2}{3}}\,t^{3/4} + \tfrac{2}{7}t + \cdots$$

$$\phi_4 = \tfrac{2}{3}t^{3/2} + \tfrac{4}{7}\sqrt{\tfrac{2}{3}}\,t^{7/4} + \tfrac{1}{7}t^2 + \cdots$$

$$= t^{3/2}\left(\tfrac{2}{3} + \tfrac{4}{7}\sqrt{\tfrac{2}{3}}\,t^{1/4} + \tfrac{1}{7}t^{1/2} + \cdots\right)$$

This, while approximate, gives a clear picture of the nature of the solution function for t very near zero. Indeed, as we will see later (Exercise 3, Section 7.4), there is a series expansion for this solution of the form

$$y(t) = t^{3/2}(c_0 + c_1 t^{1/4} + c_2(t^{1/4})^2 + c_3(t^{1/4})^3 + \cdots)$$

In the formula for ϕ_4, we have succeeded in finding the correct values for the coefficients c_0, c_1, and c_2 in this series, as well as obtaining an approximate expression for the solution that suggests strongly that there is a special series expansion of this strange form.

The remainder of this section is devoted to showing that if the function $f(t, y)$ is sufficiently smooth, the Picard process always converges to a function that solves the initial-value problem

$$y' = f(t, y), \qquad y(t_0) = y_0 \tag{3.53}$$

on a (small) neighborhood of $t = t_0$. This establishes the truth of the fundamental existence theorem discussed in Section 1.5, since the proof also shows that such a solution exists.

The mathematical arguments used in the remainder of this section are deeper than any used in other sections of the book, and depend upon some theorems of analysis that are sometimes left to a course in advanced calculus. In particular, the important concept of **uniform convergence** of series is used at one crucial step. Since the purpose of the present discussion is to provide insight into the mathematical reasons behind the convergence of the Picard iterative method for those readers for whom this is important, we choose not to develop this background in detail but refer again to the references cited at the end of Section 3.2.

We will assume that the function f is continuous in a region D about the two lines through P_0 with slopes B and $-B$, and then constructing a rectangle R centered on P_0 and having these lines as diagonals, and lying entirely in D. (See Figure 3.5.) Then, the interval I_B is specified by the width of this rectangle. The explanation of this mysterious process is as follows. We are interested in finding a solution of the equation (3.53) that passes through P_0 and lies in D. The bound on $f(t, y)$ guarantees that such a solution will have to obey $|y'(t)| \le B$, which means that the curve itself will have to lie between the two diagonal lines in Figure 3.5, and therefore will have to be inside R.

We start the Picard iteration by choosing ϕ_1 as the constant function y_0, and then define ϕ_{n+1} in terms of ϕ_n.

$$\begin{aligned} \phi'_{n+1}(t) &= f(t, \phi_n(t)) \\ \phi_{n+1}(t_0) &= y_0 \end{aligned} \tag{3.54}$$

It is clear that the graph of ϕ_1, for $t \in I_B$, lies in R. We want to prove that the same holds for each of the ϕ_n. If this is the case for ϕ_n, then the right-hand side of (3.54) is a well-defined continuous function of t for $t \in I_B$, whose values lie between $-B$ and B. When we integrate (3.54) to find ϕ_{n+1}, the result is a function whose graph goes through P_0, and whose slope lies between $-B$ and B and therefore lies in the shaded region in Figure 3.5. In particular, the graph of ϕ_{n+1} stays inside the rectangle R for all $t \in I_B$. Thus, by induction, all functions in the sequence $\{\phi_k\}$ have graphs that lie in R.

In order to prove that the functions ϕ_k converge to something, we impose an additional requirement on the function $f(t, y)$, namely, that it obeys a point $(t_0, y_0) = P_0$, and that $|f(t, y)| \le B$ for all points in D. The first step is to determine a special interval I_B about the value t_0. We do this by drawing

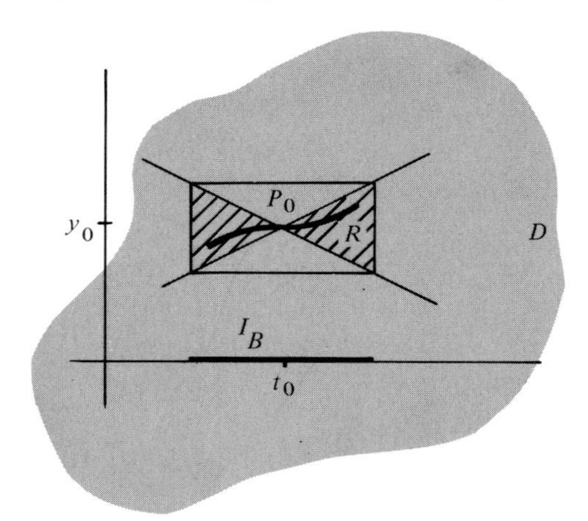

Figure 3.5

Lipschitz condition:

$$|f(t,u)-f(t,v)| \leq A\,|u-v| \tag{3.55}$$

for all choices of t, u, and v such that (t,u) and (t,v) lie in R. This property can be ensured by a simpler requirement on f, namely, that the partial derivative $f_y = \partial f/\partial y$ be continuous in D, for then it will be bounded and (3.55) follows. To see this, suppose we have $|f_y(t, y)| \leq A$. By the mean-value theorem, we can write

$$f(t,u) - f(t,v) = (u-v)\,f_y(t,\tau)$$

where τ lies between u and v. Then, taking absolute values

$$|f(t,u)-f(t,v)| = |u-v|\,|f_y(t,\tau)| \leq A\,|u-v|$$

which is (3.55).

We now state our main result.

THEOREM **3.18** *If f is continuous on the special rectangle R described in Figure* 3.5 *and there obeys the Lipschitz condition* (3.55), *then the sequence* $\{\phi_n\}$ *defined by the Picard iteration* (3.54) *converges to a function that is a solution of the initial-value problem* (3.53) *for those values of t that lie in the interval I_B and also obey $|t-t_0| < 1/A$.*

Proof Let $F_n = \phi_{n+1} - \phi_n$. Since each function ϕ_k obeys $\phi_k(t_0) = y_0$, we have $F_n(t_0) = 0$ for all n. Now, by (3.54),

$$\begin{aligned} F'_{n+1}(t) &= \phi'_{n+2}(t) - \phi'_{n+1}(t) \\ &= f(t, \phi_{n+1}(t)) - f(t, \phi_n(t)) \end{aligned}$$

so that, by the Lipschitz condition (3.55),

$$\begin{aligned} |F'_{n+1}(t)| &= |f(t,\phi_{n+1}(t))-f(t,\phi_n(t))| \\ &\leq A\,|\phi_{n+1}(t)-\phi_n(t)| \\ &\leq A\,|F_n(t)| \end{aligned} \tag{3.56}$$

holding for all t in the interval I_B.

The next step in the proof is to use this inequality to obtain a numerical bound on the size of $F_n(t)$ for all t in a subinterval I_0 of I_B determined by the number A. Choose any $\rho < 1/A$, and let I_0 be the small subinterval of I_B about t_0 consisting of those t with $|t-t_0| \leq \rho$. Let M be the maximum of the function F_1 on I_0. Then, for any $t \in I_0$, we have

$$\begin{aligned} F_2(t) &= F_2(t) - F_2(t_0) \\ &= (t-t_0)\,F'_2(\tau) \end{aligned}$$

for some point τ between t and t_0. Take absolute values, and use the inequality (3.56);

$$|F_2(t)| = |t-t_0|\,|F_2'(\tau)| \le |t-t_0|\,A\,|F_1(\tau)|$$
$$\le \rho AM$$

Hence, on I_0 the maximum of $|F_2|$ is at most ρAM.

Repeat this with F_3. For any $t \in I_0$,

$$F_3(t) = F_3(t) - F_3(t_0)$$
$$= (t-t_0)\,F_3'(\tau)$$

so that

$$|F_3(t)| = |t-t_0|\,|F_3'(\tau)|$$
$$\le |t-t_0|\,A\,|F_2(\tau)|$$
$$\le \rho A(\rho AM) = (\rho A)^2 M$$

Continuing, it is easy to see that for any t in the interval I_0,

$$|F_k(t)| \le (\rho A)^{k-1} M \tag{3.57}$$

The reason for the choice of ρ now appears, for the number $\lambda = \rho A$ is such that $\lambda < 1$ and $\lim \lambda^n = 0$. To see how we use this, we next consider the infinite series

$$\phi_1 + F_1 + F_2 + F_3 + \cdots \tag{3.58}$$

Since, for any t in I_0, $|F_n(t)| \le \lambda^{n-1} M$ and since $\sum \lambda^{n-1} M$ converges, the comparison test shows that the series (3.58) converges absolutely (and uniformly) for all $t \in I_0$. However, if we look at a typical partial sum of (3.58), we have

$$\phi_1 + (\phi_2 - \phi_1) + (\phi_3 - \phi_2) + \cdots + (\phi_n - \phi_{n-1}) = \phi_n$$

since this is a telescoping series. Thus, the sequence $\{\phi_n\}$ converges for each choice of t in I_0, and in fact it converges uniformly on the interval I_0. Denote the limit function, which is therefore the sum of the series (3.58), by ϕ. We can also apply the same argument to the series

$$\phi_1' + F_1' + F_2' + \cdots \tag{3.59}$$

Since, by (3.56) and (3.57),

$$|F_{k+1}'(t)| \le A\,|F_k(t)| \le \lambda^{k-1} AM$$

we see that the series (3.59) also is absolutely and uniformly convergent on I_0.

It follows that the integral of its limit must be the sum of the series (3.58), namely, ϕ; hence, ϕ has a derivative $\phi' = \lim \phi_n'$. Using (3.54), we conclude that $\phi' = f(t, \phi)$, which completes the proof. We have shown that the equation (3.53) has a solution and that it is the (uniform) limit of the Picard approximations.

The rate of convergence of the Picard iteration is geometric, for

$$\begin{aligned}\phi - \phi_n &= \phi_1 + F_1 + F_2 + \cdots - (\phi_1 + F_2 + \cdots + F_{n-1}) \\ &= F_n + F_{n+1} + F_{n+2} + \cdots\end{aligned}$$

and using (3.57)

$$\begin{aligned}|\phi - \phi_n| &\le |F_n| + |F_{n+1}| + \cdots \\ &\le \lambda^{n-1}M + \lambda^n M + \lambda^{n+1}M + \cdots \\ &\le \lambda^{n-1}M(1 + \lambda + \lambda^2 + \cdots) \\ &\le \lambda^{n-1}\frac{M}{1-\lambda}\end{aligned}$$

When the Lipschitz constant A is large, the number ρ must be taken quite small, and in this case the proof is able to prove convergence only on a small neighborhood of the point t_0. This is unavoidable, since in general there may not be a solution of the differential equation $y' = f(t, y)$ except on a rather small interval about the initial point t_0. For example, the equation $y' = 10y^2$, with the initial condition $y(0) = 1$, has the solution $y = 1/(1 - 10t)$; this blows up at $t = 0.1$, so that it is impossible to find a continuous solution of this equation on an interval $0 \le t \le T$ with T larger than 0.1; the Picard approximations cannot be expected to converge on an interval longer than $[0, 0.1)$.

We have given a proof of the fundamental local existence theorem for first-order equations, assuming that the defining function $f(t, y)$ is continuous and that it either has a continuous partial derivative f_y, or satisfies a Lipschitz condition (3.55). In more theoretical treatments of the subject, similar theorems are proved which require only continuity of f. However, these proofs do not seem to lead to an effective procedure for producing an approximating sequence $\{\phi_n\}$ that converges to the desired solution.

The other half of the important basic Theorem 1.1 is the uniqueness part, which asserts that an equation of the form $y' = f(t, y)$, *with appropriate conditions on* f, can have at most *one* solution passing through a given point (t_0, y_0). As suggested earlier, this has philosophical overtones connected with the concept of determinism. It also has extremely important mathematical consequences, for it is the justification for the remark that any process that leads to a function satisfying the differential equation and the initial conditions is acceptable, since there is only one solution. However, uniqueness does not hold for many simple equations in which the function f fails to obey condition (3.55). The equation

$$y' = \sqrt{y}, \qquad y(0) = 0$$

has the simple solution $y(t) \equiv 0$ as well as the solution $y = \frac{1}{4}t^2$, and in fact the solutions

$$y(t) = \begin{cases} 0 & 0 \le t \le A \\ \frac{1}{4}(t-A)^2 & A < t < \infty \end{cases}$$

and approximate procedures may yield any of these, depending on the procedure and the starting point. (See Exercise 10.)

We conclude this section by giving a proof of one form of the fundamental uniqueness theorem. It depends upon the following special theorem of analysis.

THE GRONWALL LEMMA *Let $g(t)$ be a function that is defined and has a derivative on the interval $[a,b]$ and there obeys the inequality*

$$|g'(t)| \le A\,|g(t)| \tag{3.60}$$

Let $g(t_0) = 0$ for at least one point $t_0 \in [a,b]$. Then $g(t) \equiv 0$ for all points t in $[a,b]$.

Proof Take a large integer N, and construct the special function $F(t) = \log(Ng(t)^2+1)$. Note that $F(t_0) = \log(0+1) = 0$. Moreover,

$$F'(t) = \frac{2Ng(t)g'(t)}{Ng(t)^2+1}$$

so that, using (3.60),

$$|F'(t)| \le 2\frac{N|g(t)|\,|g'(t)|}{Ng(t)^2+1}$$

$$\le 2A\frac{Ng(t)^2}{Ng(t)^2+1} \le 2A$$

for every t in $[a,b]$. Now, take any t on the interval $[a,b]$ and apply the mean-value theorem, using the fact that $F(t_0) = 0$:

$$F(t) = F(t) - F(t_0) = (t-t_0)F'(\tau)$$

where τ is between t and t_0. Take absolute values and use the bound we obtained on $[a, b]$ for the function F',

$$\begin{aligned} |F(t)| &= |t-t_0|\,|F'(\tau)| \\ &\leq 2A(b-a) \end{aligned}$$

The right-hand side of this relation does not depend on N. However, if $g(t) \neq 0$, the left-hand side, which is $\log(Ng(t)^2+1)$, could be made arbitrarily large by increasing N. Hence, $g(t) = 0$ for every t in $[a, b]$.

We now use this to obtain the following uniqueness theorem for first-order equations.

THEOREM 3.19 *Let $f(t, y)$ be defined and continuous in a region D containing the point $P_0 = (t_0, y_0)$, and bounded there by B. Construct a rectangle R, centered on P_0, as shown in Figure 3.5, and suppose that f obeys a Lipschitz condition*

$$|f(t,u)-f(t,v)| \leq A\,|u-v|$$

in R. Then, the differential equation

$$y' = f(t, y), \qquad y(t_0) = y_0 \tag{3.61}$$

can have no more than one solution defined on the interval I_B obeying the given initial conditions.

Proof Suppose that ϕ and ψ are both solutions of (3.61). As in the discussion that preceded Theorem 3.18, their graphs must lie in the rectangle R for all t in I_B. Since they solve the equation, we know that

$$\phi'(t) - \psi'(t) = f(t, \phi(t)) - f(t, \psi(t))$$

Taking absolute values and using the Lipschitz condition, we have

$$|\phi'(t)-\psi'(t)| \leq A\,|\phi(t)-\psi(t)|$$

Set $g(t) = \phi(t)-\psi(t)$, observe that $g(t_0) = y_0-y_0 = 0$, and apply the lemma to obtain $g(t) \equiv 0$, so that $\phi \equiv \psi$.

Exercises

Use the Picard process to generate approximating functions $\phi_1, \ldots, \phi_5$ for the equations of Exercises 1 and 2.

1. $y' = 3t^2 - t - y, \quad y(0) = 2$
2. $y' = \sin t + 2y, \quad y(0) = 0$

3. Find ϕ_3 for the equation $y' = t^2 + y^2$, $y(0) = 1$.

4. (a) Use the Picard process to calculate ϕ_5 with the starting choice $\phi_1(t) = e^t$ and the differential equation $y' = 2t + y$, $y(0) = 1$.
   (b) Use power series to compare ϕ_5 as obtained in part (a) with the true solution. [Note that this equation is the illustrative example (3.50) of the text.]

5. For the equation $y' = 10y^2$, $y(0) = 1$, calculate the first few Picard approximations. What do you observe about their behavior? How do you explain this?

The following is a Picard process for solving *second-order* equations. Given the equation

$$y'' = f(t, y, y'), \qquad y(t_0) = a,\ y'(t_0) = b$$

choose the function $\phi_1(t) = a + b(t - t_0)$ and define a sequence $\{\phi_k\}$ by the recursion

$$\begin{aligned} \phi''_{n+1} &= f(t, \phi_n, \phi'_n) \\ \phi_{n+1}(t_0) &= a \\ \phi'_{n+1}(t_0) &= b \end{aligned}$$

This process should be used for solving Exercises 6 and 7.

6. Use the Picard process for second-order equations to find a sequence of approximations to the solution of the equation $y'' = y'$, with $y(0) = 0$, $y'(0) = 1$. Compare this sequence with the exact solution, which can be found directly.

7. Do the same as in Exercise 6 for the equation $y'' = y$, $y(0) = 0$, $y'(0) = 1$, for which the exact solution is $y = \frac{1}{2}(e^t - e^{-t})$.

8. Consider the equation $y' = t^2 + \frac{1}{2}y^3$, with $y(0) = 1$. Use the region D consisting of points (t, y) with $|t| \le 1$, $0 \le y \le 2$, and estimate the numbers A and B that appear in Theorem 3.18. Then, obtain an explicit numerical interval about $t_0 = 0$ on which the proof of Theorem 3.18 ensures that a solution of this differential equation can be found.

9. For the equation $y' = 3y/t$ with $t_0 = 1$ and $y(1) = 1$, apply the Picard process with $\phi_1(t) = t$ to find ϕ_2, ϕ_3, and ϕ_4.

10. (a) For the equation $y' = \sqrt{y}$, with $y(0) = 0$, start with $\phi_1(t) = t$ and find ϕ_2, ϕ_3, ϕ_4. Can you guess what the sequence is converging to?
    (b) What happens if you repeat this with the choice $\phi_1(t) \equiv 0$?

The next two problems offer alternative proofs of Gronwall's Lemma.

*11. Let g and g' be defined on the interval $0 \leq t \leq 2$ and there satisfy the conditions $|g'(t)| \leq A|g(t)|$, $|g(t)| \leq M$, and $g(0) = 0$.
(a) Show that $|g(t)| \leq A\int_0^t |g(s)|\, ds$, for $t \in [0, 2]$.
(b) Show that $|g(t)| \leq AMt$, for $t \in [0, 2]$.
(c) Show that $|g(t)| \leq \frac{1}{2}A^2Mt^2$, for $t \in [0, 2]$.
(d) Show that $|g(t)| \leq A^nMt^n/n!$, for $t \in [0, 2]$.
(e) Show that $g(t) = 0$ for all t, $0 \leq t \leq 2$.

*12. Suppose that $|g'(t)| \leq A|g(t)|$ for $0 \leq t \leq 1$ and that $g(0) = 0$. Choose a small number ρ with $\rho < 1/A$, and let M be the maximum value of $|g(t)|$ for $0 \leq t \leq \rho$.
(a) Show that $|g(t)| \leq AMt$, for $0 \leq t \leq \rho$.
(b) Make a special choice of t, and conclude that $M = 0$.
(c) Show that $g(t) = 0$ for all t, $0 \leq t \leq \rho$.
(d) Show how it then follows that $g(t) = 0$ for $t \in [0, 1]$.

4

Numerical Methods

Introduction

The purpose of this chapter is to introduce a handful of numerical methods for finding approximate solutions of differential equations. While all the methods described could be done by hand, most are designed to be used in conjunction with some type of computer, especially one that can be programmed to do the dull and repetitive calculations. You are not expected to memorize the complex algorithms that describe the various methods discussed; nor are you expected to have access to a powerful computer, nor to be able to write programs yourself. Nor is the purpose to teach you to be a skilled numerical analyst, equipped to start work at a computing center. What is expected is that you will carry out some sample calculations in order to understand better the ideas behind the methods and to experience at first hand the problems that arise. In some cases, we give you vicarious experience with computers by letting you see the results of certain experiments. But your main objective should be to gain an insight into the origins and structure of some of the most frequently used numerical methods such as the Runge-Kutta, Adams-Bashforth, and predictor-corrector methods, so that you can be aware of their strengths and weaknesses.

The ordinary differential equations package that is part of a college computer-program library may provide a numerical solution (for example, a table of values) for some specific initial-value problem, but this solution may be quite incorrect for a variety of reasons, and unless the program has been written with great care and skill, the computer may not tell you this! It is fatal to depend upon the computer to obtain numerical solutions for you if you do not know enough either about differential equations or about numerical analysis.

Why do we need numerical methods at all? First, it should be clear by now that only a handful of equations can be solved completely in terms of familiar functions. For the rest, some approximate method must be used. Second, all the methods discussed in previous chapters have provided approximations that could be expected to be accurate for a very short range of t values. However, this is not the typical problem. Given an equation such as $y' = t^2 - y^2$, we are interested in the solution that obeys $y(0) = 0$, but we want the value of y for $t = 2$, or $t = 3$, or perhaps even $t = 10$. The methods of Chapter 3 might give us approximations that were useful for $t = 0.05$ or perhaps even $t = 0.2$, but none would give dependable information for a large value of t. Here is where certain numerical methods become especially important.

It should be noted in passing that there are a great many problems concerned with differential equations in which neither the theoretical nor the numerical methods help. These must wait for better methods to be developed by applied mathematicians and computer scientists in the coming generations.

While most of the exercises in this chapter can be done with pencil and paper alone, we have included a few in each section which become much less tedious if there is access to any simple calculator. The performance of pages of meticulous arithmetic may teach certain valuable virtues and skills such as patience, fortitude, and penmanship, but it is not likely to inculcate much knowledge of differential equations. For this reason, we encourage the reader to use any labor-saving gadgets that happen to be available.

Finally, there are always college students who have learned some of the simpler skills of programming, and who delight in letting a machine do as much of the work as possible. Again, we do not object to this in principle; laziness is as much a goad to research as altruism. For the benefit of such students, we have given in Appendix 2 a small collection of problems designed to be done only by students who have access to a programmable computer. These problems introduce no new mathematical ideas, and differ only in complexity from those assigned to all students.

Texts that deal with programming and with the detailed theory of numerical analysis are appearing with regularity. This book is not intended to duplicate these, so for the information of readers who wish more information in this direction, we mention the following which we happen to have enjoyed.

S. D. Conte and Carl De Boor, *Elementary Numerical Analysis: An Algorithmic Approach.* (McGraw-Hill, New York, 1965, 1972.)

C. W. Gear, *Numerical Initial Value Problems in Ordinary Differential Equations.* (Prentice-Hall, Englewood Cliffs, N.J., 1971. Automatic Computation Series.)

R. W. Hamming, *Numerical Methods for Scientists and Engineers*, 2nd. ed. (McGraw-Hill, New York, 1973.)

Jon M. Smith, *Scientific Analysis on the Pocket Calculator.* (Wiley-Interscience, New York, 1975.)

4.1 The Euler Method

Since the Euler method is the simplest numerical method for solving ordinary differential equations, it serves to introduce the basic viewpoint and notation that are used in the remainder of the chapter. As we shall show in the next section, the Euler method is capable in theory of giving approximations of arbitrarily high accuracy; however, the computation time and the practical limitations of round-off errors in the arithmetic processes make the achievement of this degree of accuracy difficult. This term "round-off" is used for any process by which a real number having an infinite number of decimal places is approximated by a rational number having a specified number of decimal places. Whenever a large number of successive arithmetic operations are carried out, rounding can introduce accumulated errors. For example, if we retain merely two decimals, the product of 0.31 and 0.37 could be calculated as 0.11, and the product of 0.23 and 0.37 as 0.09; if we then divide these, we have $(0.31)(0.37)/(0.23)(0.37) = 0.11/0.09 = 1.22$. However, the result of the calculation of $0.31/0.23$, which should be the same, is 1.35. We will not discuss the theory of round-off errors in this book, but instead assume that computations can be made with any required precision.

The basic idea behind the Euler method is that of approximating the exact solution of the given differential equation by a polygonal curve constructed by drawing a sequence of line segments tangent to the direction field of the differential equation. The geometry is easily understood.

Suppose that the equation is $y' = f(t, y)$ and we are interested in the solution which obeys the initial conditions $y(t_0) = y_0$. As in Chapter 1, we can envision this differential equation as defining a direction field in the TY-plane as in Figure 4.1, where we have shown a short line segment of slope $f(t, y)$ at each point (t, y) in the domain of f. The solutions for the differential equation are the curves that have the characteristic property that they pass

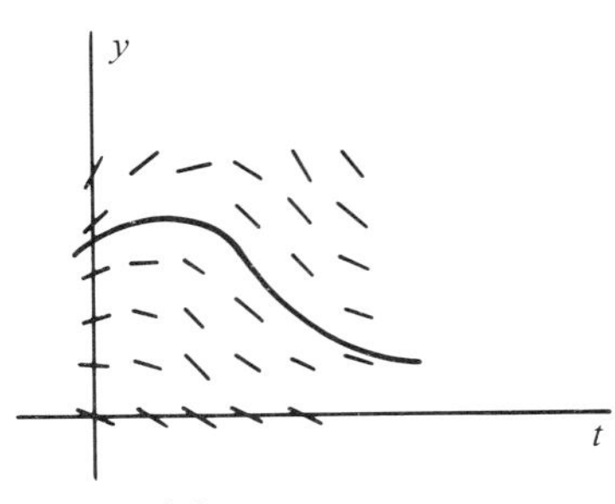

Figure 4.1

through the direction field tangent to each of these vectors at each of their points. We are interested in the special solution that passes through the point (t_0, y_0). It is possible, as we pointed out in Chapter 1, to make freehand sketches of these curves; while not accurate for detailed numerical work, these sometimes display the general trend and qualitative behavior of solutions. R. W. Hamming of Bell Laboratories (who is famous among computer scientists for his dictum: *The purpose of computing is insight, not numbers*!) makes the following remark in his excellent text on numerical analysis:

> The direction field is a simple idea, but it should not be scorned. Frequently it shows the nature of the problem so that a sound approach may be made to the more accurate numerical solution. Several times in the author's experience, the direction-field sketch has answered the original problem, and on one occasion an accurate sketch prepared on a drawing board gave sufficient accuracy to answer the problem completely.
>
> R. W. Hamming, *Numerical Methods for Scientists and Engineers*, McGraw-Hill, New York, 1962, p. 185

To see how the direction field is used in the Euler method, we start by choosing a convenient sequence of t values, $t_0 < t_1 < t_2 < t_3 < \cdots < t_n < t_{n+1} < \cdots$. Although the spacing of these does not have to be uniform, a simple choice is to put $t_n = t_0 + nh$, where h is a fixed **step size**. The object of any numerical method is to obtain estimates y_k of the numerical values $y(t_k)$, for $k = 1, 2, \cdots$, for the desired solution function $y(t)$, starting with the given value $y(t_0) = y_0$. The accuracy of any numerical method is measured by how much the estimate, usually called y_k, differs from the exact value $y(t_k)$. The error $y(t_k) - y_k$ depends upon k and the step size h, and also of course upon the particular function f that determines the differential equation. In general, this error will increase with large k, since one is estimating values for y at points t remote from the initial point t_0. One also expects this error to decrease if the step size h is decreased.

The various numerical methods for solving differential equations differ widely in their behavior. Some achieve good accuracy with a relatively large step size h but are complicated to apply. Others are simpler in their construction but require a very small step size in order to obtain a reasonable degree of accuracy. The Euler method belongs to the latter class, but has the virtue of being a very simple method and one for which it is very easy to prove that the accuracy increases as h decreases; we present the proof in the next section.

We start by examining the direction field for the given differential equation and constructing the vertical lines $t = t_0, t = t_1, t = t_2, \cdots$, as in Figure 4.2. We know that the exact solution curve passing through the point $P_0 = (t_0, y_0)$ must have there a slope equal to $m_0 = f(t_0, y_0)$. We therefore approximate the solution curve at P_0 by a straight line of slope m_0. This line meets the vertical line $t = t_1$ at a new point P_1, with coordinates (t_1, y_1). At the point P_1, we again calculate the slope of the direction field, and find $m_1 = f(P_1) = f(t_1, y_1)$. We then construct a new line through P_1 with slope m_1, and denote the point where it meets the vertical line $t = t_2$ as $P_2 = (t_2, y_2)$. Continuing in this fashion, we generate a polygonal curve made up of straight line segments; this is the Euler approximation to the true solution curve, and the numbers y_k are the estimates for the true values $y(t_k)$. In Figure 4.3 we show the relation between the Euler approximation and the exact solution. The first segment of the Euler approximation is part of the tangent line to the exact solution. However, this is not in general true for the subsequent segments.

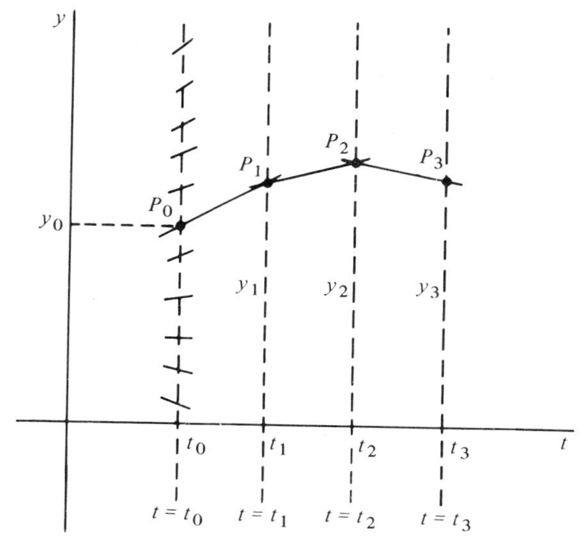

Figure 4.2

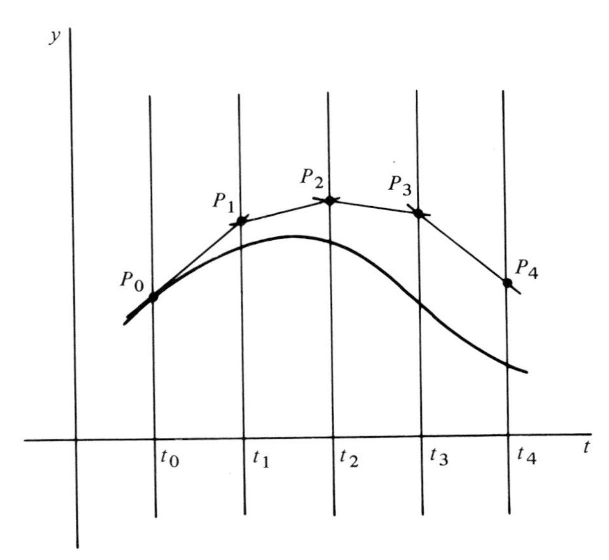

Figure 4.3

Since the construction method uses the direction field for the differential equation, each segment is tangent to *some* solution curve at an appropriate point, but after the first segment, none of these solution curves is apt to be the unique one through P_0.

Having given a geometric description of the Euler process, we next present the details of the analytic **algorithm** to show how the y_k are calculated. We start with y_0, which is $y(t_0)$. The slope of the direction field at P_0 is $m_0 = f(P_0) = f(t_0, y_0)$ so that the equation of the first line segment is

$$y - y_0 = m_0(t - t_0)$$

Setting $t = t_1 = t_0 + h$, we obtain

$$\begin{aligned} y_1 &= y_0 + hm_0 \\ &= y_0 + hf(t_0, y_0) \\ &= y_0 + hf(P_0) \end{aligned}$$

We then calculate the slope at $P_1 = (t_1, y_1)$, obtaining $m_1 = f(P_1) = f(t_1, y_1)$. The line through P_1 with slope m_1 has the equation $y - y_1 = m_1(t - t_1)$, and setting $t = t_2 = t_1 + h$, we obtain

$$\begin{aligned} y_2 &= y_1 + hf(P_1) \\ &= y_1 + hf(t_1, y_1) \end{aligned}$$

In general, we have

$$y_{n+1} = y_n + hf(t_n, y_n) \tag{4.1}$$

This formula, together with the initial data $y_0 = y(t_0)$, and the fact that $t_n = t_0 + nh$, provides a simple recursive algorithm for calculating the successive estimates y_k.

Let us try this out with a familiar equation whose solution we know. Consider the equation $y' = y$ with initial condition $y(0) = 1$. The solution of this is clearly $y = e^t$. We first apply the Euler method with $h = 0.5$ and attempt to estimate the value of y when $t = 1$ and when $t = 2$. This requires that we find y_2 and y_4, since $y_n \approx y(t_n)$ and $t_n = 0 + nh = (0.5)n$. We start with $y_0 = 1$ since this is $y(0)$. Then

$$\begin{aligned}
y_1 &= y_0 + (0.5)\,f(0, y_0) \\
&= y_0 + (0.5)\,y_0 = (1 + 0.5)\,y_0 \\
&= (1.5)(1) = 1.50 \\
y_2 &= y_1 + (0.5)\,f(0.5, y_1) \\
&= y_1 + (0.5)\,y_1 \\
&= (1.5)(1.5) = 2.25 \\
y_3 &= y_2 + (0.5)\,f(1.0, y_2) \\
&= y_2 + (0.5)\,y_2 \\
&= (1.5)(2.25) = 3.37 \\
y_4 &= y_3 + (0.5)\,f(1.5, 3.37) \\
&= (1.5)(3.37) = 5.06
\end{aligned}$$

In Table 4.1, we give these computed approximations and, for comparison, the exact values to several places. (Here, as elsewhere, we use * to indicate that additional digits have been omitted.) The accuracy of the approximation

Table 4.1

t	0	0.5	1.0	1.5	2.0
Euler					
y	1.00	1.50	2.25	3.37	5.06
exact					
y	1.00*	1.65*	2.72*	4.48*	7.39*

is certainly not impressive! To see what happened, we show in Figure 4.4 the exact exponential solution and the Euler polygonal estimate, and it is clear that the error that existed at the end of the first step was magnified by each subsequent step until we reached $t = 2$.

Things improve somewhat if we try the Euler method with a smaller value for h. For example, with step size $h = 0.01$, we have the following

$$\begin{aligned}
y_0 &= 1.0000 \\
y_1 &= y_0 + (0.01)\, f(0, y_0) \\
&= y_0 + (0.01)\, y_0 = (1 + 0.01)\, y_0 \\
&= (1.01)(1.0000) = 1.0100 \\
y_2 &= y_1 + (0.01)\, f(0.01, y_1) \\
&= y_1 + (0.01)\, y_1 \\
&= (1.01)(1.0100) = 1.0201 \\
y_3 &= (1.01)(1.0201) \\
&= 1.0303
\end{aligned}$$

And so on. (This computation is tedious to do with pencil and paper but is extremely simple with any pocket calculator.)

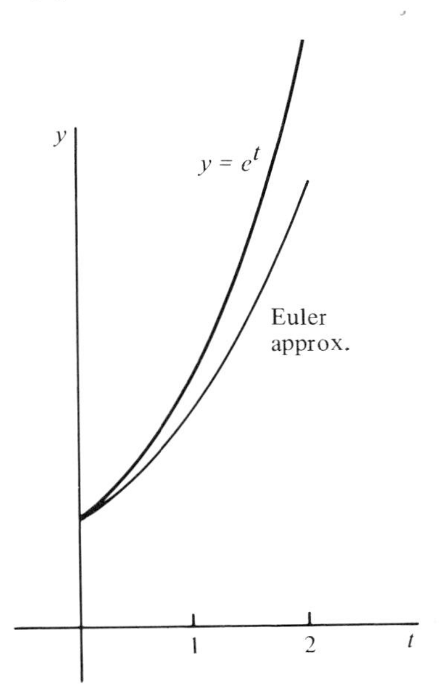

Figure 4.4

In order to obtain estimates for $y(1.0)$ and $y(2.0)$, how many terms in the sequence y_k must we calculate? Since $y_n \approx y(t_n) = y([0.01]n)$, we see that we need to know y_{100} for $y(1.0)$ and y_{200} for $y(2.0)$. The results obtained by carrying out the computation of y_n for all n from $n = 0$ to $n = 200$ is summarized in Table 4.2.

Table 4.2 *Euler method:* $h = 0.01$

t	0	0.5	1.0	1.5	2.0
Euler y	1.0000	1.6446	2.7048	4.4484	7.3160
exact y	1.0000	1.6487*	2.7183*	4.4817*	7.3891*
error	0.0000	0.0041	0.0135	0.0333	0.0731

If we decrease the step size to $h = 0.001$, the accuracy again improves. This time, since $t_n = 0 + [0.001]n$, we must calculate y_{2000} to obtain an estimate for $y(2.0)$. The results are displayed in Table 4.3. Comparison of

Table 4.3 *Euler method:* $h = 0.001$

t	0	0.5	1.0	1.5	2.0
Euler y	1.0000	1.6467	2.7115	4.4650	7.3523
error	0.0000	0.0020	0.0068	0.0167	0.0368

the data of Tables 4.2 and 4.3 shows that we have cut the errors by a half. It should seem plausible that if we were to continue to decrease h, the results would converge toward the exact values. In the next section, we prove that this is indeed the case, if all the arithmetic computations are carried out with infinite precision and no round-off. Of course, if h is *very* small, then the number of steps must be very large. This increases the total running time of the calculation; in addition, each arithmetic computation will involve round-off errors and as these accumulate, the total round-off error can become sizable. It is for these reasons that numerical analysts have devised numerical methods that are better than the Euler method and do not require as small a choice of h; we will discuss these in later sections of this chapter.

To show the usefulness of numerical methods, let us consider a more complicated differential equation, namely,

$$y' = t^2 + y^2, \qquad y(0) = 1 \tag{4.2}$$

We have worked with this before. In Section 3.4, formula (3.27) and the discussion that followed, we obtained a polynomial approximation to the solution that was useful for values of t near the initial point $t = 0$. However, this could tell us little about the nature of the solution for values of t beyond $t = 0.2$ and, in particular, for t as large as 0.5 or 1.0. In Section 3.5, formula (3.44), we obtained the first few terms in a power-series solution, but since we did not find the general term, we could not discuss convergence of the power series. Let us see what can be learned by applying Euler's method.

We take $h = 0.1$ and $y_0 = 1$. Then, the general recursion is

$$\begin{aligned} y_{n+1} &= y_n + h(t_n^2 + y_n^2) \\ &= y_n + (0.1)([0.1]^2 n^2 + y_n^2) \end{aligned}$$

Accordingly,

$$\begin{aligned} y_1 &= y_0 + (0.1)(0^2 + 1^2) \\ &= 1 + (0.1)(1.0) = 1.100 \\ y_2 &= y_1 + (0.1)([0.1]^2 + [1.100]^2) \\ &= 1.100 + 0.122 = 1.2220 \\ y_3 &= y_2 + (0.1)([0.2]^2 + [1.222]^2) \\ &= 1.222 + (0.1)(1.533) = 1.375 \end{aligned}$$

These terms so far give us estimates for $y(0.1)$, $y(0.2)$, and $y(0.3)$. If we continue, we obtain the results given in Table 4.4.

Table 4.4 *Euler method: $h = 0.1$*

t	*Euler y*	t	*Euler y*
0	1.000	0.6	2.199
0.1	1.100	0.7	2.719
0.2	1.222	0.8	3.508
0.3	1.375	0.9	4.802
0.4	1.574	1.0	7.189
0.5	1.837	1.1	12.458

If we used these values to sketch an approximate solution curve, we would have a curve that rises slowly until t reaches 0.7 and then rises very sharply between there and $t = 1.1$. Of course, we do not as yet know how accurate these values are, nor whether this picture truly represents the exact solution of the differential equation.

This illustrates one of the important problems that arise in the use of numerical methods. Since they are not exact, it is very helpful to know something about the magnitude of the error involved. Detailed error analysis is one of the main topics in numerical analysis. In the next section, we obtain a crude error bound for Euler's method of the form $|\text{error}| < Ch$, where h is the step size and C is a constant that depends on the differential equation and the interval of t values on which the solution is sought. Unfortunately, the size of C depends upon the size of y itself. In our present example, then, we may place some confidence in the numerical values for t near zero, but we should not trust those near $t = 1$ where y seems to be so large.

In such a case, it often helps to try a smaller step size. In Table 4.5, we show

Table 4.5

	$h = 0.01$	$h = 0.001$
t	*Euler y*	*Euler y*
0	1.00000	1.00000
0.1	1.11013	1.11133
0.2	1.24933	1.25264
0.3	1.43180	1.43887
0.4	1.68050	1.69450
0.5	2.03599	2.06378
0.6	2.57864	2.63712
0.7	3.49519	3.63585
0.8	5.34276	5.79053
0.9	10.80409	13.80218
1.0	90.75551	overflow

the effect of two different choices for h. (The calculations in the table were done on a small hand-held electronic calculator, and "overflow" in the second column indicates that no value could be obtained for $t = 1$; the numbers became too large for the capacity of the registers.) Comparing these results, we may conclude a number of things. First, the general agreement of the results for t in the range $0 \leq t \leq 0.6$ suggests that the computed values of y are probably fairly close to the true values of y in this range. Second, it seems clear that something strange happens when t becomes close to $t = 1$.

We shall continue to work with this example in later sections. Eventually, by an analytic method, we show that the function $y(t)$ has a vertical asymptote at a value of t between 0.9 and 1.0. The Euler method is good enough to identify the difficulty and to lead one to conjecture the correct explanation, but is not good enough to yield accurate values for y near the critical singularity without going to much smaller values of h, which in turn would require very long calculations and produce large round-off errors. The methods we discuss later in this chapter do better.

In Exercise 9, we have given an example in which something similar happens; in this case, however, it is much easier to understand what is happening since the differential equation has a simple solution, namely, the function $y = 1/(1-t)$.

Exercises

1. (a) Apply the Euler method with $h = 0.2$ to estimate the value of y for $t = 1$ and $t = 2$, where y satisfies the differential equation $y' = -2y$ with initial condition $y(0) = 1$.
   (b) Compare the numerical results with the exact value.

2. (a) Apply the Euler method with $h = 0.2$ on the interval $0 \leq t \leq 1$ to the equation $y' = t - y$, $y(0) = 1$, and estimate $y(1)$.
   (b) Compare the value for $y(1)$ computed in part (a) with the value of the exact solution.

PC3. If you have access to a pocket calculator, repeat Exercise 1 with step size $h = 0.05$, and estimate $y(1)$.

PC4. If you have access to a pocket calculator, repeat Exercise 2 with $h = 0.05$, and estimate $y(1)$.

5. (a) Sketch the direction field for $y' = \sqrt{1-y^2}$.
   (b) Check that a solution for this equation obeying the initial condition $y(0) = 0$ is given by

$$y = \begin{cases} \sin t & 0 \leq t < \frac{1}{2}\pi \\ 1 & \frac{1}{2}\pi \leq t \end{cases}$$

   (c) Without doing computations, can you predict the behavior to be expected for the Euler method with $h = 0.4$?

6. (a) There are times when it is more efficient to change the step size as you go along. Write a description for an algorithm for computing the estimates y_n in Euler's method with step sizes given as $h_1, h_2, h_3, \ldots$.
   (b) Why would you want to make the step size large when y'' is small, and small when y'' is large?

7. (a) Can you explain why the Euler estimates were consistently too small for the illustrative example $y' = y$ and for the equation $y' = -2y$ in Exercise 1?
   (b) Can you formulate a general rule which applies to any equation $y' = f(t, y)$?

8. The Euler method works just as well if h is negative. Use $h = -0.2$ to estimate the value of y at $t = -1$ for the initial-value problem $y' = y$, $y(0) = 1$.

9. (a) Apply the Euler method to the problem $y' = y^2$, $y(0) = 1$, with $h = 0.2$ on the interval [0, 1.2].
   (b) Show that the exact solution to the equation of part (a) is $y = 1/(1-t)$. Does this explain the strange numerical behavior?

PC10. If you have access to a pocket calculator, estimate $y(1)$ and $y(1.2)$ using $h = 0.05$ where $y' = y^2$ and $y(0) = 1$.

4.2 Convergence of the Euler Approximations

In this section, we will show that for sufficiently small step size h, the Euler approximation is very close to the exact solution of the differential equation. We start by examining a familiar special case.

Consider the initial-value problem $y' = y$ with $y(0) = 1$. The exact solution of this is $y = e^t$. The Euler algorithm with step size h produces a sequence $\{y_n\}$ described by

$$\begin{aligned} y_0 &= 1 \\ y_{n+1} &= y_n + hy_n \\ &= (1+h)\, y_n \end{aligned} \tag{4.3}$$

The general formula for y_n is easily seen to be

$$y_n = (1+h)^n$$

and y_n is an estimate for the exact value $y(t_n)$, where $t_n = t_0 + nh = nh$. If we take a particular value of t, say $t = T$, and choose an n and h so that $T = t_n$, then $y_n \approx y(T)$. We want to investigate the size of the approximation error $|y(T) - y_n|$ as h decreases. In our case, to obtain $t_n = T$ we must choose $h = T/N$, $n = N$, $y(T) = e^T$, and $y_N = (1 + T/N)^N$. The error at $t = T$ is then $\epsilon_N = |e^T - (1 + T/N)^N|$, and the desired convergence is established by the following result (which is sometimes a definition).

LEMMA
$$\lim_{N \to \infty} \left(1 + \frac{T}{N}\right)^N = e^T$$

This can be proved at once by L'Hospital's Rule, which implies

$$\lim_{x \to 0} \frac{\log(1+xT)}{x} = T$$

If we replace x by $1/N$, we have

$$\lim_{N \to \infty} N \log\left(1 + \frac{T}{N}\right) = T$$

and, taking exponentials, $(1+T/N)^N \to e^T$.

We want to prove that the same convergence property holds for any suitably well-behaved differential equation. We suppose that we are dealing with an initial-value problem of the usual form:

$$\begin{aligned} y' &= f(t, y) \\ y(t_0) &= y_0 \end{aligned} \tag{4.4}$$

with t in an interval I about the value t_0. We want to prove that if we apply the Euler process with step size h, calculate the value of the Euler approximations corresponding to a point $t = T$ in I, and let h tend toward zero, then these approximations will converge toward the exact solution of (4.4) at $t = T$. As in the preceding example, we choose a large integer N and then select $h = (T-t_0)/N$. This gives $T = t_0+Nh = t_N$, so that the Euler approximation is the term y_N. What we must prove is that $\lim_{N\to\infty} y_N = y(T)$, where $y = y(t)$ is the exact solution of (4.4).

To simplify the proof, we impose two conditions on the differential equation: one on the function f and one on the solution function $y(t)$. [With a little more work, it can be shown that both hold if f and its first partial derivatives are known to be continuous on some region containing the point (t_0, y_0).] These conditions are similar to those used in the existence and uniqueness theorems in Section 3.6.

THEOREM 4.1 *Assume that the following pair of conditions hold.*

$$\begin{aligned} |f(t,u)-f(t,v)| &\le B|u-v| \\ |y''(t)| &\le M \end{aligned} \tag{4.5}$$

Then, $\lim_{N\to\infty} y_N = y(T)$ *and in fact* $|y_N - y(T)| < C/N$ *for some constant* C.

Proof We choose h, and then construct the sequence $\{y_n\}$ according to the Euler algorithm; for simplicity, we assume $T > t_0$, and we will require $h = (T - t_0)/N$ later on in the proof. We have

$$\begin{aligned} y_0 &= y(t_0) \\ y_{n+1} &= y_n + hf(t_n, y_n) \end{aligned} \tag{4.6}$$

We want to compare the value of y_n and the exact solution $y(t)$ for the choice $t = t_n$, namely, $y(t_n)$. Thus, we want to study the differences $y(t_n) - y_n$.

Since the function $y(t)$ is a solution of the differential equation (4.4), we have

$$y'(t) = f(t, y(t))$$

An application of Taylor's Theorem yields

$$\begin{aligned} y(t+h) &= y(t) + hy'(t) + \tfrac{1}{2}h^2 y''(\tau) \\ &= y(t) + hf(t, y(t)) + \tfrac{1}{2}h^2 y''(\tau) \end{aligned}$$

where τ is a number lying between t and $t+h$.

In this, set $t = t_n$, and recall that $t_n + h = t_{n+1}$. Then

$$y(t_{n+1}) = y(t_n) + hf(t_n, y(t_n)) + \tfrac{1}{2}h^2 y''(\tau)$$

Subtract from this the equation describing the Euler recursion (4.6)

$$y(t_{n+1}) - y_{n+1} = y(t_n) - y_n + hf(t_n, y(t_n)) - hf(t_n, y_n) + \tfrac{1}{2}h^2 y''(\tau)$$

Taking absolute values, and using the hypotheses on f and y given in (4.5), we have

$$\begin{aligned} |y(t_{n+1}) - y_{n+1}| &\le |y(t_n) - y_n| \\ &\quad + h|f(t_n, y(t_n)) - f(t_n, y_n)| \\ &\quad + \tfrac{1}{2}h^2 |y''(\tau)| \\ &\le |y(t_n) - y_n| + hB|y(t_n) - y_n| + \tfrac{1}{2}h^2 M \end{aligned}$$

Set $\epsilon_k = |y(t_k) - y_k|$, the absolute error at the kth step in the Euler approximation; this is usually called the **local truncation** or **discretization error**. Then, we have shown that this obeys the following simple inequality:

$$\epsilon_{n+1} \le (1 + hB)\epsilon_n + \tfrac{1}{2}Mh^2$$

which has the form

$$\epsilon_{n+1} \le \alpha\epsilon_n + \beta \tag{4.7}$$

with $\alpha = 1 + hB$ and $\beta = \tfrac{1}{2}Mh^2$. In our case, $\epsilon_0 = 0$ since $y(t_0) = y_0$. Taking

$n = 1, 2, \cdots$ in (4.7), we find

$$\epsilon_1 \leq \beta$$
$$\epsilon_2 \leq \alpha\epsilon_1 + \beta \leq \alpha\beta + \beta = (1+\alpha)\beta$$
$$\epsilon_3 \leq \alpha\epsilon_2 + \beta \leq (\alpha+\alpha^2)\beta + \beta$$
$$\leq (1+\alpha+\alpha^2)\beta$$

and in general

$$\epsilon_n \leq (1+\alpha+\alpha^2+\cdots+\alpha^{n-1})\beta$$
$$\leq \frac{\alpha^n - 1}{\alpha - 1}\beta < \frac{\alpha^n}{\alpha - 1}\beta$$

Replacing α and β by their values as previously given, we have shown that

$$\epsilon_n = |y(t_n) - y_n| \leq \frac{(1+hB)^n}{(1+hB) - 1}\tfrac{1}{2}Mh^2 = h\frac{M}{2B}(1+hB)^n$$

At this point, we use an inequality whose proof is an exercise in elementary calculus:

$$1 + x < e^x \qquad \text{for all } x > 0 \tag{4.8}$$

Applying this, we have $1 + hB < e^{hB}$ so that $(1+hB)^n < e^{nhB}$. This gives us a more convenient form for the error estimate:

$$|y(t_n) - y_n| < h\frac{M}{2B}e^{nhB} \tag{4.9}$$

Finally, with $n = N$, we replace h by $(T - t_0)/N$ so that $t_N = T$, and have

$$|y(T) - y_N| < h\frac{M}{2B}e^{B(T-t_0)} = \frac{M(T-t_0)\,e^{B(T-t_0)}}{2BN} \tag{4.10}$$

The right-hand term of this has the form C/N and tends to zero as N increases. Hence, $\lim_{N\to\infty} y_N = y(T)$, which is the desired convergence property; at any choice of t in the domain where the solution exists and the bounds on f and y'' apply, the Euler approximations $\{y_N\}$ with step size $(T - t_0)/N$ converge to the value of the exact solution there.

Note that in (4.10) we have also discovered something about the rate of convergence; we have in fact shown that ϵ_N is of the order of $1/N$, meaning that $\epsilon_N \leq C/N$ for some constant C. The mathematical shorthand for this is $\epsilon_N = O(1/N)$. Such a sequence does not converge to zero very rapidly; a sequence that is $O(1/N^2)$ or $O(1/N^3)$ converges much more rapidly. While our bound for ϵ_N was only an upper bound, standard examples show that the bound $O(1/N)$ is in fact the best that can be obtained for Euler's method.

(See Exercise 3.) This explains the slow convergence that was evident in the numerical examples in Section 4.1. This fact is usually described by saying that the total accumulated truncation error of the Euler method is like h; this means that the error $|y(T)-y_{\text{Euler}}| = O(h)$. Thus, to obtain an approximation by Euler's method whose error (ignoring round-off) is 0.0001, we would have to choose a step size h that is far smaller than this, as, for example, in Exercise 6. In the next section, we discuss methods that are much better.

Exercises

1. Show by direct calculation that the Euler method, applied to the initial-value problem $y' = 2y/t$, $y(1) = 1$, converges as $h \downarrow 0$. *Hint:* First show that the recursion for y_n has the solution

$$y_n = \frac{(1+nh)(1+h+nh)}{(1+h)}$$

Then, if $T = 1 + Nh$, show that $\lim y_N = T^2$.

2. Show by elementary calculus that
   (a) $1 + x \leq e^x$ for all real x.
   (b) $e^x \leq 1 + x + x^2$ for all $x \leq 1.5$.

*3. (a) When the Euler method is applied to the equation $y' = y$ with $y(0) = 1$, the error was shown in this section to be $\epsilon_N = e^T - [1+(T/N)]^N$. Take $T = 1$ and then show that $\epsilon_N \approx e/(2N)$ for large N. *Hint:* Show that

$$\lim_{x \downarrow 0} \frac{e - (1+x)^{1/x}}{x} = \frac{e}{2}$$

PC(b) If a pocket calculator is available, compute $S_N = N\{e - [1+(1/N)]^N\}$ for increasing N and compare the results with $\frac{1}{2}e$.

4. Let $\{\epsilon_n\}$ be a positive sequence that obeys the recursive inequality $\epsilon_{n+1} \leq \epsilon_n + n$. If $\epsilon_1 = 1$, show that $\epsilon_n \leq \frac{1}{2}n^2$ for $n \geq 2$.

5. Suppose that for a particular differential equation, $t_0 = 1$, $T = 5$, $B = 1$, and $M = 2.5$. What should the step size be to ensure that $|y(T)-y_N| \leq 0.001$, and how large will N be? [*Hint:* Use (4.10).]

6. For the equation $y' = t^2 + y^2$, $y(0) = 1$, on $0 \leq t \leq 0.7$, what should h be to be sure, from formula (4.10), that the error $|y(T)-y_N|$ is less than 0.01? [Assume that $|y(t)| < 5.0$.]

4.3 RK Methods

There are a number of numerical methods for integrating differential equations that are considerably more efficient than the Euler method; some, for example, permit us to obtain surprisingly accurate approximations to the solution even when a relatively large step size is used. We will discuss several of these in this section, concentrating on a class of methods that make use only of current information available at the previous step in order to estimate the value at the next step. A number of such methods were discovered during the last century by C. Runge and M. Kutta.

Let us begin by developing a natural generalization of the Euler method. The key to the Euler method was extrapolation of the solution curve by using a tangent line; it is natural to replace the tangent line by a better-fitting polynomial curve—a parabola, for example, or more generally, a polynomial of degree $m \geq 2$. At the initial point (t_0, y_0) we know the correct slope $y'(t_0) = f(t_0, y_0)$ of the solution curve. If we also knew the correct value of $y''(t_0)$, we could approximate the solution curve in a neighborhood of the initial point by the parabola

$$y = y_0 + y'(t_0)(t-t_0) + \tfrac{1}{2}y''(t_0)(t-t_0)^2$$

and then use this to obtain an estimate for the correct value of y for $t = t_1$. (See Figure 4.5.) Since $t_1 = t_0 + h$, this yields the estimate

$$y_1 = y_0 + hy'(t_0) + \tfrac{1}{2}h^2y''(t_0)$$

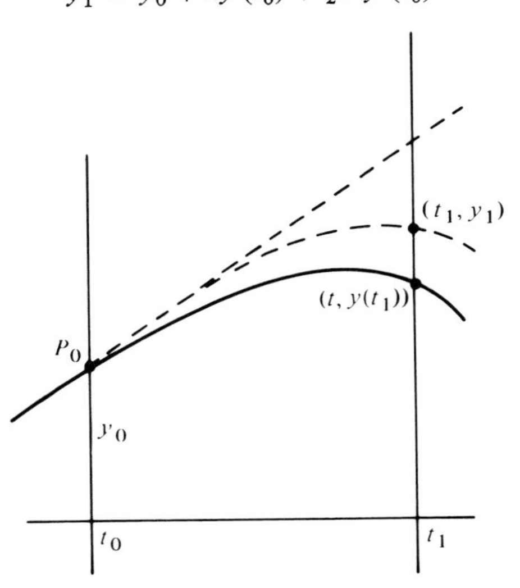

Figure 4.5

It is possible to calculate $y''(t)$ from the differential equation, since

$$y'(t) = f(t, y(t))$$
$$y''(t) = f_1(t, y(t)) + f_2(t, y(t))\, y'(t)$$

where f_1 and f_2 are the partial derivatives of f with respect to the first and second variables. Setting $t = t_0$ and $y = y_0$, we find that

$$y'(t_0) = f(t_0, y_0)$$
$$y''(t_0) = f_1(t_0, y_0) + f_2(t_0, y_0)\, f(t_0, y_0)$$

If we repeat this process at the point (t_1, y_1) and at each of the subsequent steps of the iteration, we arrive at the following algorithm for constructing an approximating sequence $\{y_n\}$ of estimates for the exact values $y(t_n)$.

The Modified Euler Method

$$t_n = t_0 + nh$$
(4.11) $$y_0 = y(t_0)$$
$$y_{n+1} = y_n + hf(t_n, y_n) + \tfrac{1}{2}h^2\,[f_1(t_n, y_n) + f_2(t_n, y_n)\, f(t_n, y_n)]$$

To see how this process works in practice, let us apply it to one of the equations we have considered before. Consider the initial-value problem

$$y' = t - y, \qquad y(0) = 1$$

whose solution is $y = t - 1 + 2e^{-t}$. To apply the modified Euler method, we need the partial derivatives $f_1 = 1$ and $f_2 = -1$. The formula for the recursion is then

$$t_n = 0 + nh = nh$$
$$y_0 = 1$$
$$y_{n+1} = y_n + h(t_n - y_n) + \tfrac{1}{2}h^2\,[1 + (-1)(t_n - y_n)]$$
$$= (1 - h + \tfrac{1}{2}h^2)\, y_n + (h - \tfrac{1}{2}h^2)\, t_n + \tfrac{1}{2}h^2$$

If we take $h = 0.2$, then $t_n = (0.2)\,n$ and

$$y_{n+1} = (0.82)\, y_n + (0.036)\, n + 0.02$$

Thus,

$$y_0 = 1$$
$$y_1 = 0.84$$
$$y_2 = 0.7448$$
$$y_3 = 0.7027$$
$$y_4 = 0.7042$$
$$y_5 = 0.7415$$

We therefore estimate $y(1.0)$ to be 0.7415; the exact value is $2e^{-1} = 0.7358*$, so the error is less than 0.006.

From this example, it seems likely that for the same step size h, the modified Euler method is more accurate than the simple Euler method. (This is confirmed by Exercise 1.) A detailed theoretical error analysis shows this same result, since the local truncation error turns out to be $O(h^3)$, producing a total accumulated error of $O(h^2)$. However, the modified Euler method has some obvious drawbacks, since at each step in the process we must calculate not only a new value of f but also values of its partial derivatives f_1 and f_2. For our illustrative example, $y' = t-y$, this is trivial. However, for a more realistic example such as

$$y' = f(t, y) = \frac{\sin(0.67t+0.43)\,e^{-0.2t^2}}{e^{yt}+t^3\cos^2(0.3y)}$$

these portions of the work might be time-consuming. Moreover, in some applications, the values of the function f are derived from empirical measurements. This means that no formula exists for f, and partial derivatives of f would have to be estimated numerically, leading to a further source for errors.

We can adopt an alternative approach which will eventually lead to the efficient RK methods. Suppose we examine the basic iteration step more carefully. We are trying to use the data available to us in the direction field to predict the next solution value, based on knowledge of the previous solution values. It will help to look at this geometrically, as in Figure 4.6. Suppose that

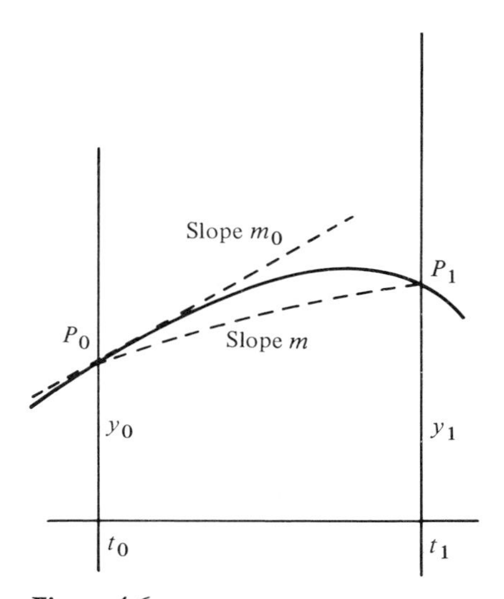

Figure 4.6

we have reached the point P_0 on the solution curve, and we want to locate the next point P_1. Clearly, we could find P_1 accurately if we knew the slope m of the chord $P_0 P_1$, for we could then write

$$y_1 = y_0 + hm$$

Many numerical methods are based on formulas that give a good estimate for the value of m. For example, the simple Euler method is based on the choice $m \approx m_0$, where m_0 is the slope of the direction field at P_0. It is clear from the diagram of Figure 4.6 that this is not a very good choice, since we are replacing the true slope of the chord by the slope of the curve at the left endpoint.

By the mean-value theorem, the slope of the chord must be the slope of the curve at some point on it between P_0 and the unknown point P_1. It seems plausible that a better choice than the endpoint would be any point that is near the middle, such as the point Q in Figure 4.7. Unfortunately, since we do not know the curve itself, we cannot find Q. However, an alternative would be to use the slope of the direction field at a point such as Q^* which is near to Q. The point Q^* is on the line tangent to the solution at P_0 and lies halfway between the two vertical lines $t = t_0$ and $t = t_1$.

We convert this geometrical description into an analytic procedure as follows. The equation of the tangent line is

$$y = y_0 + m_0(t - t_0)$$

where $m_0 = f(P_0) = f(t_0, y_0)$. Putting $t = t_0 + \frac{1}{2}h$, which is the T-coordinate

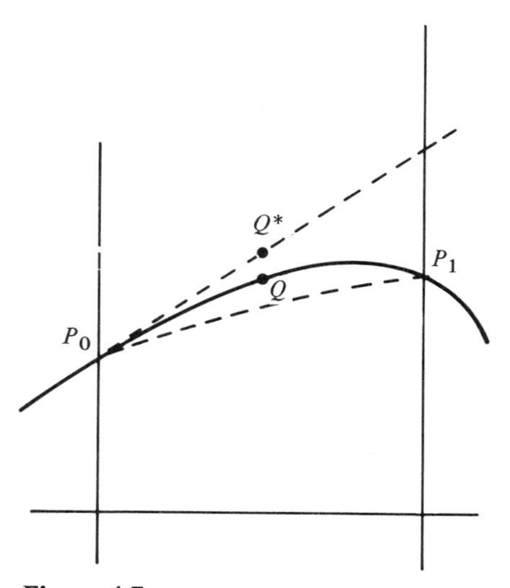

Figure 4.7

of Q^*, we find the Y-coordinate of Q^* to be

$$y^* = y_0 + \tfrac{1}{2}(m_0 h)$$

Our estimate for the true slope of the chord $P_0 P_1$ is the slope of the direction field at Q^*, namely,

$$m^* = f(Q^*) = f(t_0 + \tfrac{1}{2}h, y^*)$$

Accordingly, the Y-coordinate for the predicted point P_1 is

$$y_1 = y_0 + hm^*$$

If we iterate this process, we arrive at a useful algorithm for obtaining approximate solutions of a differential equation which does not require calculation of partial derivatives of f. In stating the algorithm, we have adopted a notation that is standard in numerical analysis and computing, and abbreviate $f(t_n, y_n)$ as f_n. (Do not confuse this with the use of numerical subscripts to denote partial derivatives!)

The Midpoint Algorithm

$$\begin{aligned} y_0 &= y(t_0) \\ t_n &= t_0 + hn \\ y_n^* &= y_n + \tfrac{1}{2}hf_n \\ m^* &= f(t_n + \tfrac{1}{2}h, y_n^*) \\ y_{n+1} &= y_n + hm^* \\ f_n &= f(t_n, y_n) \end{aligned} \tag{4.12}$$

Let us illustrate this by again using the differential equation $y' = t - y$, $y(0) = 1$, and choosing $h = 0.2$. We find

$$\begin{aligned} f_0 &= f(0, 1) = -1 \\ y_0^* &= y_0 + \tfrac{1}{2}(0.2)(-1) \\ &= 1 - 0.1 = 0.9 \\ m^* &= f(0 + \tfrac{1}{2}(0.2), 0.9) \\ &= 0.1 - 0.9 = -0.8 \\ y_1 &= y_0 + (0.2)(-0.8) \\ &= 1 - 0.16 = 0.84 \end{aligned}$$

Continuing, we find

$$
\begin{aligned}
f_1 &= -0.64, & y_1^* &= 0.776, & m^* &= f(0.3, 0.776) = -0.476 \\
& & y_2 &= 0.84 + (0.2)(-0.476) = 0.7448 \\
f_2 &= -0.3448, & y_2^* &= 0.7103, & m^* &= -0.2103 \\
& & y_3 &= 0.7027 \\
f_3 &= -0.1027, & y_3^* &= 0.6925, & m^* &= 0.0075 \\
& & y_4 &= 0.7042 \\
f_4 &= 0.0958, & y_4^* &= 0.7138, & m^* &= 0.1862 \\
& & y_5 &= 0.7415 \approx y(1.0)
\end{aligned}
$$

For comparison, recall that $y(t) = t - 1 + 2e^{-t}$ so that the exact value is $y(1.0) = 0.735758*$.

If these numbers seem familiar, turn back and compare them with the results obtained by applying the modified Euler method (4.11) to the same equation. (Also see Exercise 14.) Note that the Midpoint method used only *one* evaluation of f and involved no consideration of partial derivatives of f. Thus, it is certainly to be preferred to the modified Euler method. Its local truncation error can be shown to be $O(h^3)$ so that the total accumulated error is $O(h^2)$. The significance of this is that cutting the step size h in half will cut the error by 4 since $(\frac{1}{2}h)^2 = \frac{1}{4}h^2$.

There are numerical methods that have an even smaller truncation error. The most commonly used method of the type we have discussed is the classical Runge-Kutta method, or fourth-order RK method. It has a local truncation error $O(h^5)$ and a total accumulated error $O(h^4)$, so that cutting the step size by a half cuts the error by one-sixteenth, and going from $h = 0.1$ to $h = 0.01$ can be expected to decrease the total error by a factor of 0.0001.

The idea behind this method is again geometric. It is to estimate the true chord slope m by a complicated weighted average of slopes in the direction field at specified points. We have illustrated this in Figure 4.8. We start with the slope at the current point P_0, and use this to locate the point Q^* halfway along the P_0 tangent line. We calculate the slope of the direction field at Q^*:

$$m^* = f(Q^*)$$

This is one estimate for the slope of the solution curve at the actual midpoint Q. We next use a line through P_0 with slope m^* to locate another point Q^{**} at the halfway point, and calculate the slope of the direction field there:

$$m^{**} = f(Q^{**})$$

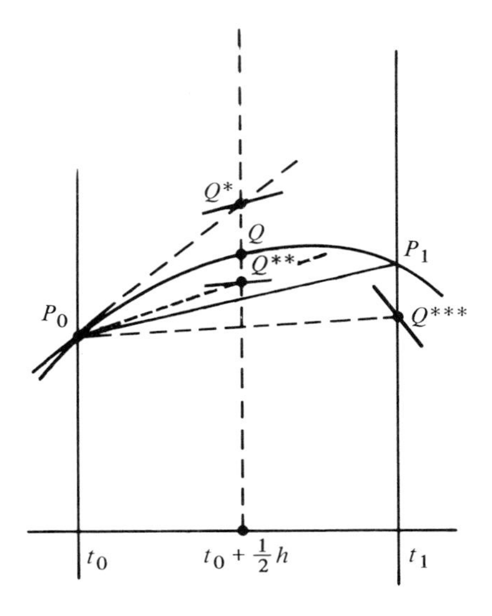

Figure 4.8

This is another estimate for the slope at Q. We then use a line through P_0 with slope m^{**} to locate a point Q^{***} on the vertical line $t = t_1$, and calculate the slope of the direction field there:

$$m^{***} = f(Q^{***})$$

This is an estimate for the slope of the solution curve at P_1. Finally, we estimate the true chord slope to be

$$m = \tfrac{1}{6}(m_0 + 2m^* + 2m^{**} + m^{***})$$

and using this, obtain the estimated value of $y(t_1)$:

$$y_1 = y_0 + hm$$

If we iterate this, we obtain a very useful algorithm.

The Classical RK Method

$$t_n = t_0 + hn$$

(4.13)
$$y_0 = y(t_0)$$

$$y_{n+1} = y_n + hm$$

where

$$
\begin{aligned}
y_n^* &= y_n + \tfrac{1}{2}hf_n \\
m^* &= f(t_n + \tfrac{1}{2}h,\ y_n^*) \\
y_n^{**} &= y_n + \tfrac{1}{2}hm^* \\
m^{**} &= f(t_n + \tfrac{1}{2}h,\ y_n^{**}) \\
y_n^{***} &= y_n + hm^{**} \\
m^{***} &= f(t_n + h,\ y_n^{***}) \\
m &= \tfrac{1}{6}(f_n + 2m^* + 2m^{**} + m^{***}) \\
f_n &= f(t_n, y_n)
\end{aligned}
$$

The proof that this method has local truncation error $O(h^5)$, and thus total accumulated error $O(h^4)$, is a long and tedious exercise in the use of Taylor's Theorem, and there are reference books that contain the detailed argument (see Bibliography reference 5). A more immediately convincing demonstration of the degree of accuracy and usefulness of the classical RK method is to see what it does with a familiar differential equation.

We take the initial-value problem $y' = t - y$, $y(0) = 1$, which we have used before, and solve it with the RK method and a rather large step size $h = 0.5$. We compare this with a solution by the Midpoint method with step size 0.1, and with Euler for steps of size 0.1 and 0.01. We also list the total number N of steps involved as well as the number of evaluations of $f(t, y)$. Table 4.6 gives the errors $\epsilon = |\text{exact value} - \text{calculated value}|$, not the values of y itself.

Table 4.6

	Tabulated error for:			
	RK	*Midpoint*	*Euler*	*Euler*
t	$h = 0.5$	$h = 0.1$	$h = 0.1$	$h = 0.01$
0.0	0	0	0	0
0.5	0.00048	0.00109	0.03208	0.00305
1.0	0.00058	0.00132	0.03840	0.00370
1.5	0.00053	0.00120	0.03448	0.00336
2.0	0.00043	0.00097	0.02752	0.00271
N	4	20	20	200
Evaluations of f	16	40	20	200

It is clear from Table 4.6 that the RK method is far superior to the Euler method and also markedly better than the Midpoint method. In *four* steps, using only 16 function evaluations, the RK method arrives at greater accuracy than the Midpoint method manages to achieve with 20 steps and 40 evaluations. Moreover, the comparison between the step size $h = 0.5$ used for the RK method, and the resultant error is striking.

Another sample may underscore this. The differential equation $y' = 3y/t$, with $y(1) = 1$, has solution $y = t^3$. In Table 4.7 we give the results of applying

Table 4.7

t	*Exact value*	*Euler*	*Midpoint*	*RK*
1.0	1	1	1	1
2.0	8	7.0000	7.9351	7.9997
3.0	27	22.5454	26.7378	26.9989

the Euler method, Midpoint method, and classical RK method, all with the same step size $h = 0.1$.

Consideration of this example points up another aspect of the choice of a numerical algorithm. Although the RK method achieved here a markedly better approximation than did the Euler or Midpoint methods, it did so at the expense of greater operating time, for in carrying out the 20 steps needed to go from $t = 1$ to $t = 3$, there were 80 evaluations of the function $f(t, y)$, compared to 40 for the Midpoint method and 20 for Euler. When the function f is simple, as in this example, this is not a crucial factor; however, if f were complex in structure so that a single evaluation of f took a significant time, then the RK method might not be the optimal choice. In the next section we will study another class of methods which have some advantages over the classical RK method.

We conclude this section by giving an argument designed to show mathematically why the RK method achieves such a high degree of accuracy. It is not a general argument, for we will discuss it only for one specific differential equation, the familiar case $y' = y$ with $y(0) = 1$. The exact solution of this is $y(t) = e^t$. For a general step size h, we have $t_n = 0 + nh = nh$, so that $y(t_n) = e^{nh}$. For later use, we record the observation that the exact values $y(t_k)$ form a sequence obeying the recursion

$$(4.14) \qquad \begin{aligned} y(t_0) &= 1 \\ y(t_{n+1}) &= e^h y(t_n) \end{aligned}$$

We now proceed to determine the recursion which defines the RK sequence $\{y_k\}$. We start from $y_0 = 1$. Then, assuming that we have determined $y_1, y_2, \ldots, y_n$, let us find the formula for y_{n+1}. Since $f(t, y) = y$, it is easy to follow the instructions in (4.13), obtaining

$$
\begin{aligned}
f_n &= f(t_n, y_n) = y_n \\
y_n^* &= y_n + \tfrac{1}{2}hf_n \\
&= (1+\tfrac{1}{2}h)\, y_n \\
m^* &= f(t_n+\tfrac{1}{2}h, y_n^*) = y_n^* \\
&= (1+\tfrac{1}{2}h)\, y_n \\
y_n^{**} &= y_n + \tfrac{1}{2}hm^* \\
&= (1+\tfrac{1}{2}h+\tfrac{1}{4}h^2)\, y_n \\
m^{**} &= f(t_n+\tfrac{1}{2}h, y_n^{**}) = y_n^{**} \\
y_n^{***} &= y_n + hm^{**} \\
&= (1+h+\tfrac{1}{2}h^2+\tfrac{1}{4}h^3)\, y_n \\
m^{***} &= f(t_n+h, y_n^{***}) = y_n^{***} \\
m &= \tfrac{1}{6}(f_n+2m^*+2m^{**}+m^{***}) \\
&= \tfrac{1}{6}(1+2[1+\tfrac{1}{2}h]+2[1+\tfrac{1}{2}h+\tfrac{1}{4}h^2]+[1+h+\tfrac{1}{2}h^2+\tfrac{1}{4}h^3])\, y_n \\
&= (1+\tfrac{1}{2}h+\tfrac{1}{6}h^2+\tfrac{1}{24}h^3)\, y_n \\
y_{n+1} &= y_n + hm \\
&= (1+h+\tfrac{1}{2}h^2+\tfrac{1}{6}h^3+\tfrac{1}{24}h^4)\, y_n
\end{aligned}
$$

For emphasis, let us place the final result in parallel display with that in (4.14).

$$
\begin{aligned}
y_0 &= 1 & \qquad\Big|\qquad y(t_0) &= 1 \\
y_{n+1} &= (1+h+\tfrac{1}{2}h^2+\tfrac{1}{6}h^3+\tfrac{1}{24}h^4)\, y_n & \qquad\Big|\qquad y(t_{n+1}) &= e^h y(t_n)
\end{aligned}
$$

Now recall the corresponding recursion which we obtained for the Euler method (4.3):

$$
\begin{aligned}
y_0 &= 1 \\
y_{n+1} &= (1+h)\, y_n
\end{aligned}
$$

It is immediately clear that the reasons for the success of the RK method when applied to this differential equation is that the coefficient of y_n is a much better approximation to the true coefficient e^h than is the corresponding coefficient in the Euler method; recall that $e^h = 1+h+h^2/2!+h^3/3!+\cdots$. This sample calculation also shows visibly the truth of the assertion that the local truncation error in the RK method is $O(h^5)$, since the agreement with the series expansion of e^h is exact through the term in h^4.

Of course, this calculation for the equation $y' = y$ in no way proves the corresponding result for the general case. However, the general argument is analogous, requiring only that one replace the various evaluations of f by Taylor expansions about the point (t, y).

Exercises

1. Apply the modified Euler method with $h = 0.2$ to the equation $y' = y$, $y(0) = 1$, to estimate the value $y(1)$. Compare with the tables of values for this equation in Section 4.1.

2. (a) Apply the Midpoint method with $h = 0.5$ to the equation $y' = -2ty$, $y(0) = 1$, to estimate the value $y(1)$.
   (b) Compare with the exact solution, which is $y = e^{-t^2}$.

3. Apply the Midpoint method with $h = 0.5$ to the equation $y' = 3y/t$, $y(1) = 1$, to estimate $y(2)$.

PC4. If you have access to a pocket calculator, apply the classical Runge-Kutta (RK) method, with $h = 0.5$, to the equation $y' = y^2$, $y(0) = 1$, to estimate $y(1)$.

PC5. Apply the classical RK method, with $h = 0.5$, to the equation $y' = -2ty$, $y(0) = 1$, to estimate $y(1)$, and compare with the exact solution (see Exercise 2).

6. Apply the Midpoint method with general h to the equation $y' = y$, $y(0) = 1$, and verify the recursion

$$y_{n+1} = (1+h+\tfrac{1}{2}h^2)\, y_n$$

7. (a) Apply the classical RK method with general h to the equation $y' = -2y$, $y(0) = 1$, and verify the recursion

$$y_{n+1} = (1-2h+2h^2-\tfrac{4}{3}h^3+\tfrac{2}{3}h^4)\, y_n$$

   (b) Can you explain why this coefficient of y_n is appropriate?

8. Sometimes one needs to estimate values of the derivatives of a function given only by a table of values. In such cases, the following formulas are sometimes used:

$$\phi'(t) \approx \frac{\phi(t+h)-\phi(t-h)}{2h}$$

$$\phi''(t) \approx \frac{\phi(t+h)-2\phi(t)+\phi(t-h)}{h^2}$$

(a) Apply these to estimate values for ϕ' and ϕ'' at $t=0.1$ and $t=0.2$, given the values in the accompanying table.

t	0	0.1	0.2	0.3
$\phi(t)$	1.2	1.5	2.1	2.9

*(b) Can you justify the formulas used in part (a), and estimate the error of the estimate?

*9. A smooth curve $y=\phi(x)$ has the shape shown in Figure 4.9. Let m_P, m_Q, m_R be the slopes of the curve at the indicated points P, Q, R, respectively. Let m be the slope of the chord PR.

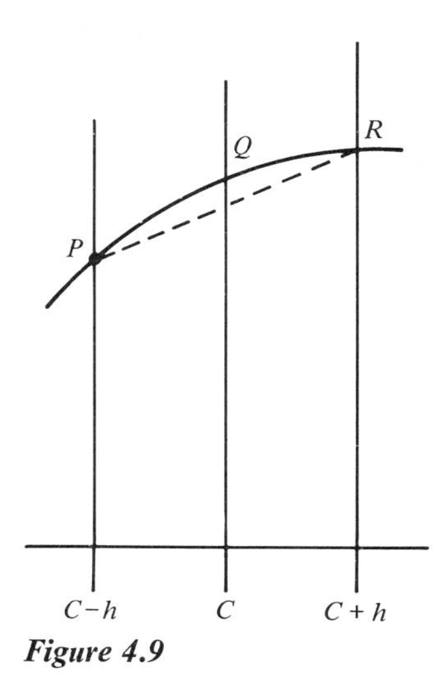

Figure 4.9

(a) Which of the following is a better estimate for m: $\frac{1}{2}(m_P+m_R)$, or m_Q?

(b) Show that $\frac{1}{6}(m_P+4m_Q+m_R)$ is a better estimate for m than either estimate in part (a).

Exercises 10–13 are more appropriately done with a pocket calculator.

PC10. (a) Use the Midpoint method with $h = 0.2$ to estimate $y(1)$ for the equation $y' = y^2$, $y(0) = 1$.
(b) Compare with the exact values at each step.

PC11. Apply the Midpoint method with $h = 0.3$ to $y' = t^2 + y^2$, with $y(0) = 1$, to estimate $y(0.9)$.

PC12. Apply the classical RK method with $h = 0.5$ to $y' = 3y/t$, with $y(1) = 1$, and estimate $y(2)$.

PC13. Apply the RK method to the problem in Exercise 11, with $h = 0.3$.

14. Show that the Modified Euler method and the Midpoint method always give the same numerical answers if the equation is of the form $y' = At + By$.

*15. If $y' = f(t, y)$, $y(t_0) = y_0$, then

$$y(t_1) = y_0 + \int_{t_0}^{t_1} f(s, y(s))\, ds$$

A standard estimate for the value of an integral $\int_{t_0}^{t_1} \phi(s)\, ds$ is $(t_1 - t_0)\,\phi[\frac{1}{2}(t_0 + t_1)]$. Show how this observation could lead to the Midpoint algorithm (4.12).

4.4 Multivalue Methods

When the classical RK method is used to solve the equation $y' = f(t, y)$, a considerable amount of calculation must be done using the values t_n and y_n in order to arrive at the next value y_{n+1} of the solution function. The time-consuming part is usually the four evaluations of the function $f(t, y)$. As an algorithm to be implemented on a computer, the RK method is inefficient in that none of the intermediate results obtained in the course of carrying out step n are of use for any later step. A better scheme would be one in which at step n, one has available and makes essential use of the values of y_k and f_k calculated during earlier steps. This should reduce the total number of complicated calculations to be performed, and thereby produce a faster and more efficient algorithm.

To show that such methods can be devised, we give a very simple example. Suppose that we have already obtained $y_0, y_1, y_2, f_0, f_1, f_2$, and we are ready to find y_3. Examine Figure 4.10, and recall the geometric basis for the Midpoint

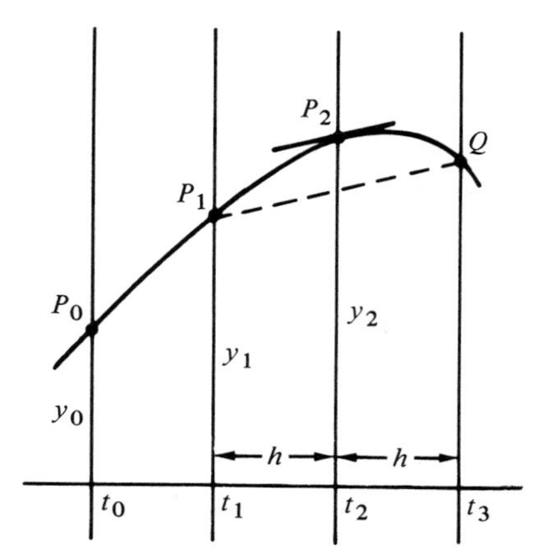

Figure 4.10

method. A useful estimate for the slope of the chord connecting P_1 to the as-yet-undetermined point Q is the slope of the curve at the middle point P_2. If the curve is the solution of $y' = f(t, y)$, then this slope is the value f_2, which is among those we have already obtained. Thus, we estimate the slope of the chord to be f_2 and obtain our desired estimate for y_3:

$$y_3 = y_1 + (t_3 - t_1) f_2$$
$$= y_1 + 2hf_2$$

When this is made into a recursive algorithm defining the sequence $\{y_k\}$, we have the following **Simple Extrapolation** (SE) method.

The SE Method

$$y_0 = y(t_0)$$

(4.15) $\qquad y_1$ obtained from other sources

$$y_{n+1} = y_{n-1} + 2hf_n$$

Note that we did not specify the value of y_1. One drawback for all the multivalue methods is that they require more information at the start than is supplied merely by the initial conditions of the differential equation. This means that to use the SE method, some other method must be used first in order to obtain a good estimate for y_1. However, once you are past this stage, the recursion generates all the remaining terms y_k.

Let us try this out on one of our standard illustrations. Consider the equation $y' = 3y/t$ with $y(1) = 1$. Recall that the exact solution for this is $y = t^3$. Let us apply Simple Extrapolation with $h = 0.2$. We have $y_0 = 1$ and we need an estimate for $y_1 \approx y(t_1) = y(1.2)$. If we apply the RK method with $h = 0.2$, we obtain $y(1.2) = 1.7275$, which we round to 1.73. [Since $y = t^3$, $y(1.2)$ is actually 1.728.] Following the instructions in (4.15), we therefore start with

$$\begin{aligned} y_0 &= 1 \\ y_1 &= 1.73 \end{aligned}$$

We will need f_1, which is

$$\begin{aligned} f_1 &= f(t_1, y_1) = f(1.2,\ 1.73) \\ &= \frac{(3)(1.73)}{1.2} = 4.32 \end{aligned}$$

Now we are ready to apply (4.15).

$$\begin{aligned} y_2 &= y_0 + 2(0.2) f_1 \\ &= 1 + (0.4)(4.32) \\ &= 2.728 = 2.73 \end{aligned}$$

Proceeding, we have

$$\begin{aligned} f_2 &= f(1.4,\ 2.73) = \frac{(3)(2.73)}{1.4} = 5.85 \\ y_3 &= y_1 + 2(0.2) f_2 \\ &= 1.73 + (0.4)(5.85) = 4.07 \\ f_3 &= f(1.6,\ 4.07) = 7.63 \\ y_4 &= y_2 + (0.4) f_3 \\ &= 2.73 + (0.4)(7.63) = 5.78 \\ f_4 &= f(1.8,\ 5.78) = 9.63 \\ y_5 &= y_3 + (0.4) f_4 \\ &= 4.07 + (0.4)(9.63) = 7.92 \end{aligned}$$

Since $t_5 = 2.0$, we have found an estimate for $y(2.0)$; it is fairly good, since the exact value is $(2)^3 = 8$. The result is comparable to that obtained by the Midpoint method—namely, 7.9351, using $h = 0.1$ (see Table 4.7)—and it used only a quarter as many evaluations of f.

There are a great variety of these multivalue extrapolation methods, often called AB methods after the mathematicians J. Adams and F. Bashforth.

They differ in the number of earlier values of y_k and f_k that are used in the calculation of y_{n+1}, in the nature of the recursion used, and of course in the degree of accuracy obtained. Although they can be carried out by hand, the repetitive nature of the calculation is better suited to a programmable calculator. To illustrate the variety of methods that have been created and used, we give a handful of examples.

The AB_1 Method

$$
\begin{aligned}
y_0 &= y(t_0) \\
&y_1, y_2 \text{ obtained from other sources} \qquad (4.16) \\
y_{n+1} &= y_{n-2} + \tfrac{1}{4}h(3f_{n-2}+9f_n)
\end{aligned}
$$

To minimize storage, it is necessary to keep only the last three values of y_k and the last two values of f_k. The value of f_n is calculated.

The AB_2 Method

$$
\begin{aligned}
y_0 &= y(t_0) \\
&y_1, y_2 \text{ obtained from other sources} \qquad (4.17) \\
y_{n+1} &= y_n + \tfrac{1}{12}h(23f_n - 16f_{n-1} + 5f_{n-2})
\end{aligned}
$$

Here, one needs only the last two values of f_k and the last value of y_k. The value of f_n is calculated.

The Hermite Method

$$
\begin{aligned}
y_0 &= y(t_0) \\
&y_1, y_2 \text{ obtained from other sources} \qquad (4.18) \\
y_{n+1} &= 10y_{n-2} + 9y_{n-1} - 18y_n + h(3f_{n-2}+18f_{n-1}+9f_n)
\end{aligned}
$$

The strange coefficients that appear in these formulas are not chosen at random but are forced by the mathematical requirements of the methods. We can show this in one case by going through the derivation of the AB_2 method. We first observe that $f_k = f(t_k, y_k)$ is an estimate for the value $y'(t_k)$ since $y(t)$ is a solution for the equation $y' = f(t, y)$. If we are looking for a method which expresses y_{n+1} in terms of y_n and the values f_{n-2}, f_{n-1}, f_n, which is the objective of AB_2, we can instead try to express $y(t_{n+1})$ in terms of $y(t_n)$, $y'(t_{n-2})$, $y'(t_{n-1})$, and $y'(t_n)$. We write $t_{n+1} = t_n + h$, $t_{n-2} = t_n - 2h$, $t_{n-1} = t_n - h$, and then drop the subscript n, writing t for t_n. Thus, our objective now is to express $y(t+h)$ in terms of $y(t)$, $y'(t-2h)$, $y'(t-h)$, and $y'(t)$.

The Taylor expansion of $y(t-h)$ starts with

$$y(t-h) = y - hy' + \tfrac{1}{2}h^2y'' - \tfrac{1}{6}h^3y^{(3)} + \cdots \tag{4.19}$$

where we have abbreviated $y(t)$ by y, $y'(t)$ by y', etc. If we differentiate this formula with respect to t, we have

$$y'(t-h) = y' - hy'' + \tfrac{1}{2}h^2y^{(3)} - \tfrac{1}{6}h^3y^{(4)} + \cdots \tag{4.20}$$

In this formula, replace h by $2h$ and obtain

$$y'(t-2h) = y' - 2hy'' + \tfrac{1}{2}(4h^2y^{(3)}) - \tfrac{1}{6}(8h^3)y^{(4)} + \cdots \tag{4.21}$$

We want to use these series to obtain an approximate relation of the form

$$y(t+h) = y(t) + h[\alpha y'(t) + \beta y'(t-h) + \delta y'(t-2h)] \tag{4.22}$$

We substitute from (4.20) and (4.21) to calculate the right-hand side of (4.22) as

$$y + h\alpha y' + h\beta[y' - hy'' + \tfrac{1}{2}h^2y^{(3)} - \cdots] + h\delta[y' - 2hy'' + \tfrac{1}{2}(4h^2y^{(3)}) - \cdots].$$
$$= y + (\alpha+\beta+\delta)\,hy' + (-\beta-2\delta)\,h^2y'' + \tfrac{1}{2}(\beta+4\delta)\,h^3y^{(3)} + \cdots$$

The Taylor series for the left-hand side of (4.22) is obtained by changing the sign of h in (4.19); it is

$$y(t+h) = y + hy' + \tfrac{1}{2}h^2y'' + \tfrac{1}{6}h^3y^{(3)} + \cdots$$

If we equate these two series in h, we obtain algebraic linear equations for the unknown coefficients α, β, δ.

$$\alpha + \beta + \delta = 1$$
$$-\beta - 2\delta = \tfrac{1}{2}$$
$$\tfrac{1}{2}\beta + \tfrac{4}{2}\delta = \tfrac{1}{6}$$

The unique solution of these turns out to be

$$\alpha = \tfrac{23}{12}, \qquad \beta = -\tfrac{16}{12}, \qquad \delta = \tfrac{5}{12}$$

This gives the form of the recursion shown in (4.17), and explains where these strange weights came from. The proof is not complete since we have not gone into the error analysis, nor have we studied the errors resulting from replacing $y(t_k)$ by y_k in deriving the form of the recursion. But at least we have shown that the form of the AB_2 method is the result of a detailed analysis and not empirical. It should also be plausible that the local truncation error is $O(h^5)$, since we obtained agreement of the Taylor series through the terms in h^4.

This suggests that the AB_2 method might be comparable in accuracy to the classical RK method. Since it requires many fewer evaluations of the function f in moving through a fixed number of steps, one might expect that it would be quite a bit faster than RK, and therefore preferable.

Unfortunately, some AB methods exhibit certain peculiarities of behavior that lead at times to large errors. This is called "numerical instability," and is particularly prone to occur with methods which use widely scattered data points and with values of h that are too large. The detailed discussion of this important matter is properly the subject of a course in numerical analysis and computing, and we therefore refer the reader to the texts cited in the introduction to this chapter. The following somewhat artificial example will show the nature of the behavior we are describing.

Suppose we use the Simple Extrapolation method (4.15) to solve the initial-value problem $y' = -y$, $y(0) = 1$, with $h = 0.5$. The exact solution is, of course, $y = e^{-t}$, which we know has a graph that approaches the horizontal axis from above in a steady way, as t increases. Compare this with the behavior of the numerical results obtained from the SE method, as shown in Table 4.8.

Table 4.8

t	*Estimated y*
0	1.00000
0.5	0.60653
1.0	0.39347
1.5	0.21306
2.0	0.18041
2.5	0.03265
3.0	0.14776
3.5	−0.11511
4.0	0.26287
4.5	−0.37798
5.0	0.64085
5.5	−1.01883
6.0	1.65968

(*Note*: To apply the SE method, we needed a value for y_1. Since $e^{-0.5} =$ 0.60653066*, we used $y_1 = 0.60653$.) It should be remarked that the SE method does much better if a smaller step size, such as $h = 0.01$, is used. Even so, erratic oscillations such as those shown by the data in the table will eventually occur.

Much depends on the choice of the method. For example, the AB_1 method applied to the same example is much more stable; with $h = 0.25$, it yields accuracy of about 0.0012 for $0 \le t \le 5$. (Other numerical illustrations will be found in the exercises.)

However, methods of the AB type are subject to other weaknesses. An initial analysis of the Hermite method, using Taylor series in the fashion that was done above for the AB_2 method, would suggest that this method ought to be rather accurate. However, the large coefficients used in the recursion formula (4.18) tend to augment any round-off errors, and the result is that it is a very undependable method. For example, suppose Hermite is used to solve the equation $y' = y^2$ with $y(0) = 1$ [the exact solution is $y = 1/(1-t)$] using $h = 0.1$ and $y_0 = 1.000000$, $y_1 = 1.111111$, and $y_2 = 1.250000$. With these initial data, y_5 turns out to be 1.978246. The exact value for $t_5 = 0.5$ is $1/(1-0.5) = 2.000000$, so the error at this stage is about 0.02. However, if we use initial data that are rounded to four decimals, so that $y_1 = 1.1111$ and $y_2 = 1.2500$, then $y_5 = 1.94093$ and the error has grown to 0.06.

One way to help avoid these problems of stability and inaccuracy is to combine the approach of the RK type with that of the AB type, using several different methods together to produce a more complicated but more dependable method. One procedure is used to obtain a trial value for y_{n+1}, which is then used to obtain a better estimate for the slope f_{n+1}, and this in turn is used to obtain a still better value for y_{n+1}. The first value of y_{n+1} is thought of as a *predicted* value, and the second as a *corrected* value, and such combined methods are termed "**predictor-corrector**" methods. We give one as a sample.

PC_1 Method

$$y_0 = y(t_0)$$

y_1, y_2 obtained from other sources

$$
\begin{aligned}
y^*_{n+1} &= y_n + \tfrac{1}{12}h(23f_n - 16f_{n-1} + 5f_{n-2}) \\
m^* &= f(t_{n+1}, y^*_{n+1}) \qquad (4.23)\\
y_{n+1} &= y_n + \tfrac{1}{24}h(9m^* + 19f_n - 5f_{n-1} + f_{n-2}) \\
f_{n+1} &= f(t_{n+1}, y_{n+1})
\end{aligned}
$$

The predicted value y^*_{n+1}, obtained by AB_2, determines m*, an estimated f_{n+1}, which is then applied to a different AB method to obtain the corrected value of y_{n+1}. Note that this uses an extra function evaluation at each step. For comparison with the other methods mentioned in this section, and in particular with the round-off weakness of the Hermite method, the following numerical data may be helpful. If we use the same equation $y' = y^2$ and apply the PC_1 method with $h = 0.1$ and with the less accurate starting values $y_0 = 1.000$, $y_1 = 1.111$, and $y_2 = 1.250$, then the calculated value for y_5 is 1.99879, which has an error of only 0.0012. (See Exercise 4.)

We conclude this section with a few remarks about the numerical solution of differential equations of order higher than one, for example, an equation such as $y'' = 2y - y'$. As it turns out, there is no need to develop numerical methods specifically designed for second-order or higher-order equations. There are two reasons why this is so.

1. Any higher-order differential equation can be converted into a system of first-order equations.

2. All of the numerical methods that we have discussed for a *single* first-order equation have a parallel version which can be used for a first-order *system*.

To see why the first of these statements is true, consider the general second-order equation $y'' = f(t, y, y')$. Introduce two new variables u and v by the equations $u = y$, $v = y'$. Then

$$\begin{cases} \dfrac{du}{dt} = v \\ \dfrac{dv}{dt} = f(t, u, v) \end{cases}$$

is a first-order system which is completely equivalent to the original single second-order equation. To see why the second statement above holds would take somewhat longer; its truth will become very plausible later, when we discuss the theory of such systems in more detail. At that time, we shall also give an example of the use of some of these numerical methods in their modified form as they apply to systems.

Nevertheless, there is no difficulty in constructing numerical methods directly for higher-order equations. The following illustration may clarify this point. Suppose that we want a method we can apply to second-order equations in the standard form $y'' = f(t, y, y')$ with initial conditions of the form $y(t_0) = a$, $y'(t_0) = b$. Having first chosen a step size h, we want an algorithm that will generate two sequences $\{y_k\}$ and $\{y_k'\}$ which will be numerical estimates for the true values $y(t_k)$ and $y'(t_k)$, respectively. Since $y''(t)$ can be obtained from f and the values of $y(t)$ and $y'(t)$, the number $f_n = f(t_n, y_n, y_n')$ will serve as an estimate for $y''(t_n)$ that we can use to move from y_n' to y_{n+1}'. For example, we could use an analog of the simple Euler method, and define y_{n+1}' as $y_n' + hf_n$. A more accurate formula is obtained by using the analog of the Midpoint method. We first obtain crude estimates for y and y' for $t = t_n + \frac{1}{2}h$, and next use these to calculate an estimate for y'', using the function f. We then use this to obtain y'_{n+1} as in the Midpoint method. To obtain y_{n+1}, we use a formula which takes advantage of the available information about y''. The result is the following useful but simple algorithm.

Method for Second-Order Equations

$$
\begin{aligned}
y_0 &= y(t_0) \\
y'_0 &= y'(t_0) \\
f_n &= f(t_n, y_n, y'_n) \\
{}^*y_n &= y_n + \tfrac{1}{2}hy'_n \\
{}^*y'_n &= y'_n + \tfrac{1}{2}hf_n \\
m &= f(t_n+\tfrac{1}{2}h, {}^*y_n, {}^*y'_n) \\
y'_{n+1} &= y'_n + hm \\
y_{n+1} &= y_n + \tfrac{1}{3}h(2y'_n+y'_{n+1}) + \tfrac{1}{6}h^2 f_n
\end{aligned}
\tag{4.24}
$$

To see exactly how this is carried out, let us solve the equation $y'' = 2y - y'$ with initial conditions $y(0) = 2$, $y'(0) = -1$. (As can be checked easily, the solution for this is $y = e^t + e^{-2t}$.) To apply the algorithm, we have $f(t, y, y') = 2y - y'$, $y_0 = 2$, $y'_0 = -1$. We take $h = 0.1$ as a convenient step size. Then,

$$
\begin{aligned}
f_0 &= 2y_0 - y'_0 = 2(2) - (-1) = 5 \\
{}^*y_0 &= y_0 + \tfrac{1}{2}(0.1)\, y'_0 \\
&= 2 + (0.05)(-1) = 1.95 \\
{}^*y'_0 &= y'_0 + \tfrac{1}{2}(0.1)\, f_0 \\
&= -1 + (0.05)(5) = -0.75 \\
m &= f(0+0.05,\ 1.95, -0.75) \\
&= 2(1.95) - (-0.75) = 4.65 \\
y'_1 &= y'_0 + (0.1)\, m \\
&= -1 + 0.465 = -0.535 \\
y_1 &= y_0 + \tfrac{1}{3}(0.1)(2y'_0 + y'_1) + \tfrac{1}{6}(0.01)\, f_0 \\
&= 2 + \tfrac{1}{3}(0.1)(2[-1] + [-0.535]) + \tfrac{1}{6}(0.01)(5) \\
&= 1.9238
\end{aligned}
$$

This completes one step in the iteration. To continue,

$$\begin{aligned}
f_1 &= 2y_1 - y'_1 \\
&= 2(1.9238) - (-0.535) \\
&= 4.3826 \\
{}^*y_1 &= y_1 + \tfrac{1}{2}(0.1)\, y'_1 \\
&= 1.9238 + (0.05)(-0.535) \\
&= 1.8971 \\
{}^*y'_1 &= y'_1 + \tfrac{1}{2}(0.1)\, f_1 \\
&= -0.535 + (0.05)(4.3826) \\
&= -0.3159 \\
m &= f(0.1+0.05,\ 1.8971,\ -0.3159) \\
&= 2(1.8971) - (-0.3159) \\
&= 4.1101 \\
y'_2 &= y'_1 + (0.1)\, m \\
&= -0.535 + (0.1)(4.1101) \\
&= -0.1240 \\
y_2 &= y_1 + \tfrac{1}{3}(0.1)(2y'_1 + y'_2) + \tfrac{1}{6}(0.01)\, f_1 \\
&= 1.9238 + \tfrac{1}{3}(0.1)(2[-0.535] + [-0.1240]) + \tfrac{1}{6}(0.01)(4.3826) \\
&= 1.8913
\end{aligned}$$

etc.

If this is carried on to find y_5, one finds

$$y_5 = 2.0143$$

which accordingly is the estimated value of $y(0.5)$. From the formula $y = e^t + e^{-2t}$, the exact value is 2.0166∗, so that the error in the estimate is 0.0023. If the step size is decreased to $h = 0.005$, then to find $y(0.5)$, we must calculate y_{100} which turns out to be 2.0165948, with an error of only 0.000006. No one would do this calculation by hand since it is the repetitive type of loop which a computer—even one of the very small hand-held programmable machines—does so well. Furthermore, there are special ways of obtaining more accurate answers without doing as much work as this, one of which ("bootstrapping") is explained in the next section.

Exercises

1. Use the Simple Extrapolation method (4.15) with $h = 0.2$ to estimate $y(1)$ for the equation $y' = t - y$, $y(0) = 1$, starting with $y_1 = 0.84$. (The exact value is $2e^{-1}$.)

2. Apply method AB_1 [the algorithm (4.16)] to $y' = y^2$, $y(0) = 1$, using $h = 0.1$, and find an estimate for $y(0.5)$. Use $y_1 = 1.111$, $y_2 = 1.250$.

PC3. If a pocket calculator is available, use the AB_2 method of algorithm (4.17) with $h = 0.1$ to obtain an estimate for $y(0.5)$ where $y' = y^2$, $y(0) = 1$. Use $y_1 = 1.111$ and $y_2 = 1.250$ for starting estimates.

4. By applying PC_1 (4.23) with $h = 0.1$ to the equation $y' = y^2$, $y(0) = 1$, using $y_1 = 1.111$, $y_2 = 1.250$, and rounding to four decimals, the numerical results shown in Table 4.9 were obtained. Verify them.

Table 4.9

		$k = 3$	$k = 4$
	y_k^*	1.4266	1.6627
$f_1 = 1.2343$	m_k^*	2.0351	2.7646
$f_2 = 1.5625$	y_k	1.4285	1.6663
	f_k	2.0405	

5. Derive the AB_1 algorithm in a manner similar to that used in the text for the AB_2 algorithm.

6. Can you explain the choice of the coefficients in the last line of the algorithm in (4.24)?

PC7. For the differential equation $y' = t^2 - y^2$, $y(0) = 1$, estimate $y(1)$ using the AB_1 method with $h = 0.2$, and using the Midpoint method to first find y_1 and y_2. (You will probably want to use a pocket calculator, and retain only four decimals.)

8. Use the second-order method (4.24) and $h = 0.2$ to solve $y'' = 2y'$ with initial conditions $y(0) = y'(0) = 2$, verifying Table 4.10. Also, compare the results with the exact solution, which is $y = e^{2t} + 1$. [*Note:* The computations can be done with relative ease if you first obtain the general recursion defining the sequences $\{y_k\}$ and $\{y_k'\}$ for this problem.]

Table 4.10

t	y	y'
0	2.0000	2.0000
0.2	2.4907	2.9600
0.4	3.2168	4.3808
0.6	4.2916	6.4836
0.8	5.8822	9.5957
1.0	8.2363	14.2016

4.5 The Art of Computing

The scientific study of the design and use of high-speed computers is still in its beginning stages, and much of what has been discovered remains in the realm of skills to be acquired by experience rather than formal procedures to be learned in class. For example, it does not seem possible as yet to prescribe a "best" method for getting numerical solutions of a differential equation. Each method seems to have some advantages and some drawbacks, and the right selection depends upon the equation under consideration and the ultimate objectives of the person who wants the solution. How accurate must the answer be? Over how large an interval do you want the answers tabulated? What kind of computer can you use? How much computer time do you have?

Once a method has been selected, the step size h must be chosen. If h is too small, too many steps are required and the total time and round-off error go up. If h is too large, the truncation error is apt to be large. A different method may give the same accuracy but allow you to use a larger h. However, as with the classical RK method, it may require many more evaluation steps. If the function f is very complicated, this may slow down the procedure considerably. Furthermore, there may even be effects that come from the type of computer being used. Some computers round numbers by "chopping," so that 3.1415926 might become 3.1415 rather than 3.1416. Others might permit the use of "interval arithmetic." All of this is quite important for anyone who is directly concerned with numerical analysis, but it is not within the scope of the present book, and we leave it for others to present.

As has been emphasized before, a numerical answer is of little practical value unless one has some idea as to its accuracy. Unfortunately, error control is one of the aspects of computing that is less well understood. The theoretical bounds are entirely too large, sometimes by several orders of magnitude, when compared with experience. (Errors sometimes behave like random noise and can cancel each other in a statistical way.) People who do computing have devised tricks to help estimate errors. One simple trick is to carry out all arithmetic to ten places or so, keeping only four or five places in the final answer. This may help avoid some of the round-off errors.

Another device is to repeat a calculation with half the step size to see what effect this has on the final answers. For example, if we were solving $y' = y$ with $y(0) = 1$, and we used the classical RK method with $h = 0.2$ and then with $h = 0.1$, the results would be as given in Table 4.11. These suggest that

Table 4.11

t	y $(h = 0.2)$	y $(h = 0.1)$
0	1.00000	1.00000
1.0	2.71825	2.71828
2.0	7.38889	7.38905
3.0	20.08486	20.08549

we would not be wrong to give $y(2) = 7.389$ and $y(3) = 20.085$. In fact, the exact values are $y(2) = e^2 = 7.389056*$ and $y(3) = e^3 = 20.085537*$.

In these circumstances, a special mathematical procedure is sometimes used to improve the accuracy without going to still smaller step size. We call it "bootstrapping," because it uses the data already obtained to decrease the error. Suppose that a particular numerical procedure is intended to provide an estimate for the exact answer A_E, and uses step size h. We denote this approximate answer by $A(h)$, since it depends in a complicated way upon h. In many cases, it is possible to represent $A(h)$ with a formula such as

$$A(h) = A_E + Bh^r + \text{small error term} \tag{4.25}$$

where r is an integer like 1, 2, or 3, and A_E, B, and the error term are unknown. It is then possible to eliminate the effect of the term Bh^r and thus obtain a better estimate for A_E by comparing the value $A(h)$ with the value $A(\frac{1}{2}h)$ obtained by repeating the calculation with half the step size. An example may make this convincing.

The Euler method corresponds to $r = 1$ so that if $A(h)$ is the value obtained for $y(T)$ with step size h, then

$$A(h) = A_E + Bh + \text{small error term}$$

Then

$$A(\tfrac{1}{2}h) = A_E + \tfrac{1}{2}Bh + \text{small error term}$$

and we have

$$2A(\tfrac{1}{2}h) - A(h) = A_E + \text{small error terms} \tag{4.26}$$

As an illustration, if we use the Euler method to solve $y' = -y$ with $y(0) = 1$ to estimate $y(2)$, then with step size $h = 0.1$ we obtain $A(0.1) = 0.12158$, whereas with $h = 0.05$, we have $A(0.05) = 0.12851$. Bootstrapping these, we obtain

$$\begin{aligned} 2A(0.05) - A(0.1) &= 2(0.12851) - (0.12158) \\ &= 0.13544 \end{aligned}$$

The exact value is $e^{-2} = 0.13534*$, showing how we have improved the accuracy.

As a second illustration, the Midpoint method corresponds to $r = 2$, so that we have the approximate formula

$$A(h) = A_E + Bh^2 + \text{error term}$$

Replacing h by $\frac{1}{2}h$, we have

$$A(\tfrac{1}{2}h) = A_E + \tfrac{1}{4}Bh^2 + \text{error term}$$

Accordingly,

$$\frac{4A(\frac{1}{2}h) - A(h)}{3} = A_E + \text{error terms} \tag{4.27}$$

If we apply the Midpoint method to solve $y' = 3y/t$ with $y(1) = 1$, the resulting estimates for $y(2)$ with several different step sizes are given in Table 4.12. If we bootstrap, using formula (4.27), we obtain

$$\begin{aligned} \frac{4A(0.1) - A(0.2)}{3} &= \frac{4(7.93506) - (7.77587)}{3} \\ &= 7.988123 \end{aligned}$$

The exact solution of the equation is $y = t^3$, so $y(2) = 8.000$. The improvement in accuracy is striking. (Other examples are to be found in the exercises.)

Table 4.12

h	$y(2)$ *estimate*
0.2	7.77587
0.1	7.935060

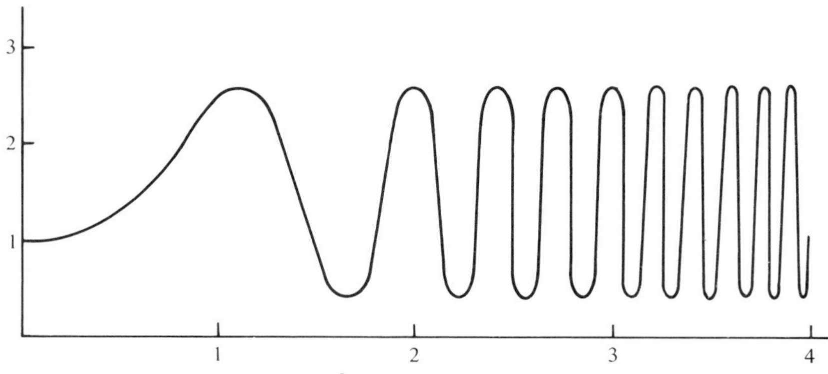

Figure 4.11 $y = \exp(\sin t^3)$

The ability to use a packaged routine to solve differential equations on a computer does not answer all problems. The differential equation itself may have built-in properties that make it difficult to solve numerically. Sometimes this is evident from the equation. Suppose we wanted to solve $y' = 3yt^2 \cos(t^3)$ with $y(0) = 1$, and wanted to find $y(4)$ or $y(5)$. As t increases, the term $\cos(t^3)$ will oscillate between 1 and -1, and the term t^2 will become large. Hence, the slope y' of the solution curve must swing wildly between large positive values and large negative values, doing so more and more rapidly as you move to the right. Such a behavior is bound to be hard to follow numerically. It may even be hard to detect if you did not know in advance that it was there.

The exact solution of this equation is $y = e^{\sin(t^3)}$, whose graph is shown in Figure 4.11. However, a routine use of one of the numerical methods gives the data and approximate curve shown in Figure 4.12. This starts out correctly but does not show the detailed behavior near $t = 4$.

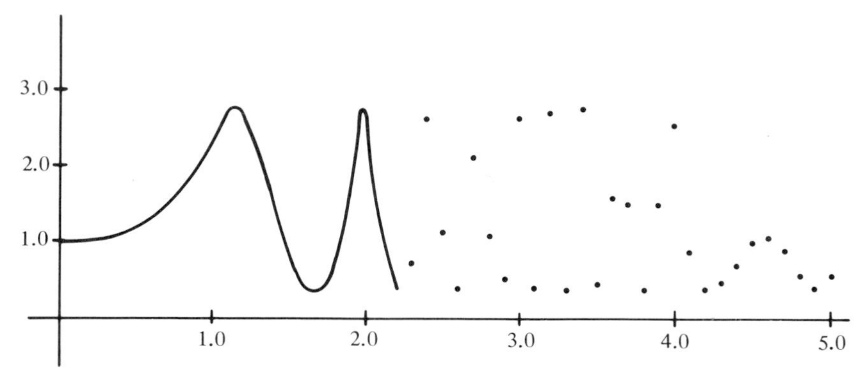

Figure 4.12

Another example may be even more striking. Consider the equation

$$y' = 2ty - t^2, \qquad y(0) = C$$

This is a linear equation whose solution was discussed in Section 2.2, where it appeared as formula (2.19). There, we stated that a mathematical analysis of the solution showed that the solutions had the form $y = F(t) + Ce^{t^2}$, where F was a function which was asymptotically $\frac{1}{2}t$. How does this equation respond to numerical solution? The result is shown in Figure 4.13, where solution curves are sketched for various choices of C. Note how different the solutions

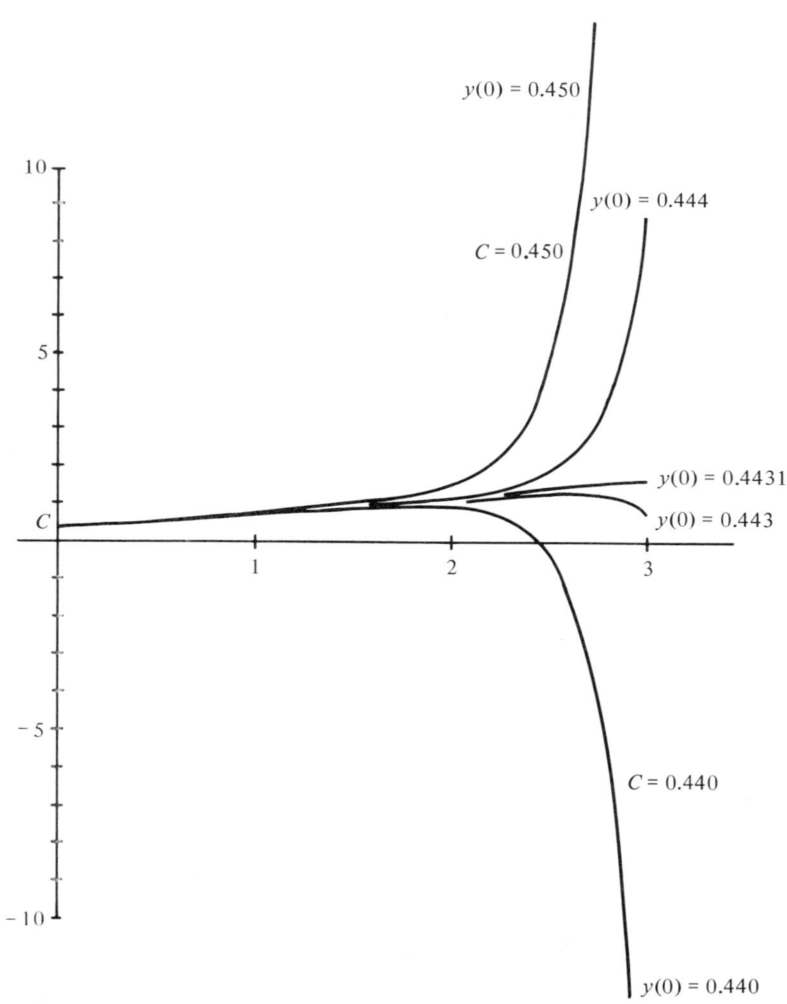

Figure 4.13 $y = 2ty - t^2$

are for $C = 0.440$ and $C = 0.450$. This illustrates a different type of instability, one that is not associated with the numerical algorithm, but that lies in the nature of the differential equation itself. Small changes in the initial conditions produce very large differences in the actual values of $y(3)$. The reason for this is in the nature of the terms Ce^{t^2}, which grow very rapidly as t increases, and which can be either positive or negative, depending on the sign of the constant C. When you try to solve such an equation numerically without knowing precisely the value $y(0)$, the computer may generate a different solution curve than the one intended.

Of course, such a differential equation may in fact be a good model for a physical system. There are processes in which a very small change in one parameter—here represented by C—can produce a very large change in another variable—here represented by the value $y(3)$.

Finally, we give an example to support the statement in the introduction to this chapter that it is fatal to trust the computer unless you know enough about both numerical analysis and differential equations. In Table 4.13, we

Table 4.13

	Estimated y value		
t	$h = 0.1$	$h = 0.05$	$h = 0.025$
1.00	2.00000	2.00000	2.00000
1.05		1.92823	1.92823
1.10	1.84370	1.84372	1.84372
1.15		1.73989	1.73990
1.20	1.59969	1.59996	1.60000
1.25		1.28668	1.27643
1.30	2.97083	0.28246	0.81714
1.35		0.30580	0.93396
1.40	2.85260	0.33023	1.09442
1.45		0.35577	1.38907
1.50	2.70918	0.38245	1.75792

give the results of three different attempts to solve the differential equation

$$y' = \frac{2yt}{t^2 - y^2}, \qquad y(1) = 2 \tag{4.28}$$

In each case, the classical RK method was used with the step sizes shown. It is clear from the data in the table that we seem to start out in good shape, but something certainly begins to go wrong between $t = 1.2$ and $t = 1.3$ since all three "solutions" are wildly different there. Which do we trust, if any? (See Exercise 6.)

Recall again the Hamming dictum: "The purpose of computing is insight, not numbers!" The information obtained by doing a numerical solution will often suggest questions which can be answered only by a mathematical study of the equation and its solutions. Answers obtained by a computer may be totally incorrect for reasons that stem from the nature of the equation and its solution, or from the nature of the particular algorithm used, but may be helpful in studying the equation.

Exercises

1. The Euler method applied to $y' = y$, $y(0) = 1$, with two step sizes yields the data in Table 4.14.

Table 4.14

	t	0	1	2	3
$(h = 0.01)$	y	1.00	2.7048	7.3160	19.7885
$(h = 0.005)$	y	1.00	2.7115	7.35233	19.93596

Use these data to obtain better estimates for $y(2)$ and $y(3)$ and compare those estimates with the exact values.

2. For the classical RK method, the exponent r in the bootstrapping procedure (4.25) is 3. Develop a formula for bootstrapping and apply it to the data in Table 4.11.

3. Two applications of the Midpoint method to the equation $y' = 3y/t$, $y(1) = 1$, yield the data in Table 4.15. Can you use these data to obtain a better estimate for $y(3)$?

Table 4.15

$h = 0.25$	$y(3) \approx 25.6705$
$h = 0.10$	$y(3) \approx 26.7378$

4. (a) The following bootstrapping formula is sometimes used to improve Euler's method.

$$A_E \approx \tfrac{1}{3}[8A(\tfrac{1}{4}h) - 6A(\tfrac{1}{2}h) + A(h)]$$

Justify this when $A(h)$ has the form $A_E + Bh + Ch^2 +$ error term.

(b) Apply this formula to the following data to obtain a better estimate for $y(2)$ where $y' = y$, $y(0) = 1$.

$$\text{Euler estimate of } y(2) = \begin{cases} 6.727500 & \text{when } h = 0.1 \\ 7.039989 & \text{when } h = 0.05 \\ 7.209568 & \text{when } h = 0.025 \end{cases}$$

(c) How many evaluations of f were required to obtain this final estimate? Was this an efficient use of computer time?

5. In solving the equation $y' = 36t^3y^2$ with $y(0) = 1$, the data in Table 4.16 were obtained, using the RK method with $h = 0.1$.
(a) What do you think explains the data?
(b) Verify your conjecture by mathematical methods.

Table 4.16

t	*Estimated y*	t	*Estimated y*
0	1.0000	0.5	2.2776
0.1	1.0009	0.6	78.2717
0.2	1.0146	0.7	1.6×10^{25}
0.3	1.0786	0.8	overflow
0.4	1.2994		

6. For the illustrative example (4.28), $y' = 2ty/(t^2 - y^2)$, sketch the direction field and give an explanation for the wide discrepancies in the numerical "solutions" given in the text. Alternatively, the equation is one of the types discussed in Section 2.1; can you use the methods of that section to obtain a formula for $y(t)$ and thus explain the data in the Table 4.13?

5

Linear Equations

Introduction

In the preceding chapters, we have discussed certain special types of differential equations which can be solved readily, and looked at methods which enable us to solve more general equations, sometimes exactly but usually only approximately. It is a mathematical fact that most differential equations do not have solutions that can be described in simple terms, so that numerical methods such as those treated in Chapter 4 are the ultimate refuge for the applied mathematician.

However, there is one class of differential equations for which a satisfactory general theory has been developed, and this is the topic of the present chapter. Linear equations are important for a variety of reasons. First, there *is* a coherent mathematical theory and structure to the subject, one which has led to many significant mathematical concepts and insights. Second, linear differential equations supply some of the most frequently used models in all branches of science. Finally, because *any* differential equation is linear to a first-order approximation, knowledge of the linear theory often suggests how one should study a nonlinear problem.

We also use this chapter to introduce, in Section 5.2, some of the basic ideas of linear analysis such as linear dependence and independence, basis and dimension. This will be review for those who have already studied some linear algebra. We emphasize that it is not our intention to discuss the general theory of the subject; only those topics are introduced which have an immediate connection with the solution of linear differential equations. The goal is to provide a treatment of the solution of linear differential equations with constant coefficients, although we also discuss some aspects of the solution of more general linear equations. (We also call attention to the appendix on the use of the Laplace Transform, Appendix 1, which provides an alternative method for solving many of the same equations.)

We devote Section 5.5 to a detailed study of one equation, that for the motion of a damped harmonic oscillator such as a mass attached to a simple spring. Our purpose is to discuss in a parallel fashion the physical behavior of the system and the corresponding mathematical equivalents. This discussion can be sharpened and accelerated by having a suitable demonstraiton model at hand.

5.1 Linearity

It will be easier if we adopt a modern approach at the start and discuss linear spaces of functions and linear operators. The term "space of functions" is used to refer to a specified collection of functions when it is looked at from a geometric viewpoint. We shall use two special notations to indicate the spaces that will mostly concern us. The letter $\mathcal{C}$ will be used for the collection of all continuous functions f defined for $-\infty < t < \infty$, and $\mathcal{C}^\infty$ will denote those functions f that have derivatives there of all orders. Functions in $\mathcal{C}^\infty$ are also said to be **infinitely differentiable**. If we want to restrict our discussion to the functions that are defined only on a specific interval I, $a < t < b$, then we will denote the corresponding spaces of functions by $\mathcal{C}[I]$ and $\mathcal{C}^\infty[I]$.

Thus, $\mathcal{C}$ contains functions such as $3t^4 - t^3 + 1$, $\sin(3t)$, and $te^{t^2} + t^3$; it does not contain $\tan t$, $1/t$, or $\sin[1/(1-t)]$, since these fail to be continuous at certain values of t. The first three functions also belong to $\mathcal{C}^\infty$, since each can be differentiated as often as one wishes. The function $f(t) = |t|$ is continuous everywhere and so belongs to the space $\mathcal{C}$, but it does not belong to $\mathcal{C}^\infty$ since it does not have a derivative at $t = 0$. Since all functions in $\mathcal{C}^\infty$ must be continuous, $\mathcal{C}^\infty$ is a subset of $\mathcal{C}$; we usually say that $\mathcal{C}^\infty$ is a subspace of $\mathcal{C}$, and write $\mathcal{C}^\infty \subset \mathcal{C}$.

We will also include complex-valued functions such as

$$f(t) = (2-3i)t^2 + 5it^3 - 3\sin(t)$$

although we will not make explicit use of these until later in this chapter when they arise naturally in discussing the exponential function $e^{\gamma t}$ for complex values of γ.

The spaces $\mathcal{C}$ and $\mathcal{C}^\infty$ are examples of an important category of mathematical structures usually called **linear spaces** or **vector spaces.**

DEFINITION **5.1** A (nonempty) collection $\mathcal{E}$ of functions defined on the same domain is a **function space** (or **linear space of functions**) if it has the following two properties.

1. If ϕ and ψ belong to $\mathcal{E}$, so does $\phi-\psi$.
2. If ϕ belongs to $\mathcal{E}$, so does $c\phi$ for any real or complex constant c.

The pair of conditions are often combined into one, namely,

3. If ϕ and ψ belong to $\mathcal{E}$, so does the function $c_1\phi+c_2\psi$, for any choice of the constants c_1 and c_2.

An immediate implication of this definition is the property that if $\phi_1, \phi_2, \ldots, \phi_n$ are each functions in $\mathcal{E}$, then the function

$$\psi = \sum_1^n c_i\phi_i = c_1\phi_1 + c_2\phi_2 + \cdots + c_n\phi_n$$

also belongs to $\mathcal{E}$.

The fact that $\mathcal{C}$ and $\mathcal{C}^\infty$ are examples of function spaces is an immediate consequence of basic properties of continuity and differentiability; if f and g are continuous (differentiable), so is the function $c_1 f+c_2 g$.

Linear function spaces are special cases of the more general concept of an **abstract linear space** (or **vector space**) over a scalar field k. In the most common cases, k is the field **R** of real numbers or the field **C** of complex numbers, and one speaks of a "real linear space" or a "complex linear space," respectively. Although we shall have little need for the general concept, we give the formal definition for the latter.

A *complex linear space* consists of a collection $\mathcal{E}$ of objects which we call *vectors*, and two algebraic operations which we call *vector addition* and *scalar multiplication*. The following axioms are to hold.

1. There is a rule that assigns to each pair of vectors v_1 and v_2 a unique vector v_1+v_2.
2. There is a rule that assigns to each pair consisting of a complex number β and a vector v, a vector βv.

3. For any vectors v_1, v_2, v_3 and complex numbers α, β, the following identities hold.

$$
\begin{aligned}
v_1 + (v_2+v_3) &= (v_1+v_2) + v_3 \\
v_1 + v_2 &= v_2 + v_1 \\
\beta(v_1+v_2) &= \beta v_1 + \beta v_2 \\
(\alpha+\beta) v_1 &= \alpha v_1 + \beta v_1 \\
(\alpha\beta) v_1 &= \alpha(\beta v_1) \\
(1) v_1 &= v_1
\end{aligned}
$$

4. For any vectors v_1 and v_2, there is a unique vector w such that $v_1 + w = v_2$.

In algebra courses, it is customary to prove that certain useful additional properties follow from these axioms. For example, there must exist a vector denoted by $\mathbf{0}$ with the property that $v+\mathbf{0} = \mathbf{0}+v = v$, and $\mathbf{0}$ is obtained by multiplying any vector by the scalar number zero. Again, the operation of subtraction is defined for vectors by writing

$$v_1 - v_2 = v_1 + (-1) v_2$$

and the usual laws of algebra are obeyed, except that there is usually no process for multiplying two vectors.

It is convenient to introduce another special term.

DEFINITION 5.2 A function f is said to be a **linear combination** of the functions $\phi_1, \phi_2, \ldots, \phi_n$ if there are constants c_i such that

$$f = c_1 \phi_1 + c_2 \phi_2 + \cdots + c_n \phi_n$$

The constants used can be real or complex; for example, the function $2t^2+t-3$ is a linear combination of the functions $t-1$ and t^2-t, since

$$3(t-1) + 2(t^2-t) = 2t^2 + t - 3$$

The following result is an immediate consequence of Definitions 5.1 and 5.2.

THEOREM 5.1 *Let $\phi_1, \phi_2, \ldots, \phi_n$ be a collection of n functions, and let M be the set of all functions f that are linear combinations of the ϕ_i. Then, M is a linear space. Furthermore, if $\mathcal{E}$ is another linear space to which each of the ϕ_i belongs, then M is a subspace of $\mathcal{E}$.*

Proof We must first prove that the set M satisfies Definition 5.1. Let f and g be functions that belong to M. Then, there are constants a_i and b_i so that

$$
\begin{aligned}
f &= a_1 \phi_1 + a_2 \phi_2 + \cdots + a_n \phi_n \\
g &= b_1 \phi_1 + b_2 \phi_2 + \cdots + b_n \phi_n
\end{aligned}
$$

Accordingly,

$$f-g = (a_1-b_1)\phi_1 + (a_2-b_2)\phi_2 + \cdots + (a_n-b_n)\phi_n$$

which belongs to M since it too is a linear combination of the ϕ_i. Similarly,

$$Cf = Ca_1\phi_1 + Ca_2\phi_2 + \cdots + Ca_n\phi_n$$

is a linear combination of the ϕ_i and therefore belongs to M. The last statement in the theorem is also immediate. Any f belonging to M has the form $f = c_1\phi_1 + c_2\phi_2 + \cdots + c_n\phi_n$ for some choice of the constants c_i. If each of the ϕ_i belongs to the linear space $\mathcal{E}$, then f itself belongs to $\mathcal{E}$, and M is thus a subspace of $\mathcal{E}$.

DEFINITION **5.3** The linear space M obtained by taking all the linear combinations of a specified set of functions ϕ_i is said to be the **linear span** of the $\phi_1, \phi_2, \ldots, \phi_n$, or the space **generated** by the set $\{\phi_1, \phi_2, \ldots, \phi_n\}$.

To illustrate these definitions further, consider the three functions $1, t, t^2$. The space generated by them is clearly the space of all polynomials of degree at most two, given in general by

$$f(t) = a + bt + ct^2$$

This space is a subspace of the space of polynomials of degree at most four, which in turn can be generated by the five functions 1, t, t^2, t^3, and t^4. Both of these are subspaces of the spaces $\mathcal{C}$ and $\mathcal{C}^\infty$, since polynomials are both continuous and infinitely differentiable.

We note that the space of polynomials of degree at most two can also be generated by the three functions $t-1$, t^2-t, and t^2+t+1. This can be verified in several ways. We can show that every polynomial f of degree at most two can be expressed as a linear combination of these three specific functions. If we set $\phi_1 = t-1$, $\phi_2 = t^2-t$, and $\phi_3 = t^2+t+1$, then

$$\begin{aligned} c_1\phi_1 + c_2\phi_2 + c_3\phi_3 &= c_1(t-1) + c_2(t^2-t) + c_3(t^2+t+1) \\ &= (c_3-c_1) + (c_1-c_2+c_3)t + (c_2+c_3)t^2 \end{aligned}$$

If we want this result to be the general polynomial of degree two, then we equate it to $f = A + Bt + Ct^2$. Comparing coefficients, we have

$$\begin{aligned} c_3 - c_1 &= A \\ c_1 - c_2 + c_3 &= B \\ c_2 + c_3 &= C \end{aligned}$$

These can be solved for the c_i, obtaining

$$c_1 = \tfrac{1}{3}(B+C-2A)$$

$$c_2 = \tfrac{1}{3}(2C-A-B)$$

$$c_3 = \tfrac{1}{3}(A+B+C)$$

A second way of showing this is to recall that the space M of polynomials is spanned by the three functions 1, t, and t^2. If we can show that each of these is a linear combination of ϕ_1, ϕ_2, and ϕ_3, then it will follow that M is generated by the ϕ_i. Direct observation shows that $\phi_1+\phi_2 = t^2-1$ and $\phi_2+\phi_3 = 2t^2+1$, so that $3t^2 = \phi_1+2\phi_2+\phi_3$; if we divide by 3, we have expressed t^2 as a linear combination of the ϕ_i. We also have $t = t^2-(t^2-t) = t^2-\phi_2$, so that t is also a linear combination of the ϕ_i. Finally, $1 = (t^2+t+1)-t-t^2 = \phi_3-t-t^2$, and substituting for t and t^2 the linear combinations already found, we have expressed 1 as a linear combination of the ϕ_i. (See Exercise 4.)

Having digressed to introduce some terminology from linear algebra, let us move back toward the study of differential equations. We start from the observation that the process of differentiation carries the space $\mathcal{C}^\infty$ into itself. If f is any member of $\mathcal{C}^\infty$, then the function f' is also in $\mathcal{C}^\infty$. If we denote the differentiation operation d/dt by D, we can write this in the form: $D\{f\} \in \mathcal{C}^\infty$ for every $f \in \mathcal{C}^\infty$. More generally, if we write D^2 for the operation of taking the second derivative, D^3 for the third derivative, and so on, then we observe that $D^n\{f\}$ is in $\mathcal{C}^\infty$ for every $f \in \mathcal{C}^\infty$. We say that the linear space $\mathcal{C}^\infty$ is **closed** under the operation of differentiation. We could also say that D^n is a function which is defined on the space $\mathcal{C}^\infty$ and whose values also lie in $\mathcal{C}^\infty$; however, because the members of the space $\mathcal{C}^\infty$ are already functions, we change to the common terminology and call D^n an **operator**, mapping $\mathcal{C}^\infty$ into $\mathcal{C}^\infty$.

More generally, the term *operator* denotes any transformation that can be applied to each function f in $\mathcal{C}^\infty$ and that sends it into some member of $\mathcal{C}^\infty$ in a unique way. Here are descriptions of four specific operators.

T_1 sends f into the function $3f''-2f'+5f$

T_2 sends f into the function g, where

$$g(t) = f''(t) - (3t+1)f'(t) + 5f(t) \tag{5.1}$$

T_3 sends f into the function $f''+ff'+f^2$

T_4 sends f into the function g, where $g(t) = f(-2t)$

For example, if the operator T_1 is applied to t^3, then

$$\begin{aligned} T_1\{t^3\} &= 3(6t) - 2(3t^2) + 5(t^3) \\ &= 18t - 6t^2 + 5t^3 \end{aligned}$$

If T_2 is applied to t^3, then

$$\begin{aligned} T_2\{t^3\} &= 6t - (3t+1)(3t^2) + 5(t^3) \\ &= -4t^3 - 3t^2 + 6t \end{aligned}$$

If T_3 is applied to t^3, then

$$\begin{aligned} T_3\{t^3\} &= 6t + (t^3)(3t^2) + (t^3)^2 \\ &= 6t + 3t^5 + t^6 \end{aligned}$$

Finally,

$$T_4\{t^3\} = (-2t)^3 = -8t^3$$

We are not going to be concerned in this chapter with very general operators, but only with a very small and rather special subclass of operators called the **linear differential operators** of **finite order**. A linear operator may be defined as follows.

DEFINITION **5.4** An operator T on $\mathcal{C}^\infty$ is said to be **linear** if and only if it has the following two properties

$$\begin{aligned} T\{f+g\} &= T\{f\} + T\{g\} \\ T\{Cf\} &= CT\{f\} \end{aligned}$$

for any constant C and any functions f and g in $\mathcal{C}^\infty$.

Of the preceding examples, T_1, T_2, and T_4 are linear operators; T_3 is not. To verify the last statement, one need only check that $T_3\{5t\} \neq 5T_3\{t\}$:

$$\begin{aligned} T_3\{5t\} &= 0 + (5t)(5) + (5t)^2 \\ &= 25t + 25t^2 \\ 5T_3\{t\} &= 5[0+(t)(1)+(t)^2] \\ &= 5t + 5t^2 \end{aligned}$$

Operators can be combined by addition, subtraction, and multiplication. Addition and subtraction of operators are defined by the formula

$$(T_1 \pm T_2)\{f\} = T_1\{f\} \pm T_2\{f\}$$

and multiplication (or **composition**) of operators is defined by

$$(T_1 T_2)\{f\} = T_1\{T_2\{f\}\}$$

Multiplication of operators is not in general **commutative**, and it can easily happen that $T_1 T_2$ and $T_2 T_1$ are different. For example, using D for d/dt, consider the pair of operators

$$T_1 = D^2 - 2D + 5 \tag{5.2}$$

$$T_2 = D^2 - (3t+1)D + 5 \tag{5.3}$$

We can show that $T_1 T_2 \neq T_2 T_1$ by calculating $T_1 T_2\{t\}$ and $T_2 T_1\{t\}$ and noting that we obtain different answers. (See also Exercise 13.)

We first calculate $T_1 T_2\{t\}$. We start with

$$\begin{aligned} T_2\{t\} &= [D^2 - (3t+1)D + 5]\{t\} \\ &= D^2\{t\} - (3t+1)D\{t\} + 5\{t\} \\ &= 0 - (3t+1)(1) + 5t \\ &= 2t - 1 \end{aligned}$$

Then

$$\begin{aligned} T_1 T_2\{t\} &= T_1\{2t-1\} \\ &= D^2\{2t-1\} - 2D\{2t-1\} + 5\{2t-1\} \\ &= 0 - 2(2) + 10t - 5 \\ &= 10t - 9 \end{aligned}$$

On the other hand,

$$\begin{aligned} T_1\{t\} &= D^2\{t\} - 2D\{t\} + 5\{t\} \\ &= 0 - 2(1) + 5t \\ &= 5t - 2 \end{aligned}$$

and

$$\begin{aligned} T_2 T_1\{t\} &= T_2\{5t-2\} \\ &= D^2\{5t-2\} - (3t+1)D\{5t-2\} + 5\{5t-2\} \\ &= 0 - (3t+1)(5) + 25t - 10 \\ &= 10t - 15 \end{aligned}$$

Since $T_1 T_2\{t\} = 10t-9$, but $T_2 T_1\{t\} = 10t-15$, we see that it is not in general true that $T_1 T_2$ and $T_2 T_1$ are the same.

[Strictly speaking, we should have written the last terms in both (5.2) and (5.3) as $5I$ rather then merely 5, using I to represent the identity operator that sends any function f into itself. However, for simplicity, we have followed customary usage and dropped the I.]

There are important cases in which the multiplication of differential operators is always commutative. One such case is for the class of differential polynomials with constant coefficients. Thus, the operators $T_1 = D^2-2D+5$ and $T_2 = D^2+3D-7$ will commute, and their product turns out to be what would be obtained by following the usual rules for multiplying polynomials, namely,

$$\begin{aligned} T_1 T_2 &= (D^2-2D+5)(D^2+3D-7) \\ &= D^4 + (3-2)D^3 + (5-6-7)D^2 + (15+14)D - 35 \\ &= D^4 + D^3 - 8D^2 + 29D - 35 \end{aligned}$$

The reason for this is that in calculating $T_1 T_2\{f\}$ we must first calculate $T_2\{f\}$, which is in this case $(D^2+3D-7)\{f\} = f''+3f'-7f$. Since the coefficients of T_2 were constants, we do not end up with an expression in which a derivative of f is multiplied by another function of t, and thus further differentiation will never involve the use of the product formula for differentiation as it did in working with the operators given in (5.2) and (5.3).

What does all this have to do with differential equations? To answer this question, let us begin with an example. Suppose we want to solve the differential equation

$$y'' = -3ty + t^2y' + 5t^2 - \sin t$$

We first rewrite this as

$$y'' - t^2y' + 3ty = 5t^2 - \sin t \tag{5.4}$$

Then, let T be the differential operator

$$T = D^2 - t^2D + 3t$$

We see that we can rewrite (5.4) as the equation

$$T\{y\} = 5t^2 - \sin t$$

or if we let $y = \phi(t)$ be the desired solution of (5.4), then we want to find a function $\phi \in \mathcal{C}$ such that

$$T\{\phi\} = 5t^2 - \sin t$$

In other words, we have an operator T, and we want to find all the functions ϕ in $\mathcal{C}$ which T sends into the specific function $5t^2 - \sin t$.

This analysis has recast the differential equation (5.4) in a new form. In this form our equation fits a general pattern that occurs in many areas of mathematics. The pattern is the following. We have a linear space $\mathcal{E}$, a particular member $g \in \mathcal{E}$, and an operator T, and we want to find all the ϕ such that $T\{\phi\} = g$. The approach of the research mathematician is to study this general problem and develop an abstract theory for the solution of problems having this pattern that works for *any* linear space $\mathcal{E}$, *any* g, and *any* operator T. The general theory is then specialized to apply to special cases such as the one from which we started. The general theory is called *functional analysis*, and its development is one of the major advances in the mathematics of the twentieth century. We cannot go into it much further, but we shall look at several simple aspects, particularly studying the relationship to differential equations.

The next section reviews some of the basic concepts of linear spaces, including independence and dependence.

Exercises

1. The symbol $\mathcal{C}^1$ is used for the space of functions that are continuous on $-\infty < t < \infty$ and have continuous first derivatives. Which of the following functions belong to $\mathcal{C}^1$?

   (a) $t^2|t|$ *(b) $t^{-1}\sin t$ (c) $|\cos t|$

2. Prove that the single condition (3) in Definition 5.1 is equivalent to the pair of conditions (1) and (2).

3. Show that the space $\mathcal{C}^1$ as defined in Exercise 1 is a linear space.

4. Show that the space of all polynomials of degree at most 2 can be generated by the functions $2t+1$, t^2+2, and $3t+t^2$.

5. Show that the function $\sin(3t+5)$ is a linear combination of $\sin 3t$ and $\cos 3t$.

6. Show that the function $(2t+1)/(t^3-t)$ is a linear combination of the functions $1/t$, $1/(t-1)$, and $1/(t+1)$.

7. Describe the linear spaces generated by the following sets of functions.

(a) $\{t^2-2t,\ t,\ t^3-t\}$
(b) $\{t^3-t,\ 1+t,\ t^2+t\}$
(c) $\{t^3-t,\ 1+t^2,\ 2-t,\ t^3+t^2\}$

8. Show that Definition 5.4 is equivalent to the following: T is linear on $\mathcal{C}^\infty$ if and only if $T\{c_1 f_1+c_2 f_2\} = c_1\, T\{f_1\}+c_2\, T\{f_2\}$, for all constants c_i and all f_i in $\mathcal{C}^\infty$.

9. For each of the operators T_i in (5.1), find $T_i\{3t^2+t\}$.

10. Prove that T_1, T_2, and T_4 in (5.1) are linear operators.

11. Define three operators T_i by $T_1\{f\} = ff''$; $T_2\{f\} = 3f+tf'$; and $T_3\{f\} = g$, where $g(t) = f(t^2)$.

(a) Find $T_i\{t^3\}$ for $i = 1, 2, 3$.
(b) Which of these operators are linear?

12. (a) Verify that

$$D(D+t)\{e^{2t}\} = (5+2t)\,e^{2t}$$

$$D(D+t)\{t^3\} = 6t + 4t^3$$

(b) By calculating $D(D+t)\{f\}$ for a general f, show that $D(D+t) = D^2+tD+1$.

13. Show that

$$(D+t)(D-t) = D^2 - (t^2+1)$$

whereas

$$(D-t)(D+t) = D^2 - (t^2-1)$$

*14. Let $\mathcal{E}$ be a linear space of functions that has the special property that it contains the square of any function in it. Prove that if f and g belong to $\mathcal{E}$, so does fg.

15. Show that $\sin^2(3t+5)$ is a linear combination of the set $\{1, \sin 6t, \cos 6t\}$.

16. Show that

$$D^2 - 3D - 28 = (D-7)(D+4)$$

$$= (D+4)(D-7)$$

17. Show that $(D+g)^2 = D^2+2gD+g^2+g'$.

18. Calculate $[D+(1/t)][D-(1/t)]$.

5.2 Linear Independence

We will say that a linear space M is of finite dimension if it can be generated by a finite number of functions $\phi_1, \phi_2, \ldots, \phi_N$, so that every member of M is of the form $\sum_1^N c_i \phi_i$. The space $\mathscr{P}_2$ of polynomials of degree at most 2 is finite-dimensional since it can be generated by 1, t, and t^2. The space $\mathscr{P}$ of all polynomials is not finite-dimensional, since it cannot be generated by a finite set. For example, no linear combination of a finite set could be a polynomial whose degree is higher than that of any member of the set.

DEFINITION **5.5** The **dimension** of a linear space M, written $\dim(M)$, is the smallest number of elements in any generating set for M.

As we will see shortly, the space $\mathscr{P}_2$ has dimension 3; more generally, the polynomial space $\mathscr{P}_N$ has dimension $N+1$. The calculation of dimension is made possible by the following important concept.

DEFINITION **5.6** A set of distinct functions $\phi_1, \phi_2, \ldots, \phi_N$ is said to be **linearly dependent** in $\mathcal{C}^\infty$ if there is a set of constants c_i, not all zero, such that $\sum_1^N c_i \phi_i(t) = 0$ for all t, $-\infty < t < \infty$. The set is **linearly independent** in $\mathcal{C}^\infty$ if the only choice of c_i for which $\sum_1^N c_i \phi_i(t)$ is identically zero is $c_1 = c_2 = \cdots = c_N = 0$. (If the values of t are restricted to a specific interval I, then one says that the set of functions is dependent or independent in $\mathcal{C}^\infty[I]$.)

For example, $\{1, t, t^2\}$ is an independent set in $\mathcal{C}^\infty$ since $c_1(1) + c_2 t + c_3 t^2$ cannot be zero identically without having $c_1 = c_2 = c_3 = 0$. (A quadratic polynomial can have at most two roots.) However, the set $\{t+3,\ 2t+5,\ 3t-7\}$ is linearly dependent since

$$29(t+3) - 16(2t+5) + (1)(3t-7) = 0$$

for all t.

It is easy to tell if a set of two functions $\{\phi_1, \phi_2\}$ is linearly dependent since the only way this can happen is to have one of the functions a constant multiple of the other. Thus, $\sin t$ and $\cos t$ are independent, as are the pair $\{e^t, e^{2t}\}$. On the other hand, $\{e^{t+1}, e^{t-2}\}$ is a linearly dependent pair; the quotient $\phi_1(t)/\phi_2(t)$ is $e^{t+1}/e^{t-2} = e^3 = \text{const}$. However, for three or more functions it may be less apparent that a set is linearly independent. For example,

$$\{t-1,\ t^2+2,\ 2t^2+3t+1\}$$

is dependent, whereas

$$\{t-1,\ t^2+2,\ 2t^2+3t+2\}$$

is independent. (See Exercises 4 and 5.) The following determinant test for independence is sometimes very useful. We state it for three functions but it can be generalized at once to any number.

THEOREM 5.2 *Let g_1, g_2, g_3 belong to $\mathcal{C}^\infty[I]$. Form the determinant function*

$$W(t) = \det\begin{bmatrix} g_1(t) & g_2(t) & g_3(t) \\ g_1'(t) & g_2'(t) & g_3'(t) \\ g_1''(t) & g_2''(t) & g_3''(t) \end{bmatrix} \tag{5.5}$$

Then if the function $W(t)$ is not identically zero on the interval I, the set $\{g_1, g_2, g_3\}$ is linearly independent in $\mathcal{C}^\infty[I]$. [The function $W(t)$ is called the **Wronskian** of $\{g_1, g_2, g_3\}$.]

Before proving this result, we give several illustrations. If the set of functions is $\{1, t, t^2\}$, then $W(t)$ is

$$W(t) = \det\begin{bmatrix} 1 & t & t^2 \\ 0 & 1 & 2t \\ 0 & 0 & 2 \end{bmatrix} = 2 \neq 0$$

Thus, we confirm that $\{1, t, t^2\}$ is a linearly independent set in $\mathcal{C}^\infty$. For the set $\{t^2, t^3, t^5\}$, $W(t)$ is

$$\begin{aligned} W(t) &= \det\begin{bmatrix} t^2 & t^3 & t^5 \\ 2t & 3t^2 & 5t^4 \\ 2 & 6t & 20t^3 \end{bmatrix} \\ &= t^2 \det\begin{bmatrix} 3t^2 & 5t^4 \\ 6t & 20t^3 \end{bmatrix} - 2t\det\begin{bmatrix} t^3 & t^5 \\ 6t & 20t^3 \end{bmatrix} + 2\det\begin{bmatrix} t^3 & t^5 \\ 3t^2 & 5t^4 \end{bmatrix} \\ &= (t^2)(30t^5) - (2t)(14t^6) + (2)(2t^7) \\ &= 6t^7 \end{aligned}$$

Since this is not *identically* zero, the set $\{t^2, t^3, t^5\}$ is linearly independent on $(-\infty, \infty)$, or any other interval.

The proof of Theorem 5.2 is based on the following familiar theorem about systems of linear algebraic equations.

THEOREM 5.3 *The system of m linear equations in m unknowns*

$$\begin{aligned} a_{11}x_1 + a_{12}x_2 + \cdots + a_{1m}x_m &= 0 \\ a_{21}x_1 + a_{22}x_2 + \cdots + a_{2m}x_m &= 0 \\ &\vdots \\ a_{m1}x_1 + a_{m2}x_2 + \cdots + a_{mm}x_m &= 0 \end{aligned}$$

has only the zero solution $x_1 = x_2 = \cdots = x_m = 0$ *if the coefficient determinant* $\det[a_{ij}]$ *is not zero. Conversely, if* $\det[a_{ij}] = 0$, *there is a solution* $x_1, x_2, \ldots, x_m$ *with at least one of the* x_i *different from zero.* (Bibliography, reference 12.)

To see how this yields the proof of Theorem 5.2, suppose that there are constants c_1, c_2, c_3 such that $\sum_1^3 c_i g_i(t) = 0$ for all t in an interval I. If we differentiate this identity twice and then set $t = t_0 \in I$, we obtain the following linear equations for the unknown c_i.

$$\begin{aligned} c_1 g_1(t_0) + c_2 g_2(t_0) + c_3 g_3(t_0) &= 0 \\ c_1 g_1'(t_0) + c_2 g_2'(t_0) + c_3 g_3'(t_0) &= 0 \qquad (5.6) \\ c_1 g_1''(t_0) + c_2 g_2''(t_0) + c_3 g_3''(t_0) &= 0 \end{aligned}$$

The coefficient determinant of this system is exactly the number $W(t_0)$, defined by (5.5). Thus, if there is some choice of t_0 in I for which $W(t_0) \neq 0$, then the only solution of the system (5.6) is $c_1 = c_2 = c_3 = 0$. Accordingly, the functions g_1, g_2, g_3 are linearly independent.

The converse of Theorem 5.2 is not true: unfortunately, as Exercise 11 shows, it is possible to have a set of functions that is linearly independent and still have $W(t)$ identically zero. Thus, the vanishing of $W(t)$ does not guarantee that the set g_i is linearly *dependent*. However, in the special case we will examine in the next section, connected with differential equations, there is a valid converse which can be used. (See Exercise 24 in Section 5.3 and Exercise 12 of the present section.)

Let us now return to the topic of dimension of linear spaces.

DEFINITION 5.7 A **basis** for a linear space M is a linearly independent set that generates M.

For example, the set $\{1, t, t^2\}$ is a basis for $\mathscr{P}_2$. We shall show shortly that if M has dimension n, then every basis for M consists of n elements. We first need some preliminary results.

THEOREM 5.4 *If $\phi_1, \phi_2, \ldots, \phi_N$ generate M and are linearly dependent, then there is one of the ϕ_i that can be deleted, and the rest will still generate M.*

Proof By hypothesis, there are constants c_i, not all zero, such that

$$c_1 \phi_1 + c_2 \phi_2 + \cdots + c_N \phi_N = 0$$

Suppose that c_k is one of the nonzero coefficients. Then, we can solve for ϕ_k, obtaining

$$\begin{aligned} \phi_k &= -\frac{c_1}{c_k}\phi_1 - \frac{c_2}{c_k}\phi_2 - \cdots - \frac{c_N}{c_k}\phi_N \\ &= b_1 \phi_1 + \cdots + b_{k-1}\phi_{k-1} + b_{k+1}\phi_{k+1} + \cdots + b_N \phi_N \end{aligned}$$

Any function f in M is a linear combination of the ϕ_i such as

$$f = a_1 \phi_1 + a_2 \phi_2 + \cdots + a_k \phi_k + \cdots + a_N \phi_N$$

and in this expression, we can replace ϕ_k by its representation in terms of the other ϕ_i, obtaining

$$\begin{aligned} f &= a_1 \phi_1 + \cdots + a_k \left(\sum_{i \neq k} b_i \phi_i \right) + \cdots + a_N \phi_N \\ &= (a_1 + a_k b_1)\phi_1 + \cdots + (a_N + a_k b_N)\phi_N \end{aligned}$$

which is seen to be a linear combination of the ϕ_i with ϕ_k absent. Thus, $\{\phi_1, \phi_2, \ldots, \phi_{k-1}, \phi_{k+1}, \ldots, \phi_N\}$ continues to generate M.

We can obtain two important facts immediately from this.

COROLLARY 1 *If $\{\phi_1, \phi_2, \ldots, \phi_N\}$ generates M, then this set contains a subset that is a basis for M.*

COROLLARY 2 *Every finite-dimensional linear space has a basis.*

To prove Corollary 1, we observe that if the generating set $\{\phi_1, \phi_2, \ldots, \phi_N\}$ is not a basis, it must be a dependent set. One of the ϕ_i must then be a linear combination of the rest and is therefore redundant; if we drop it, those that remain still generate M. Repeating this, we eliminate as many as possible until we arrive at a generating set for M in which none is redundant. Those that remain are then linearly independent, and therefore form a basis for M.

The next step is apparently only a slightly strengthened form of Theorem 5.4; however, it is the key to the final result of this section.

LEMMA 5.1 *If $\{\phi_1, \phi_2, \ldots, \phi_N\}$ is a dependent set and none of the members are* **0**, *then there is a choice of $k \geq 2$ and constants c_i such that*

$$\phi_k = c_1 \phi_1 + c_2 \phi_2 + \cdots + c_{k-1} \phi_{k-1}$$

Proof Merely repeat the proof of Theorem 5.4, choosing k to be the index of the last nonzero coefficient c_i. Since the c_i for $i > k$ will then be zero, none of the ϕ_i for $i > k$ will appear in the representation for ϕ_k.

THEOREM 5.5 *If M is a linear space of dimension n, then M does not contain any set of $n+1$ linearly independent elements. Accordingly, every basis for M has n members.*

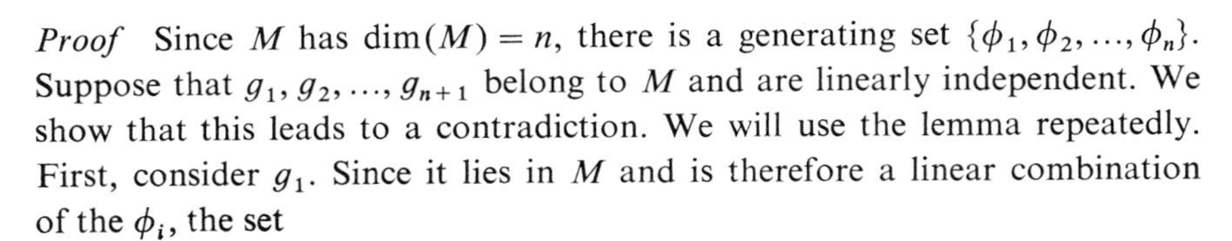

Proof Since M has $\dim(M) = n$, there is a generating set $\{\phi_1, \phi_2, \ldots, \phi_n\}$. Suppose that $g_1, g_2, \ldots, g_{n+1}$ belong to M and are linearly independent. We show that this leads to a contradiction. We will use the lemma repeatedly. First, consider g_1. Since it lies in M and is therefore a linear combination of the ϕ_i, the set

$$\{g_1, \phi_1, \phi_2, \ldots, \phi_n\}$$

is a dependent set that generates M. By the lemma, one of these functions is a linear combination of those in the list that precede it; since $k \geq 2$, it is not g_1. Thus, one of the ϕ_i is a combination of the functions that precede it in the list. We can renumber and rearrange the ϕ_i so that this one is ϕ_n. As in Theorem 5.4, we can delete this function and the remaining functions

$$\{g_1, \phi_1, \phi_2, \ldots, \phi_{n-1}\}$$

continue to generate M.

Now consider g_2, which also lies in M and is therefore a linear combination of the functions in our current list. Hence, the set

$$\{g_2, g_1, \phi_1, \phi_2, \ldots, \phi_{n-1}\}$$

is dependent and generates M. Apply the lemma again; one of these functions in the list is a linear combination of those that precede it in the list. It cannot be g_2 since g_1 and g_2 are linearly independent. Thus, it must be one of the ϕ_i. renumbering, we can assume that it is ϕ_{n-1}, and deleting ϕ_{n-1} we discover that

$$\{g_2, g_1, \phi_1, \phi_2, \ldots, \phi_{n-2}\}$$

generates M.

Proceeding in this fashion, we arrive at the stage where we know that $\{g_{n-1}, g_{n-2}, \ldots, g_2, g_1, \phi_1\}$ is a set of generators for M. We then know that g_n is in their span, so that the set

$$\{g_n, g_{n-1}, \ldots, g_2, g_1, \phi_1\}$$

is dependent and also generates M. Apply the lemma again. Since the g_i are independent, the only candidate for the special element produced by the lemma is ϕ_1, which must therefore be a combination of those that precede it. Deleting it, we have arrived at the conclusion that

$$\{g_n, g_{n-1}, \ldots, g_2, g_1\}$$

generates M. However, $g_{n+1} \in M$, and must therefore be a linear combination of the g_i with $i \leq n$, contradicting the assumed independence of the entire set of g_i.

The last statement in the theorem is an immediate deduction. If $f_1, f_2, \ldots, f_m$ and $h_1, h_2, \ldots, h_r$ were two different basis sets for M, then each would generate M and each would be a linearly independent set. By the theorem, we could not have $m < r$ nor could we have $r < m$; thus, $m = r$.

This theorem has a very important corollary, whose proof we leave as an exercise.

COROLLARY *If M is a linear space of dimension n, then any set of n linearly independent functions in M forms a basis.*

In the next section, we apply these results to the solution of differential equations.

Exercises

1. Can a set consisting of one function be dependent? Can it be independent?

2. Show that every nonempty subset of an independent set is independent.

3. If $\{\phi_1, \phi_2, \ldots, \phi_N\}$ is a basis for M, show that every function in M has a *unique* representation as a linear combination of the ϕ_i.

Test for the linear dependence or independence of the sets of functions in Exercises 4–9, using the Wronskian test of Theorem 5.2 involving $W(t)$ as in (5.5).

4. $\{t-1,\ t^2+2,\ 2t^2+3t+2\}$

5. $\{t-1,\ t^2+2,\ 2t^2+3t+1\}$

6. $\{\sin t, \cos t\}$

7. $\{e^{\alpha t}, e^{\beta t}\}$

8. $\{\sin t, \sin 2t\}$

9. $\{te^{\alpha t}, e^{\beta t}\}$

10. Show that the set of functions $\{1, t, t^2, \ldots, t^n\}$ is an independent set and forms a basis for $\mathscr{P}_n$.

11. Apply the Wronksian test to the following pair of functions:

$$g_1(t) = t^3 \qquad \text{for all } t$$

$$g_2(t) = \begin{cases} 2t^3 & \text{for } t > 0 \\ 3t^3 & \text{for } t \leq 0 \end{cases}$$

Prove that $W(t) = 0$ for all t, but that g_1 and g_2 are independent in $\mathcal{C}^\infty$.

*12. Prove that if g_1, g_2, g_3 are in $\mathcal{C}^\infty$, and $W(t) = 0$ for all t, then for each t_0 there are constants c_i, not all zero, such that $\sum_1^3 c_i g_i = 0$ at t_0. Why doesn't this show that the g_i are dependent?

13. If the set $\{\phi, \psi\}$ is linearly independent when considered as continuous functions on the interval $(-\infty, \infty)$, must they also be independent as functions on $-1 < t < 1$? What about the case for dependent functions?

14. Show that if M is a linear space of dimension n, then any set of n linearly independent functions in M forms a basis.

15. Let $p_k(t)$ be a polynomial of degree exactly k. Show that $\{p_0, p_1, \ldots, p_N\}$ is a basis for $\mathscr{P}_N$.

5.3 Null Spaces of Differential Operators

The following result is central to all the work of this chapter.

THEOREM 5.6 *Let T be a linear operator on $\mathcal{C}^\infty$, and $g \in \mathcal{C}^\infty$. Let $\mathcal{N}_T$ be the set of all solutions of $T\{y\} = 0$. Then,*

1. *$\mathcal{N}_T$ is a linear subspace of $\mathcal{C}^\infty$ (called the* **null space** *of T).*

2. *Let* ϕ_0 *be one solution of* $T\{y\} = g$. *Then, all solutions of* $T\{y\} = g$ *have the form*

$$y = \phi_0 + f \tag{5.7}$$

where f *is any member of* $\mathcal{N}_T$.

Proof Property (1) is an immediate consequence of the linearity of T: if f_1 and f_2 are both members of the null space $\mathcal{N}_T$, then $T\{f_1\} = 0$ and $T\{f_2\} = 0$. Take any constants C_1 and C_2 and construct the function $f = C_1 f_1 + C_2 f_2$. Because T is linear,

$$\begin{aligned} T\{f\} &= C_1 T\{f_1\} + C_2 T\{f_2\} \\ &= C_1(0) + C_2(0) = 0 \end{aligned}$$

Thus, f belongs to $\mathcal{N}_T$, proving that $\mathcal{N}_T$ is a linear space. To prove property 2, assume that ϕ_0 is a solution of $T\{y\} = g$, and let f be any function in the null space $\mathcal{N}_T$. Because T is linear,

$$T\{\phi_0 + f\} = T\{\phi_0\} + T\{f\} = g + 0 = g$$

Hence, every function of the form (5.7) is a solution. Conversely, let y be any solution of $T\{y\} = g$. Then,

$$T\{y - \phi_0\} = g - g = 0$$

Setting $f = y - \phi_0$, we see that f lies in the null space $\mathcal{N}_T$ so that $y = \phi_0 + f$ and has the required form (5.7).

Theorem 5.6 reveals the general abstract pattern to the solution of all linear operator equations. In order to solve for y in the equation $T\{y\} = g$, three steps are involved

Step 1 Find at least one solution ϕ_0 *for the equation.*

Step 2 Determine the null space $\mathcal{N}_T$ *for* T.

Step 3 Then, all solutions, including those that obey any suitable list of initial conditions or other requirements, can be obtained from ϕ_0 *by adding members of* $\mathcal{N}_T$.

This pattern is applicable to the solution of algebraic linear equations and to the solution of linear differential equations.

Before we can apply this technique we need to know more about the nature of the null space of a linear operator. We will deal only with differential operators of the form

$$T = D^n + a_1(t) D^{n-1} + \cdots + a_{n-1}(t) D + a_n(t) \tag{5.8}$$

where the coefficients $a_i(t)$ are in $\mathcal{C}^\infty$ or $\mathcal{C}^\infty[I]$. The central result is the following.

THEOREM 5.7 *If T is the operator (5.8), then its null space $\mathcal{N}_T$ has dimension n.*

Restated in terms of differential equations, this means that the set of all solutions of the so-called **homogeneous equation**

$$\frac{d^n y}{dt^n} + a_1(t)\frac{d^{n-1}y}{dt^{n-1}} + \cdots + a_{n-1}(t)\frac{dy}{dt} + a_n(t)\,y = 0 \tag{5.9}$$

is a linear space of dimension n which can therefore be generated by a set of n linearly independent solutions. (Observe that the word *homogeneous* as used here has a meaning different from that given in Definition 2.1.)

The proof of Theorem 5.7 for general n depends upon the existence and uniqueness theorem for differential equations. However, we can give a simple proof when $n = 1$. The operator T will have the form $T = D + a_1(t)$, and the equation $T\{y\} = 0$ becomes merely

$$\frac{dy}{dt} + a_1(t)\,y = 0$$

This is a separable differential equation, so we rewrite it as

$$\frac{dy}{y} = -a_1(t)\,dt$$

and obtain the general solution

$$y = Ce^{-A(t)} \tag{5.10}$$

where $A'(t) = a_1(t)$. It is evident from (5.10) that the null space $\mathcal{N}_T$ has dimension 1, and that $e^{-A(t)}$ is a basis for $\mathcal{N}_T$.

Suppose now that T is of second order so that $n = 2$. The equation $T\{y\} = 0$ becomes

$$y'' + a_1(t)\,y' + a_2(t)\,y = 0$$

and the key fact we shall use (from Theorems 3.18 and 3.19, Section 3.6) is that there is a unique solution of this equation obeying any assigned pair of initial conditions of the form $y(t_0) = A$, $y'(t_0) = B$. In particular, we can choose two solutions ϕ_1 and ϕ_2 corresponding to the following initial conditions:

$$\phi_1\colon\quad \phi_1(t_0) = 1 \qquad \phi_1'(t_0) = 0$$

$$\phi_2\colon\quad \phi_2(t_0) = 0 \qquad \phi_2'(t_0) = 1$$

The functions ϕ_1 and ϕ_2 are linearly independent. For example, we can use the Wronskian test and calculate

$$W(t) = \det\begin{bmatrix} \phi_1(t_0) & \phi_2(t_0) \\ \phi_1'(t_0) & \phi_2'(t_0) \end{bmatrix} = \det\begin{bmatrix} 1 & 0 \\ 0 & 1 \end{bmatrix}$$

$$= 1 \neq 0$$

We next show that ϕ_1 and ϕ_2 form a basis for $\mathcal{N}_T$ by showing that they generate $\mathcal{N}_T$. Let $f \in \mathcal{N}_T$, and compute the values $A = f(t_0)$, $B = f'(t_0)$. Then, construct the function $g = A\phi_1 + B\phi_2$, which is also a member of $\mathcal{N}_T$ since $\mathcal{N}_T$ is a linear space. Calculating $g(t_0)$ and $g'(t_0)$ we have

$$g(t_0) = A\phi_1(t_0) + B\phi_2(t_0) = A$$

$$g'(t_0) = A\phi_1'(t_0) + B\phi_2'(t_0) = B$$

Thus, g is a member of $\mathcal{N}_T$ that obeys the same initial conditions as does f. By the uniqueness theorem, $g = f$ so that f has been expressed as a linear combination of ϕ_1 and ϕ_2. This shows that $\mathcal{N}_T$ has dimension 2.

The proof of the general case follows the same pattern. Choose a particular set of solutions ϕ_i of (5.9) that obey special initial conditions as follows:

$$\phi_i\colon\quad \phi_i^{(k)}(t_0) = \begin{cases} 1 & \text{if } k = i-1 \\ 0 & \text{for all other } k = 0, 1, \ldots, n-1 \end{cases}$$

Because $W(t_0) = 1 \neq 0$, the functions $\{\phi_1, \phi_2, \ldots, \phi_n\}$ are linearly independent and lie in $\mathcal{N}_T$. To show that they form a basis, choose any $f \in \mathcal{N}_T$, and calculate the numbers $C_k = f^{(k)}(t_0)$. Then, construct the function

$$g = C_0\phi_1 + C_1\phi_2 + \cdots + C_{n-1}\phi_n$$

One discovers that $g^{(k)}(t_0) = C_k$ for $k = 0, 1, 2, \ldots, n-1$ so that g and f obey the same initial conditions. Since both are solutions of (5.9), they are identical. Thus, f has been represented as a linear combination of the ϕ_i.

Theorem 5.7 reduces the problem of finding all solutions of (5.9) to the problem of finding n linearly independent solutions. By virtue of the results proved in Section 5.2, it is not necessary to find the special functions $\{\phi_i\}$ used in the proof of Theorem 5.7, since *any* set of n independent solutions will be a basis for $\mathcal{N}_T$.

While the theory is transparent, its practical application is somewhat limited since there is no simple way to obtain even one solution of the general equation (5.9), except by power-series methods such as those we discussed in

Section 3.5. However, there is one special case where a complete treatment can be given. It is also a case that has considerable practical importance, since many differential equations fall into this class. These are the operators T of the standard form (5.9) but where the coefficients $a_i(t)$ are merely constants. For example, we might have $T = D^3 - D^2 - D + 1$. In such cases, it is possible to describe a basis for $\mathcal{N}_T$ directly from the form of T. For example, here a basis is $\{e^{-t}, e^t, te^t\}$. The purpose of the next two theorems is to show how one determines such a basis.

THEOREM 5.8 *With any constant-coefficient operator*

$$T = D^n + a_1 D^{n-1} + a_2 D^{n-2} + \cdots + a_{n-1} D + a_n \tag{5.11}$$

we associate the ordinary polynomial P of degree n

$$P(s) = s^n + a_1 s^{n-1} + \cdots + a_n$$

Then, if γ is a root of P [so that $P(\gamma) = 0$], $e^{\gamma t}$ lies in $\mathcal{N}_T$.

Proof For any choice of γ, we have

$$\begin{aligned} T\{e^{\gamma t}\} &= (D^n + a_1 D^{n-1} + \cdots + a_n)\{e^{\gamma t}\} \\ &= \gamma^n e^{\gamma t} + a_1 \gamma^{n-1} e^{\gamma t} + \cdots + a_n e^{\gamma t} \\ &= P(\gamma)\, e^{\gamma t} \end{aligned}$$

Thus, if $P(\gamma) = 0$, $T\{e^{\gamma t}\} = 0$ and $e^{\gamma t}$ lies in $\mathcal{N}_T$.

COROLLARY *If the polynomial $P(s)$ has n distinct roots $\gamma_1, \gamma_2, \gamma_3, \ldots, \gamma_n$, then a basis for $\mathcal{N}_T$ consists of the functions $\{e^{\gamma_1 t}, e^{\gamma_2 t}, \ldots, e^{\gamma_n t}\}$.*

To conclude this, it is only necessary to know that these n functions are linearly independent. (See Exercise 6.)

As an illustration, suppose we want to find all solutions of the equation $y'' - 2y' - 3y = 0$. The polynomial $P(s)$ is $s^2 - 2s - 3$, which factors as $(s-3)(s+2)$. Thus, the null space of T is generated by $\{e^{3t}, e^{-2t}\}$ and the set of all solutions of the differential equation consists of the functions $y = c_1 e^{3t} + c_2 e^{-2t}$.

What happens if the **associated polynomial** $P(s)$ has two or more equal roots? Let us look at the operator

$$T = D^2 - 2D + 1$$

The associated polynomial is $P(s) = s^2 - 2s + 1 = (s-1)^2$, whose roots are 1 and 1. By Theorem 5.8 we know that e^t is in the null space of T. But the pair of roots yield only one function, and this function cannot be a basis for this space, which must have dimension 2. Where can we find another function ϕ satisfying $T(\phi) = 0$?

We can guess the correct answer by looking at an even simpler operator, D^2. Here, the associated polynomial is s^2, and the roots are 0 and 0. Theorem 5.8 gives us only the function $e^{0t} = 1$ as a member of the null space. However, we know that the complete null space of D^2 is generated by the functions 1 and t. Likewise, the associated polynomial for the operator D^3 has roots $0, 0, 0$, and its null space is generated by 1, t, and t^2. We are thus led to conjecture that the null space of the operator $T = D^2 - 2D + 1$ contains both e^t and te^t. This is easily checked:

$$\begin{aligned} T\{te^t\} &= \frac{d^2}{dt^2}\{te^t\} - 2\frac{d}{dt}\{te^t\} + te^t \\ &= \frac{d}{dt}\{e^t + te^t\} - 2(e^t + te^t) + te^t \\ &= e^t + e^t + te^t - 2e^t - 2te^t + te^t \\ &= 0 \end{aligned}$$

Since e^t and te^t are linearly independent, the null space of T consists of all the functions $C_1 e^t + C_2 te^t = (C_1 + C_2 t) e^t$. This example extends to the general case.

THEOREM 5.9 *If the associated polynomial $P(s)$ for an operator T has a factor of the form $(s-\gamma)^m$, then the null space of T contains the functions $e^{\gamma t}, te^{\gamma t}, t^2 e^{\gamma t}, \ldots, t^{m-1} e^{\gamma t}$.*

Proof In Theorem 5.8 we established the fact that for any γ, $T\{e^{\gamma t}\} = P(\gamma) e^{\gamma t}$. If this formula is differentiated with respect to γ, we obtain

$$T\{te^{\gamma t}\} = P'(\gamma) e^{\gamma t} + P(\gamma) te^{\gamma t} \tag{5.12}$$

If γ were a double root of $P(s)$, then $P(\gamma)$ and $P'(\gamma)$ would both be zero, so that from (5.12) we would have

$$\begin{aligned} T\{te^{\gamma t}\} &= (0) e^{\gamma t} + (0) te^{\gamma t} \\ &= 0 \end{aligned}$$

Thus, when $P(\gamma) = P'(\gamma) = 0$, $e^{\gamma t}$ and $te^{\gamma t}$ both lie in $\mathcal{N}_T$; they are, of course, linearly independent. (See Exercise 9, Section 5.2.) The same approach works in general. If we differentiate the relation (5.12) again with respect to γ (a device which works since the operators d/dt and $d/d\gamma$ commute), we obtain

(5.13) $$T\{t^2 e^{\gamma t}\} = P''(\gamma)\, e^{\gamma t} + 2P'(\gamma)\, te^{\gamma t} + P(\gamma)\, t^2 e^{\gamma t}$$

Thus, if γ is a triple root of $P(s)$, $P(\gamma) = P'(\gamma) = P''(\gamma) = 0$ and we have $T\{t^2 e^{\gamma t}\} = 0$. In this case, the functions $e^{\gamma t}$, $te^{\gamma t}$, and $t^2 e^{\gamma t}$ all lie in $\mathcal{N}_T$ and are linearly independent.

We can put the results of Theorems 5.8 and 5.9 together to yield a complete description of the null space of an operator of the form (5.11). Given T, we form the polynomial $P(s)$ and write it as a product of factors of the form $(s-\gamma)$. Each distinct γ corresponds to a function $e^{\gamma t}$, and a factor that is repeated m times corresponds to the set of m functions $e^{\gamma t}, te^{\gamma t}, t^2 e^{\gamma t}, \ldots, t^{m-1} e^{\gamma t}$. All of the functions so obtained form a set of n linearly independent functions, and together these form a basis for the null space $\mathcal{N}_T$.

As usual, practical questions arise as soon as one starts to use a theoretical result. We shall illustrate this with a brief discussion of some second-order equations.

Let us consider first the equation

$$y'' - 3y' - 2y = 0$$

The operator is $D^2 - 3D - 2$, and $P(s) = s^2 - 3s - 2$. The roots of P are $\frac{1}{2}(3 \pm \sqrt{9+8})$, or 3.56 and -0.56, to two decimal places. Hence, the solutions consist of all the functions

$$y = C_1 e^{3.56t} + C_2 e^{-0.56t}$$

Now let us solve the equation

(5.14) $$y'' + 2y' + 5y = 0$$

The operator is $D^2 + 2D + 5$, and $P(s) = s^2 + 2s + 5$. The roots of P are $\frac{1}{2}(-2 \pm \sqrt{4-20}) = -1 \pm 2i$. Hence, the solutions consist of all the functions

(5.15) $$y = C_1 e^{(-1+2i)t} + C_2 e^{(-1-2i)t}$$

Although it is entirely correct and often easier to leave the solutions in this form, it is also useful to know how to choose another basis not containing complex functions such as $e^{(-1+2i)t}$ and $e^{(-1-2i)t}$. There are two identities satisfied by the exponential function which make this possible. Together they constitute the following lemma.

LEMMA 5.2 *(a) For any real or complex numbers u and v,*

$$e^{u+v} = e^u e^v$$

(b) For any real number θ,

$$e^{i\theta} = \cos\theta + i\sin\theta$$

These were discussed and proved in Section 3.3.

Returning to (5.15), we apply the lemma to obtain

$$e^{(-1+2i)t} = e^{-t+2it} = e^{-t}e^{2it}$$
$$= e^{-t}(\cos 2t + i\sin 2t)$$

and in the same way,

$$e^{(-1-2i)t} = e^{-t}(\cos(-2t) + i\sin(-2t))$$
$$= e^{-t}(\cos 2t - i\sin 2t)$$

Hence, the null space of T, which is also the set of solutions of (5.14), can be described as the set of all functions

$$C_1 e^{-t}(\cos 2t + i\sin 2t) + C_2 e^{-t}(\cos 2t - i\sin 2t)$$

If we take $C_1 = C_2 = \frac{1}{2}$, then this expression becomes $e^{-t}\cos 2t$; if we take $C_1 = -i/2$ and $C_2 = i/2$, it becomes $e^{-t}\sin 2t$. Hence, the null space of T also has a basis consisting of the two functions $e^{-t}\cos 2t$ and $e^{-t}\sin 2t$, and comprises all the functions of the form $g(t) = C_1 e^{-t}\cos 2t + C_2 e^{-t}\sin 2t$. One direct advantage of the use of this basis is that it is often easier to graph real functions such as $e^{-t}\cos 2t$ and to understand their properties than to do this for complex functions such as $e^{(-1+2i)t}$.

This type of reduction is possible whenever a pair of roots of the associated polynomial $P(s)$ are conjugate complex numbers of the form $a+bi$, $a-bi$. In complex form, the corresponding functions in the basis are $e^{(a+bi)t}$ and $e^{(a-bi)t}$, and a similar calculation based on Lemma 5.2 shows that these functions may be replaced by the real functions $e^{at}\cos bt$ and $e^{at}\sin bt$. More generally, any *real* linear differential equation of the form $T\{y\} = 0$ which has a complex solution of the form $y = u(t) + iv(t)$ will also have the real functions $u(t)$ and $v(t)$ as solutions. For, if $0 = T\{u+iv\}$, which by linearity is $T\{u\} + iT\{v\}$, then $T\{u\} = 0$ and $T\{v\} = 0$.

We can give a complete treatment of second-degree operators $T = D^2 + pD + q$ with constant *real* coefficients. The nature of the roots of the associated polynomial $P(s) = s^2 + ps + q$ depends upon the discriminant $p^2 - 4q$:

1. *If* $p^2-4q>0$, *the roots of* P *are real and distinct, and the null space of* T *has for a basis two exponential functions of the form* $e^{\gamma_1 t}, e^{\gamma_2 t}$.
2. *If* $p^2-4q=0$, P *has a double root* γ, *and the null space of* T *has* $e^{\gamma t}$ *and* $te^{\gamma t}$ *as a basis*
3. *If* $p^2-4q<0$, *then the roots of* P *are two conjugate complex numbers* $a \pm bi$, *and a basis for* $\mathcal{N}_T$ *is the pair of functions* $e^{at}\cos bt, e^{at}\sin bt$.

In Section 5.5, we explore the physical meaning of these various cases in the context of a specific model for a physical system, the harmonic oscillator.

These methods also work for linear equations of order higher than 2 if the associated polynomial $P(s)$ can be factored easily. Consider as an illustration the equation

$$y^{(4)} - 3y'' - 6y' + 8y = 0 \tag{5.16}$$

Inspection will show that the polynomial $P(s)$ has $s=1$ and $s=2$ as roots, and that it can be factored in the form

$$P(s) = (s-1)(s-2)(s^2+3s+4)$$

Factoring the quadratic by solving $s^2+3s+4=0$, we arrive at four distinct roots γ_i: $\{1,\ 2,\ -\frac{3}{2}\pm\frac{1}{2}i\sqrt{7}\}$. The set of solutions of (5.16) is therefore the linear space of dimension 4 generated by the basis

$$\{e^t,\ e^{2t},\ e^{-1.5t}\cos(0.5\sqrt{7}\,t),\ e^{-1.5t}\sin(0.5\sqrt{7}\,t)\}$$

In a more typical practical case, $P(s)$ will not be factorable by inspection, and one must resort to some numerical procedure for finding its roots. In such cases, we will be able to find only approximate values for the γ_i, and Theorem 5.9 will then produce functions that are themselves only approximate solutions of the differential equation. A theoretical analysis, which is outside the scope of this text, shows that if the roots are fairly accurate, so are the basis functions obtained from them.

So far we have used these methods to determine the complete null space of an operator T. How do we in turn use the null space to solve an initial-value problem? One example will demonstrate the procedure. Suppose we wish to find a solution of the equation

$$y'' + 3y' - 4y = 0 \tag{5.17}$$

obeying the initial conditions $y(0)=1$, $y'(0)=2$. By the method described above we find that a basis for the null space is $\{e^t, e^{-4t}\}$ so that the most general solution of (5.17) is

$$y(t) = C_1 e^t + C_2 e^{-4t}$$

Differentiate this and obtain

$$y'(t) = C_1 e^t - 4C_2 e^{-4t}$$

If we impose the required conditions by setting $t = t_0 = 0$, we have

$$1 = C_1 + C_2$$

$$2 = C_1 - 4C_2$$

which yields $C_1 = 1.2$, $C_2 = -0.2$, and gives us the desired solution $y = (1.2)e^t - (0.2)e^{-4t}$.

In the next section, we return to the more general equation of the form $T\{y\} = g$, the so-called inhomogeneous linear equation.

Exercises

1. (a) Prove that the null space of D^3 consists of the polynomials of degree at most 2.
   (b) What is the null space of D^n?
2. By solving a separable differential equation, find the null space of the operator $D + 2t$.
3. Verify (5.12) and (5.13) directly when $T = D^3$.
4. Write out the proof of Theorem 5.7 for a general third-order operator T.
5. Prove that $e^{\alpha t}, te^{\alpha t}, t^2 e^{\alpha t}$ are linearly independent.

*6. Show that the functions $e^{\alpha t}, e^{\beta t}, e^{\gamma t}$ are linearly independent if α, β, γ are distinct. [*Hint:* If an identity is differentiated, it remains an identity.]

7. Verify that the null space of $T = D^3 + 3D^2 + 3D + 1$ has a basis consisting of $e^{-t}, te^{-t}, t^2 e^{-t}$.
8. Find the null space of each of the operators:

   (a) $D^2 + 3D - 10$ (b) $D^2 + 9$ (c) $D^2 - 4D + 4$
9. Find the null space of each of the operators:

   (a) $D^2 - 2D - 1$ (b) $D^2 + 2D + 3$ (c) $D^2 + 3D + 18$
10. Find all solutions of $y^{(4)} - 3y'' + 2y' = 0$.
11. Find all solutions of $d^3x/dt^3 = 5\, dx/dt - 4x$.

12. Find the solution of $y''+2y'-3y=0$, where $y(0)=1$ and $y'(0)=1$.

13. Solve: $x''+2x'+3x=0$, where $x(0)=1$, $x'(0)=2$.

14. Solve: $y''=y'+3y$, where $y(0)=0.3$ and $y'(0)=2.4$.

Find a basis for the null space of the operators in Exercises 15–17.

15. $T=D^5+4D^3$

16. $T=D^4+2D^2+1$

17. $T=D^5+D^4-2D^3-2D^2+D+1$

18. Show that the null space of the operator $D^2-(3/t)\,D+(3/t^2)$ is generated by t and t^3.

19. Show that the null space of the operator $D^2-(1/t)\,D+(1/t^2)$ is generated by t and $t\log t$.

20. Show that e^{t^2} and e^{-t^2} form a basis for the null space of $D^2-(1/t)\,D-4t^2$.

21. Show that the null space of $D^2-(2/\sin 2t)\,D-\frac{1}{4}(\tan^2 t)$ is generated by $\sqrt{\cos t}$ and $1/\sqrt{\cos t}$.

*22. If $T=D^2+a(t)\,D+b(t)$ and $A(t)=\int_{t_0}^{t} a(s)\,ds$ and ϕ is in $\mathcal{N}_T$, prove that

$$\phi(t)\int_{t_0}^{t}\frac{e^{-A}}{\phi^2}$$

is also in $\mathcal{N}_T$. Check this for $T=D^2-2D-3$ and $\phi(t)=e^{-t}$.

23. Verify the result of Theorem 5.6 directly for the linear operator $T=D+a(t)$ applied to $T\{y\}=g$ by using the method of Section 2.2 for solving first-order linear equations.

*24. Let ϕ_1, ϕ_2, and ϕ_3 be functions in the null space of a linear operator of the form (5.8) of order 3. Suppose the Wronskian

$$W(t)=\det\begin{bmatrix}\phi_1(t) & \phi_2(t) & \phi_3(t)\\ \phi_1'(t) & \phi_2'(t) & \phi_3'(t)\\ \phi_1''(t) & \phi_2''(t) & \phi_3''(t)\end{bmatrix}$$

is zero for some $t=t_0$, $-\infty<t_0<\infty$.

(a) Prove that $\{\phi_1,\phi_2,\phi_3\}$ is a linearly dependent set.
(b) Prove that $W(t)=0$ for all t.

25. Show that no operator T on $\mathcal{C}^\infty$ of order 2 of form (5.8) can have t and e^t in its null space $\mathcal{N}_T$. What happens if T is an operator of order 3?

If you have access to a pocket calculator, find all solutions of Exercises 26–29.

PC26. $y'' - (4.2)\,y' + (3.6)\,y = 0$

PC27. $x'' - (2.4)\,x' + (3.6)\,x = 0$

PC28. $x''' - 5x' + 3x = 0$

PC29. $y''' - 2y' + 3y = 0$

5.4 Solution of $T\{y\} = g$

Let us see how the results we have obtained can now be used to solve a linear differential equation for which the right-hand side is not zero. Suppose we wish to find all solutions of the equation

$$y'' - 2y' - 3y = 3t^2 + t - 4 \tag{5.18}$$

This has the form $T\{y\} = g$ where $g(t) = 3t^2 + t - 4$ and $T = D^2 - 2D - 3$. Using the methods of the previous section, we have $P(s) = s^2 - 2s - 3 = (s+1)(s-3)$. Thus, the null space of T is generated by $\{e^{-t}, e^{3t}\}$.

By Theorem 5.6, in order to solve (5.18) we need to find some particular solution ϕ_0. Suppose we search for it among the class of all polynomials, and try to find a solution of the form $y = \phi_0(t) = A_0 + A_1 t + A_2 t^2$. Substituting this into (5.18), we find that

$$\begin{aligned} T\{\phi_0\} &= D^2\{A_0 + A_1 t + A_2 t^2\} - 2D\{A_0 + A_1 t + A_2 t^2\} \\ &\quad - 3(A_0 + A_1 t + A_2 t^2) \\ &= 2A_2 - 2(A_1 + 2A_2 t) - 3A_0 - 3A_1 t - 3A_2 t^2 \\ &= -3A_2 t^2 + (-4A_2 - 3A_1)t + (2A_2 - 2A_1 - 3A_0) \end{aligned}$$

Equating this to the given polynomial g, we compare coefficients of powers of t, and must therefore choose the A_i so that

$$\begin{aligned} -3A_2 &= 3 \\ -4A_2 - 3A_1 &= 1 \\ 2A_2 - 2A_1 - 3A_0 &= -4 \end{aligned}$$

Solving these, we find $A_2 = -1$, $A_1 = 1$, $A_0 = 0$. Hence, by something like a trial-and-error method, we have arrived at a particular solution for (5.18), namely, $\phi_0(t) = t - t^2$. Theorem 5.6 tells us that the general solution of (5.18) is then found by adding to this the functions in the null space of T. This gives us

$$y = t - t^2 + C_1 e^{-t} + C_2 e^{3t} \tag{5.19}$$

If the original problem had been to find the solution of (5.18) which satisfied the initial conditions $y(0) = 1$, $y'(0) = 0$, we would impose these on (5.19) and determine the correct values of C_1 and C_2. From (5.19) we would have

$$y' = 1 - 2t - C_1 e^{-t} + 3C_2 e^{3t}$$

Setting $t = 0$ in both this and (5.19), we would have

$$y(0) = 1 = 0 - 0 + C_1 + C_2$$

$$y'(0) = 0 = 1 - 0 - C_1 + 3C_2$$

These yield $C_1 = 1$, $C_2 = 0$, so the required solution of the initial-value problem would be $y = t - t^2 + e^{-t}$.

This simple example illustrates the general procedure for solving a linear equation of the form $T\{y\} = g$ in the case when T is an operator with constant coefficients. The general case would follow the same pattern, but it would be harder both to find a particular solution ϕ_0 and to determine a basis for the null space $\mathcal{N}_T$. Sometimes the only way of doing either of these is to use the method of power series as explained in Section 3.5. There is a general method, known as *variation of parameters*, which in theory permits one to find a particular solution ϕ_0 for any linear equation $T\{y\} = g$ if the null space of T is known. We will explain this method at the end of the section.

When T is a constant-coefficient operator, there are a number of simple cases in which it is easy to find a particular solution ϕ_0. The first has already been illustrated; it suggests the following theorem.

THEOREM 5.10 *If $T = D^n + a_1 D^{n-1} + \cdots + a_n$, with $a_n \neq 0$, and if $g(t)$ is a polynomial of degree N, then $T\{y\} = g$ has a solution $y = \phi_0(t)$ that is a polynomial of degree N.*

Proof This is illustrated by the solution of equation (5.18): there, g was a polynomial of degree 2 and $T\{y\} = g$ had a solution $y = t - t^2$. We give a proof of Theorem 5.10 for $n = 3$ which can be immediately generalized to any n. Suppose $T = D^3 + a_1 D^2 + a_2 D + a_3$ with $a_3 \neq 0$. Construct a sequence of polynomials $p_i(t)$ by $p_i(t) = T\{t^i\}$. Thus

$$
\begin{aligned}
p_0 &= T\{1\} &&= a_3 \\
p_1 &= T\{t\} &&= a_2 + a_3 t \\
p_2 &= T\{t^2\} &&= 2a_1 + 2a_2 t + a_3 t^2 \\
p_3 &= T\{t^3\} &&= 6 + 6a_1 t + 3a_2 t^2 + a_3 t^3 \\
p_4 &= T\{t^4\} &&= 24t + 12a_1 t^2 + 4a_2 t^3 + a_3 t^4
\end{aligned}
$$

etc.

Since $a_3 \neq 0$, $p_i(t)$ is a polynomial of degree exactly i for each $i = 0, 1, 2, \ldots$. Accordingly, $\{p_0, p_1, \ldots, p_N\}$ form a basis for the space $\mathscr{P}_N$. (See Exercise 15, Section 5.2.) Since $g \in \mathscr{P}_N$, there are constants c_i such that

$$g = c_0 p_0 + c_1 p_1 + \cdots + c_N p_N$$

But, $p_i = T\{t^i\}$ so that

$$
\begin{aligned}
g &= \sum_0^N c_i T\{t^i\} \\
&= T\{c_0 + c_1 t + c_2 t^2 + \cdots + c_N t^N\} \\
&= T\{\phi_0\}
\end{aligned}
$$

and we have found a polynomial solution for $T\{y\} = g$.

Another simple case is covered by the following result.

THEOREM 5.11 *If T is the constant-coefficient operator $D^n + a_1 D^{n-1} + \cdots + a_n$ with associated polynomial $P(s)$, and $P(\beta) \neq 0$, then the equation $T\{y\} = Ae^{\beta t}$ has a solution of the form $y = Ce^{\beta t}$. If $P(\beta) = 0$ but $P'(\beta) \neq 0$, then it has a solution of the form $Cte^{\beta t}$.*

Proof As in the proof of Theorem 5.8, we have, for any γ,

$$T\{e^{\gamma t}\} = P(\gamma) e^{\gamma t}$$

Accordingly, for any constant C,

$$
\begin{aligned}
T\{Ce^{\beta t}\} &= CT\{e^{\beta t}\} \\
&= CP(\beta) e^{\beta t}
\end{aligned}
$$

and this is equal to $Ae^{\beta t}$ if we choose $C = A/P(\beta)$. If $P(\beta) = 0$, we recall formula (5.12):

$$T\{te^{\gamma t}\} = P'(\gamma) e^{\gamma t} + P(\gamma) te^{\gamma t}$$

Hence, using this with $\gamma = \beta$,

$$\begin{aligned} T\{Cte^{\beta t}\} &= CP'(\beta)\,e^{\beta t} + CP(\beta)\,te^{\beta t} \\ &= CP'(\beta)\,e^{\beta t} \end{aligned}$$

This yields $Ae^{\beta t}$ if we choose $C = A/P'(\beta)$.

Finally, we record a simple consequence of linearity that often goes by the title of the **principle of superposition**.

THEOREM 5.12 *If ϕ_k is a solution of the equation $T\{y\} = g_k$, then $y = \phi_1 + \phi_2 + \cdots + \phi_m$ is a solution of the equation*

$$T\{y\} = g_1 + g_2 + \cdots + g_m$$

The last three results are often combined in solving a specific differential equation. Suppose we want the complete solution of the equation

$$y'' - 4y = t - 2 + 5e^{-t} \tag{5.20}$$

Here, $T = D^2 - 4$. We know that $T\{y\} = t - 2$ has a polynomial solution ϕ_1 of degree 1, and that $T\{y\} = 5e^{-t}$ has an exponential solution ϕ_2 of the form Ce^{-t}. These turn out to be $\phi_1 = \frac{1}{2} - \frac{1}{4}t$ and $\phi_2 = -\frac{5}{3}e^{-t}$. Invoking Theorem 5.12 and adding the null space of T, we obtain the general solution of (5.20):

$$y = \tfrac{1}{2} - \tfrac{1}{4}t - \tfrac{5}{3}e^{-t} + C_1 e^{2t} + C_2 e^{-2t}$$

Complex exponentials are often replaced by the corresponding trigonometric functions, and vice versa. For example, suppose we want to find the complete solution for

$$y'' + 9y = 4e^{-2t} + \sin t - 3\cos 2t \tag{5.21}$$

Here $T = D^2 + 9$ and the roots of $P(s)$ are $3i$, $-3i$. If the trigonometric functions are replaced by their equivalents as illustrated in Section 5.3, the right-hand side of (5.21) would become a linear combination of e^{-2t}, e^{it}, e^{-it}, e^{2it}, and e^{-2it}, where $i = \sqrt{-1}$. Since the exponents of all these differ from $3i$ and $-3i$, Theorem 5.11 applies and we know that (5.21) has a particular solution which is also a linear combination of these five exponential functions.

We can also use the trigonometric forms instead. Thus, we first look for a solution of $T\{y\} = 4e^{-2t}$ by trying a function of the form $\phi_1 = Ce^{-2t}$, and easily find that $\phi_1 = \frac{4}{13}e^{-2t}$. We then look for a solution of $T\{y\} = \sin t$ among the functions of the form $\phi_1 = A\sin t + B\cos t$, to cover both e^{it} and e^{-it}. Substituting this, we have

$$T\{A\sin t + B\cos t\} = 8A\sin t + 8B\cos t$$

and equating this to $\sin t$, we have $\phi_1 = \frac{1}{8}\sin t$. We next look for a solution of $T\{y\} = -3\cos(2t)$ among the functions $\phi_3 = A\sin(2t) + B\cos(2t)$. We have

$$T\{\phi_3\} = 5A\sin(2t) + 5B\cos(2t)$$

so that the desired solution is $\phi_3 = -\frac{3}{5}\cos(2t)$. Finally, the complete solution of (5.21) is

$$y = \tfrac{4}{13}e^{-2t} + \tfrac{1}{8}\sin t - \tfrac{3}{5}\cos(2t) + C_1\sin(3t) + C_2\cos(3t)$$

Further examples of this technique will be found in the exercises. Note in particular Exercises 7, 8, and 20, which suggest what should be done when the exponential or trigonometric functions on the right-hand side have functions in common with the null space of T.

We now take up two closely related general methods which use functions in the null space of T to find particular solutions of $T\{y\} = g$ and which can, in theory, be applied to linear equations with nonconstant coefficients. Both of these methods are generalizations of the method for solving first-order linear equations explained in Section 2.2. This means that the final steps in the solution will involve indefinite integration which may not be possible as a practical matter if the integrals cannot be expressed in terms of standard functions.

The first method works only for second-order equations.

THEOREM 5.13 *Let $T = D^2 + a(t)D + b(t)$ and let $\phi(t)$ be any nontrivial solution of $T\{y\} = 0$, namely, a function in $\mathcal{N}_T$ different from the zero function. Then, a method exists for constructing a solution of $T\{y\} = g$ for any $g \in \mathcal{C}^\infty$ requiring only integration.*

Proof We look for a solution with the special form $y = u\phi$, where $u = u(t)$ is a function to be determined. Substituting this into the equation, we have

$$\begin{aligned} g(t) = T\{u\phi\} &= D^2\{u\phi\} + a(t)D\{u\phi\} + b(t)u\phi \\ &= u''\phi + 2u'\phi' + u\phi'' + a(t)(u'\phi + u\phi') + bu\phi \\ &= u''\phi + 2u'\phi' + u[\phi'' + a\phi' + b\phi] + a(t)u'\phi \end{aligned}$$

The expression in brackets is exactly $T\{\phi\}$, which is zero since ϕ lies in $\mathcal{N}_T$. Thus, we obtain a differential equation that must be satisfied by the function u, namely,

$$\phi u'' + [2\phi' + a(t)\phi]u' = g \tag{5.22}$$

This is still a second-order differential equation, but it is simpler than the original equation since u is missing from (5.22). If we put $v = u'$, then (5.22) becomes a first-order linear equation

$$v' + \left[\frac{2\phi'}{\phi} + a(t)\right]v = \frac{g}{\phi} \tag{5.23}$$

This can be solved by the method explained in Section 2.2. Let $A(t) = \int a(t)\, dt$. Then, setting

$$\psi(t) = \exp\left\{\int\left[\frac{2\phi'}{\phi} + a(t)\right]dt\right\} = \phi^2 e^{A(t)}$$

and multiplying both sides of (5.23) by ψ yields the equivalent equation

$$\frac{d}{dt}\{\phi^2 e^{A(t)} v\} = \phi(t)\, e^{A(t)} g(t)$$

Integrating this, we have

$$\phi^2 e^{A(t)} v = \int_{t_0}^{t} \phi(s)\, e^{A(s)} g(s)\, ds$$

Solving for v, which we recall is u', we have

$$\frac{du}{dt} = \frac{e^{-A(t)}}{[\phi(t)]^2}\int_{t_0}^{t} \phi(s)\, e^{A(s)} g(s)\, ds \tag{5.24}$$

Integrating again, we obtain a formula for $u(t)$ and then for y.

$$y = \phi(t)u(t) = \phi(t)\int \frac{e^{-A(t)}}{[\phi(t)]^2}\left(\int_{t_0}^{t} \phi(s)\, e^{A(s)} g(s)\, ds\right) dt$$

The second procedure for obtaining a particular solution of $T\{y\} = g$ works for linear equations of any order, but requires knowledge of a complete basis for $\mathcal{N}_T$. For historical reasons this procedure is called *variation of parameters*. For simplicity, we give the proof only for second-order operators, but suggest how it can be generalized.

THEOREM 5.14 *Let $T = D^2 + a(t)D + b(t)$, and suppose that $\{\phi, \psi\}$ is a basis for its null space $\mathcal{N}_T$. Let $u(t)$ and $v(t)$ be two functions whose derivatives satisfy the pair of algebraic equations*

$$\begin{aligned} u'(t)\phi(t) + v'(t)\psi(t) &= 0 \\ u'(t)\phi'(t) + v'(t)\psi'(t) &= g(t) \end{aligned} \tag{5.25}$$

Then, the function $f(t) = u(t)\phi(t) + v(t)\psi(t)$ is a solution of $T\{y\} = g$.

Proof The basic idea is to choose u and v so that $f = u\phi + v\psi$ is a solution of $T\{y\} = g$. Since $T\{f\} = f'' + af' + bf$, we start by calculating f'.

$$f' = u'\phi + u\phi' + v'\psi + v\psi' \tag{5.26}$$

If the first equation in (5.25) holds, then (5.26) simplifies to

$$f' = u\phi' + v\psi'$$

Differentiating again, we have

$$f'' = u'\phi' + u\phi'' + v'\psi' + v\psi''$$

This then yields

$$\begin{aligned} T\{f\} &= u'\phi' + u\phi'' + v'\psi' + v\psi'' + a(u\phi' + v\psi') + b(u\phi + v\psi) \\ &= u'\phi' + v'\psi' + u(\phi'' + a\phi' + b\phi) + v(\psi'' + a\psi' + b\psi) \\ &= u'\phi' + v'\psi' + uT\{\phi\} + vT\{\psi\} \end{aligned}$$

Since ϕ and ψ lie in $\mathcal{N}_T$, $T\{\phi\} = T\{\psi\} = 0$. We have thus shown that

$$T\{f\} = u'\phi' + v'\psi'$$

If the second equation in (5.25) holds, then this reduces merely to $T\{f\} = g$, and we have shown that $f = u\phi + v\psi$ is indeed a solution.

The equations (5.25) always have a solution for $u'(t)$ and $v'(t)$. In fact, the determinant of this pair of equations is

$$W(t) = \det\begin{bmatrix} \phi(t) & \psi(t) \\ \phi'(t) & \psi'(t) \end{bmatrix}$$

This is the Wronskian of ϕ and ψ and it is nonzero everywhere, since ϕ and ψ are linearly independent solutions of the equation $T\{y\} = 0$. (See Exercise 24, Section 5.3.) If (5.25) is solved for $u'(t)$ and $v'(t)$, then u and v are obtained by integration.

$$u(t) = -\int \frac{g(t)\psi(t)}{W(t)}\,dt \qquad v(t) = \int \frac{g(t)\phi(t)}{W(t)}\,dt$$

A similar procedure works for third- or higher-order operators T. Thus if T is order 3 and $\{\phi_1, \phi_2, \phi_3\}$ is a basis for $\mathcal{N}_T$, choose three functions u_1, u_2, u_3 whose derivatives satisfy the system of algebraic equations

$$\begin{aligned} u_1'\phi_1 + u_2'\phi_2 + u_3'\phi_3 &= 0 \\ u_1'\phi_1' + u_2'\phi_2' + u_3'\phi_3' &= 0 \\ u_1'\phi_1'' + u_2'\phi_2'' + u_3'\phi_3'' &= g(t) \end{aligned} \tag{5.27}$$

Then, $f = u_1\phi_1 + u_2\phi_2 + u_3\phi_3$ is a solution of $T\{y\} = g$. Note that the determinant of the equations (5.27) is again the Wronskian $W(t)$ of the functions ϕ_1, ϕ_2, ϕ_3.

We illustrate Theorems 5.13 and 5.14 with the same example, the differential equation

$$y'' - \frac{1}{t}y' - 4t^2 y = t^4, \qquad t_0 = 1 \tag{5.28}$$

The operator T was examined in Exercise 20, Section 5.3, where it was shown that the null space $\mathcal{N}_T$ was generated by $\{e^{t^2}, e^{-t^2}\}$. Suppose we first use Theorem 5.13 to obtain a particular solution of (5.28). We need one nontrivial member of $\mathcal{N}_T$. It does not matter which one we use; we choose $\phi = e^{t^2}$. Then, comparing (5.28) with the form of T given in Theorem 5.13, we see that $a(t) = -1/t$ so that $A(t)$, which is the integral of $a(t)$, is $-\log t$, and $e^{A(t)} = t^{-1}$. In our case $g(t) = t^4$ so that if we make all the required substitutions in (5.24), we have

$$\begin{aligned}\frac{du}{dt} &= \frac{t}{(e^{t^2})^2}\int_1^t e^{s^2}\frac{1}{s}s^4\,ds \\ &= te^{-2t^2}\int_1^t s^3 e^{s^2}\,ds\end{aligned}$$

Now, $\int s^3 e^{t^2}\,ds = \frac{1}{2}(s^2 - 1)e^{s^2} + \text{const}$, so that we find

$$\begin{aligned}\frac{du}{dt} &= te^{-2t^2}\tfrac{1}{2}(t^2 - 1)e^{t^2} \\ &= \tfrac{1}{2}t^3 e^{-t^2} - \tfrac{1}{2}te^{-t^2}\end{aligned}$$

Integrating, we have

$$\begin{aligned}u &= \tfrac{1}{2}\int t^3 e^{-t^2}\,dt - \tfrac{1}{2}\int te^{-t^2}\,dt \\ &= -\tfrac{1}{4}(t^2+1)e^{-t^2} + \tfrac{1}{4}e^{-t^2} = -\tfrac{1}{4}t^2 e^{-t^2}\end{aligned}$$

Then, the desired solution is obtained by multiplying by ϕ.

$$\begin{aligned}y &= u\phi = -\tfrac{1}{4}t^2 e^{-t^2}e^{t^2} \\ &= -\tfrac{1}{4}t^2\end{aligned} \tag{5.29}$$

This is easily checked in (5.28).

Let us now see how the method of variation of parameters, given in Theorem 5.14, applies in this example. This time, we need the full null space of T, and use $\phi = e^{t^2}$, $\psi = e^{-t^2}$. Following the instructions in the theorem, we write the pair of linear algebraic equations

$$\begin{aligned} u'e^{t^2} + v'e^{-t^2} &= 0 \\ u'(2te^{t^2}) + v'(-2te^{-t^2}) &= t^4 \end{aligned} \tag{5.30}$$

Solving these for u', we find $u' = \frac{1}{4}t^3e^{-t^2}$. If we use this in the first equation of (5.30), we find $v' = -\frac{1}{4}t^3e^{t^2}$. Integrating to find u and v we have

$$u = -\tfrac{1}{8}(t^2+1)e^{-t^2}$$

$$v = -\tfrac{1}{8}(t^2-1)e^{t^2}$$

Then, according to Theorem 5.14, a particular solution of (5.28) is given by

$$\begin{aligned} y = u\phi + v\psi &= -\tfrac{1}{8}(t^2+1) + (-\tfrac{1}{8})(t^2-1) \\ &= -\tfrac{1}{4}t^2 \end{aligned}$$

The general solution of (5.28) is then

$$y = -\tfrac{1}{4}t^2 + C_1e^{t^2} + C_2e^{-t^2}$$

As this example shows, the method of variation of parameters is limited in its practical usefulness by the fact that it often leads to integrals that cannot be done in terms of the usual tabulated functions. The illustrative example we have just worked was carefully chosen so that the integrals could be carried out. Several other examples of this sort are given in the exercises.

Exercises

1. Find the solution of $y'' - y = 2e^{-2t}$ such that when $t = 0$, $y = 0$, $y' = 0$.

2. For $y' - y = 3e^t$, find the solution such that $y = 1$ when $t = 0$.

3. Find the solution of the equation $y'' + y = 3e^{2it}$ which satisfies the initial conditions $y = 1$, $y' = 0$, when $t = 0$.

4. (a) Find all solutions of the differential equation

$$x'' + 3x' + 2x = 1 - 4t - 2t^2$$

   (b) Find the particular solution so that $x = 1$, $x' = 0$, when $t = 0$.

5. (a) Find the general solution of the equation $y'' + y' = e^{-2t}$.
   (b) Find the solution such that $y = 2$, $y' = 2$, when $t = 0$.

6. Show that if $T = D^n + a_1 D^{n-1} + \cdots + a_n$ and $P(\pm\lambda i) \neq 0$, then $T\{y\} = \sin\lambda t$ has a solution of form $A \sin(\lambda t) + B \cos(\lambda t)$.

7. Show that if $P(\pm\lambda i) = 0$ but $P'(\pm\lambda i) \neq 0$, then $T\{y\} = \sin(\lambda t)$ has a solution of form $At \sin(\lambda t) + Bt \cos(\lambda t)$.

8. Show that if $P(\beta) = P'(\beta) = 0$ and $P''(\beta) \neq 0$, then $T\{y\} = Ae^{\beta t}$ has a solution of the form $Ct^2 e^{\beta t}$. [*Hint:* Use a proof similar to that of Theorem 5.9.]

9. Show that if $T = D^n + a_1 D^{n-1} + \cdots + a_{n-1} D$, and $a_{n-1} \neq 0$, and g is a polynomial of degree N, then $T\{y\} = g$ has a solution that is a polynomial of degree $N+1$. [*Hint:* Let $u = y'$.]

Find the general solutions for the equations of Exercises 10–19.

10. $x' + 2x = \sin t$

11. $y'' + 2y' - 3y = \cos t$

12. $\dfrac{d^2y}{ds^2} - 4y = s^2 + 5s$

13. $\dfrac{d^2x}{ds^2} + 4\dfrac{dx}{ds} = 2s - 1$

14. $y'' - 4y = 3e^{2t}$

15. $x'' + 4x = 6 \sin 2t + 5e^{-t}$

16. $x'' + 2x' + 5x = 2 \sin t - 3e^{-t}$

17. $y'' + y' - 2y = 2s + e^{-s} + 3$

18. $y''' - y'' - 6y' = 3t^2 + 2$

19. $x'' - 4x' + 4x = s^2 + e^{2s}$

20. (a) Let P be the polynomial associated with the linear operator T with constant coefficients. Suppose $P(\beta) = 0$, $P'(\beta) \neq 0$. Show that $T\{y\} = te^{\beta t}$ has a solution of the form $y = (At + Bt^2)e^{\beta t}$.
    (b) If $P(\beta i) = 0$, $P'(\beta i) \neq 0$, show that $T\{y\} = t \sin(\beta t)$ has a solution of the form $y = (At + Bt^2) \sin(\beta t) + (Ct + Dt^2) \cos(\beta t)$.

Find general solutions for the equations of Exercises 21–28.

21. $y'' - y = te^{-t}$

22. $y'' + 4y' + 3y = (1+t)e^{-3t}$

23. $x'' + x = 3t \sin t$

24. $y'' + 6y' + 13y = 13s^2$

25. $x'' - 8x' + 25x = \frac{1}{2} \sin t$

26. $x''' + x'' + 8x' - 10x = e^{-s}$

27. $y'' - 3y' + 2y = 2 \cosh(2t)$

28. $y^{(4)} + 4y'' = 3s^3 + s$

29. For $T\{y\} = y'' - \frac{1}{2}ty' + [1 - (2/t^2)]\, y$:
    (a) Verify that t^2 lies in the null space of the operator T.
    (b) Use Theorem 5.13 to find a particular solution of the differential equation, $T\{y\} = 1/t$.

30. Solve $y'' - 2t^{-2}y = 3te^{-t}$ knowing that the null space of the operator $[D^2 - (2/t^2)]$ contains a function of the form t^m for some m.

31. Solve the equation $y'' + y = 2 \tan t$.

32. Solve by the method of variation of parameters (Theorem 5.14): $y''' + y' = \sec t$.

5.5 The Harmonic Oscillator

In this section, we use the techniques developed in the rest of the chapter to discuss in detail one second-order linear differential equation with constant coefficients. It is an equation that appears as the mathematical model for many different physical systems. We are interested in the interplay between the mathematical analysis and its physical interpretation, between theory and reality. We will see that useful and significant results are obtained by applying even a very simple variety of mathematics, and that one can be led to formulate conjectures about predicted behaviors of a physical system that can then be checked in the laboratory.

The physical system we shall use, shown in Figure 5.1, consists of a rod R which supports a spring S to which is attached a mass M. We will present a mathematical study of four laboratory experiments:

1. Set the mass into motion by giving it an upward or downward push.
2. Set the mass into motion by moving the supporting rod up and down in a periodic manner.
3. Place the mass M in a jar of liquid, and then start the mass moving as in (1).
4. Place the mass in the jar, and set it moving by vibrating the rod as in (2).

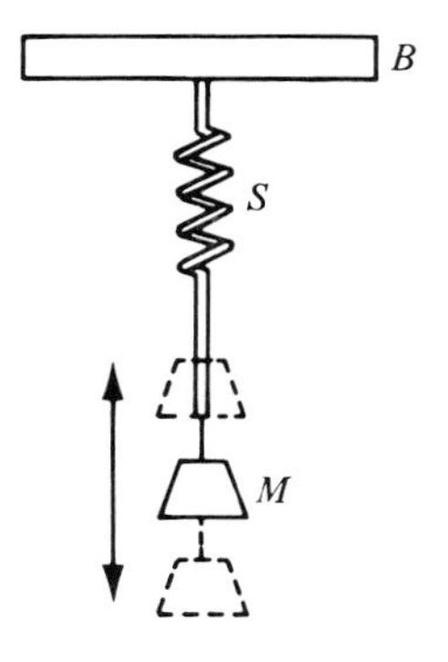

Figure 5.1

The last three of these are suggested by Figures 5.2, 5.3, and 5.4. In each case, we want to see if we can "understand" the resulting behavior of the system by showing that a model can be created which predicts the result observed. (*Attention:* To the reader! We would like you to stop reading at this point and spend some time setting down a description of what you think the motion of the mass will be for each of the four cases; for (3) and (4), be sure to discuss the differences in the motion you would expect to see depending on whether the jar is filled with alcohol or water, or with a thick, heavy oil.)

We shall not go into the reasoning which leads to an appropriate model for this system. It can be found in any adequate text on mechanics; the result is the equation

$$M\frac{d^2x}{dt^2} = -A\frac{dx}{dt} - Bx + f(t) \tag{5.31}$$

Here, x specifies the location of the moving mass at time t and is the vertical distance from the mass (regarded as a point mass) to the equilibrium point or rest position of the mass. The number B is positive and measures the strength of the spring; when the mass is displaced from equilibrium, the spring acts to try to restore equilibrium, so that the force exerted by the spring on the mass must be negative when x is positive and positive when x is negative. This explains the presence of the negative sign in front of Bx in formula (5.31). The number A is also positive and measures the resistance effect of the liquid on the moving mass. Again, the associated force must oppose the direction of motion, so that it must act downward when dx/dt is positive and upward when dx/dt is negative. This explains why this term, too, appears with a negative sign in (5.31). The assumption that the resistance is proportional to the speed is convenient mathematically; it also matches experience for many types of fluids and for speeds that are not too great. Finally, the function $f(t)$ describes an external force and reflects the manner in which the supporting rod R is moved.

We can now describe each of the separate cases (5.31) we want to study. For case (1), with no external forcing motion of the bar and no frictional forces on the moving mass, we take $A = 0$ and $f(t) = 0$. For case (3) we still have $f(t) = 0$, but now we allow $A > 0$; increasing the value of A corresponds to increasing the viscosity of the liquid, for example, changing from alcohol to water to light oil to heavy oil. For case (2) we again set $A = 0$, but this time we take for $f(t)$ a periodic function such as $\sin\beta t$. Finally, in case (4), we leave $f(t) = \sin(\beta t)$, but take $A > 0$.

We are not interested in deriving general formulas, so we take special numerical values for some of the parameters to simplify the calculations.

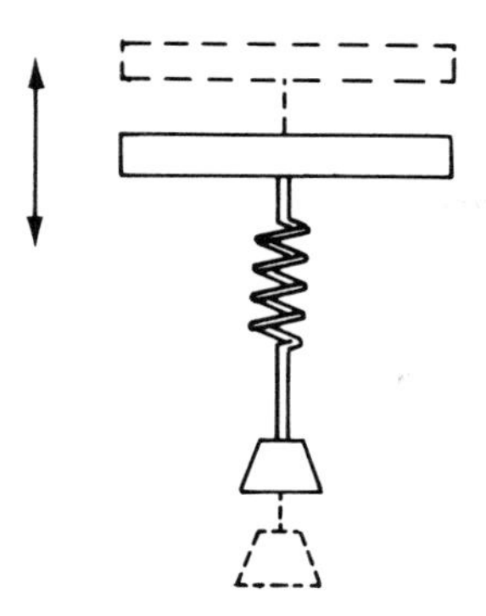

Figure 5.2

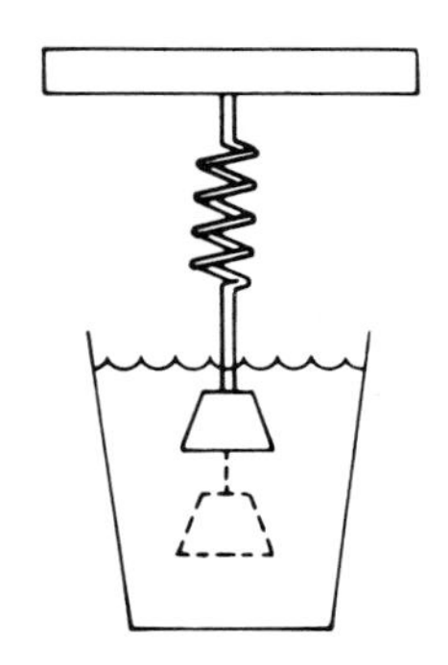

Figure 5.3

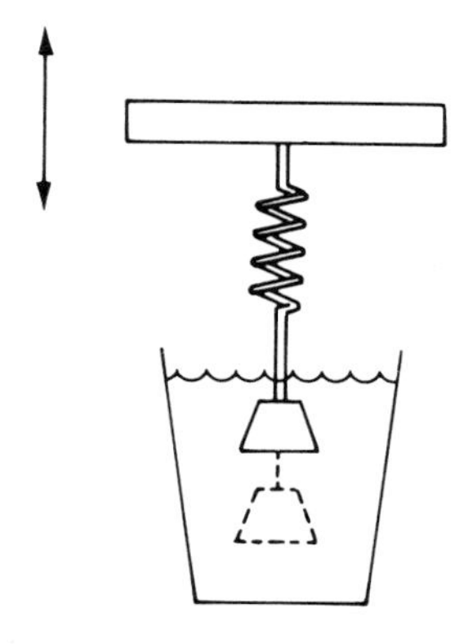

Figure 5.4

Case (1). Solve the equation $d^2x/dt^2 = -Bx$. If we set $B = \omega^2$, this equation becomes

$$(D^2+\omega^2)\{x\} = 0 \tag{5.32}$$

and by the methods of the preceding section, the general solution is

$$x = C_1 \sin(\omega t) + C_2 \cos(\omega t)$$

If we use Exercise 5 in Section 5.1, we note than an alternative form is

$$x = C \sin(\omega t+\alpha)$$

where C and α are now arbitrary constants. In either form, the resulting solution shows that the motion of the mass is periodic with a constant amplitude determined by the initial conditions. The length of the period is $2\pi/\omega$, since that is the time required for the expression $\omega t+\alpha$ to increase by exactly 2π. A typical graph of $x = \phi(t)$ is shown in Figure 5.5. We note that the period becomes shorter as ω (and therefore B) increases; this means that the stiffer the spring, the shorter the period of the up-and-down motion of the moving mass and the higher the frequency $\omega/2\pi$.

Case (2). We take the equation to be

$$\frac{d^2x}{dt^2} = -Bx + \sin\beta t$$

where, as before, $B > 0$. If we set $B = \omega^2$, then this equation becomes

$$(D^2+\omega^2)\{x\} = \sin(\beta t) \tag{5.33}$$

The roots of the associated polynomial are ωi and $-\omega i$. Recalling that $\sin(\beta t)$ can be expressed in terms of $e^{i\beta t}$ and $e^{-i\beta t}$, we have two cases to consider, that in which $\beta \neq \omega$, and that in which $\beta = \omega$. In the first case, we know that (5.33) has a solution $x = a \sin(\beta t)+b \cos(\beta t)$. Substituting this in (5.33),

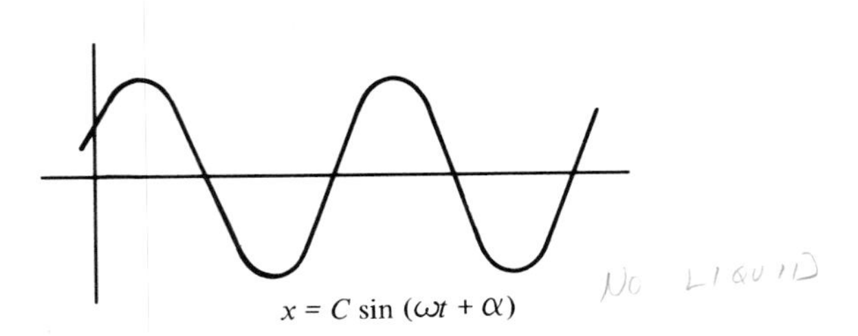

Figure 5.5

we want to have

$$\begin{aligned}\sin(\beta t) &= (D^2+\omega^2)\{a\sin(\beta t)+b\cos(\beta t)\}\\ &= -a\beta^2\sin(\beta t)+a\omega^2\sin(\beta t)-b\beta^2\cos(\beta t)+b\omega^2\cos(\beta t)\\ &= a(\omega^2-\beta^2)\sin(\beta t)+b(\omega^2-\beta^2)\cos(\beta t)\end{aligned}$$

To achieve this we must choose $b=0$ and $a=1/(\omega^2-\beta^2)$. The general solution of (5.33) is then

$$\begin{aligned}x &= C_1\sin(\omega t)+C_2\cos(\omega t)+\frac{1}{\omega^2-\beta^2}\sin(\beta t)\\ &= C\sin(\omega t+\alpha)+\frac{1}{\omega^2-\beta^2}\sin(\beta t)\end{aligned}$$

This is a combination of two sine curves with different periods. Its shape depends on the numerical relationship between ω and β. If we assume that the moving weight is at rest (in equilibrium) at $t=0$, so that when $t=0$ we have $x=dx/dt=0$, then we can solve for the constants C_1 and C_2, obtaining the special solution

(5.34) $$x = \frac{1}{\omega^2-\beta^2}\left[\sin(\beta t)-\frac{\beta}{\omega}\sin(\omega t)\right]$$

A portion of the graph of this motion, for particular values of ω and β, is given in Figure 5.6.

When $\beta=\omega$, equation (5.33) becomes

(5.35) $$(D^2+\omega^2)\{x\} = \sin(\omega t)$$

This does not have a solution of the form $x=a\sin(\omega t)+b\cos(\omega t)$ since this function in fact belongs to the null space of $D^2+\omega^2$. Instead, since $P(\pm\omega i)=0$, we can (by Theorem 5.11) expect a particular solution of the form $x_0=ate^{i\omega t}+bte^{-i\omega t}$. We prefer the trigonometric form, and so look for a solution of (5.35) with $x_0=at\sin(\omega t)+bt\cos(\omega t)$. We have

$$\begin{aligned}D\{x_0\} &= a\sin(\omega t)+a\omega t\cos(\omega t)+b\cos(\omega t)-b\omega t\sin(\omega t)\\ D^2\{x_0\} &= 2a\omega\cos(\omega t)-2b\omega\sin(\omega t)-a\omega^2 t\sin(\omega t)-b\omega^2 t\cos(\omega t)\end{aligned}$$

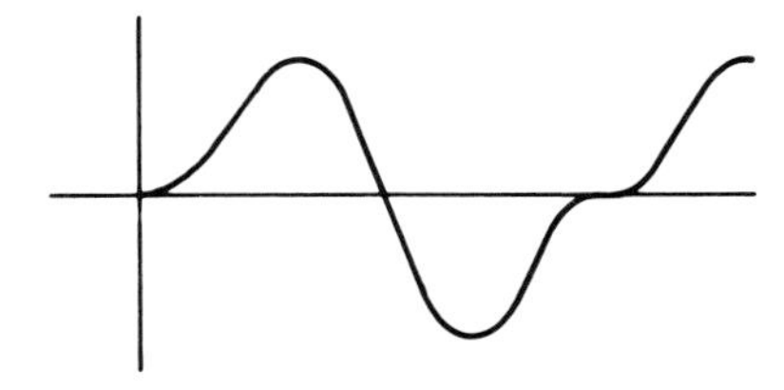

Figure 5.6

so that

$$(D^2 + \omega^2)\{x_0\} = 2a\omega \cos(\omega t) - 2b\omega \sin(\omega t)$$

In order for this to equal $\sin(\omega t)$ as required by (5.35), we must choose $a = 0$ and $b = -1/2\omega$. This gives the desired particular solution $x_0 = (-1/2\omega)\, t \cos(\omega t)$. Adding the functions in the null space of $D^2 + \omega^2$, we arrive at the general solution of (5.33) in the case $\beta = \omega$:

$$x = C_1 \sin(\omega t) + C_2 \cos(\omega t) - \frac{1}{2\omega} t \cos(\omega t)$$

If we again take the initial conditions to be $t = 0$, $x = 0$, $dx/dt = 0$, then the corresponding solution is

$$x = \frac{1}{2\omega^2} \sin(\omega t) - \frac{1}{2\omega} t \cos(\omega t) \tag{5.36}$$

A portion of the graph is shown in Figure 5.7.

Let us now return to the physical system and ask what these mathematical solutions say about the behavior of the system. We are still considering experiment (2). The mass suspended from the spring is initially at rest, and we begin to move the supporting rod up and down in a periodic way associated with the function $f(t) = \sin(\beta t)$. The period of this motion is $2\pi/\beta$. Energy is transferred to the mass-spring system, which also starts to move. However, the mathematical solutions suggest that there is a profound difference in the behavior between the case when $\beta = \omega$ and the case when $\beta \neq \omega$. We have seen that the number ω, which is derived from the physical properties of the spring and which is large if the spring is stiff, determines the "natural" frequency and period of the mass-spring system. If β and ω are different, so that the equation of motion of the mass is (5.34), then we see that the mass oscillates up and down in a complex way. The maximum and minimum positions can be found from (5.34). If we observe that each of the sine terms can take on

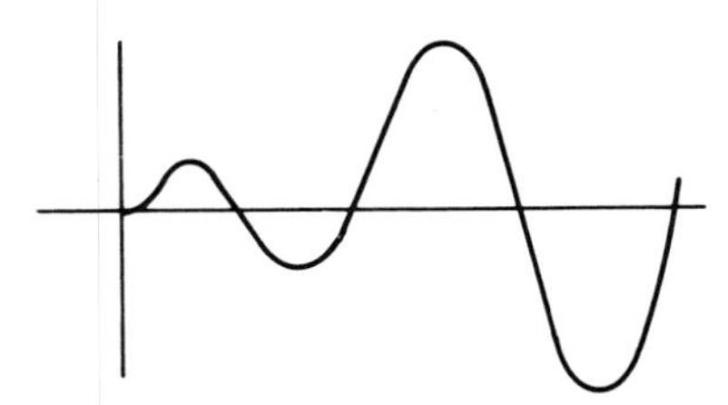

Figure 5.7

only values between -1 and 1, then we know that for all t,

$$|x| \leq \frac{1+\beta/\omega}{|\omega^2-\beta^2|} = \frac{1}{\omega\,|\omega-\beta|} \tag{5.37}$$

On the other hand, when $\beta = \omega$ and the corresponding solution is given by (5.36), we see that the cosine term is multiplied by t, and accordingly the function describing the motion of the moving mass is unbounded. The values of x would quickly exceed the distances allowed by the geometry of the physical system itself.

Do these mathematical conclusions correspond to observable physical phenomena? If we were to carry out the actual experiment and gradually allow the frequency β of the external driving force to approach the natural frequency ω of the mass-spring system, would there be a change in behavior? The answer is affirmative; as suggested by (5.37), as β approaches ω and $1/|\omega-\beta|$ becomes large, the extent of oscillation of the moving mass becomes wider and wider. When the frequencies β and ω coincide, the phenomenon called **resonance** occurs, sometimes with destructive effects. In practice, of course, no physical system is entirely without resistance, and no external supplied force can have exactly the form $f(t) = \sin(\beta t)$ with β exactly the same real number as ω. However, the relatively crude mathematical model has predicted the existence of a very important aspect of the physical system.

Case (3). We now include the effect of a simple resistance law by studying the equation

$$\frac{d^2x}{dt^2} = -2k\frac{dx}{dt} - x$$

The number k is positive and measures the amount of resistance to the motion of the moving weight: When $k = 0$, the resistance is absent and we have the undamped oscillator of case (1), with period 2π. Rewrite the general equation as

$$(D^2+2kD+1)\{x\} = 0 \tag{5.38}$$

and examine the associated polynomial $P(s) = s^2+2ks+1$. Its roots are

$$\tfrac{1}{2}(-2k\pm\sqrt{4k^2-4}) = -k\pm\sqrt{k^2-1}$$

There are three special cases to be studied: (a) $0 < k < 1$, (b) $k = 1$, (c) $k > 1$.

In case (a), let $\gamma = \sqrt{1-k^2}$. Then, the roots of the associated polynomial are $-k\pm i\gamma$ and the corresponding solutions of the equation (5.38) will be

$$x = C_1 e^{-kt}\sin\gamma t + C_2 e^{-kt}\cos\gamma t$$

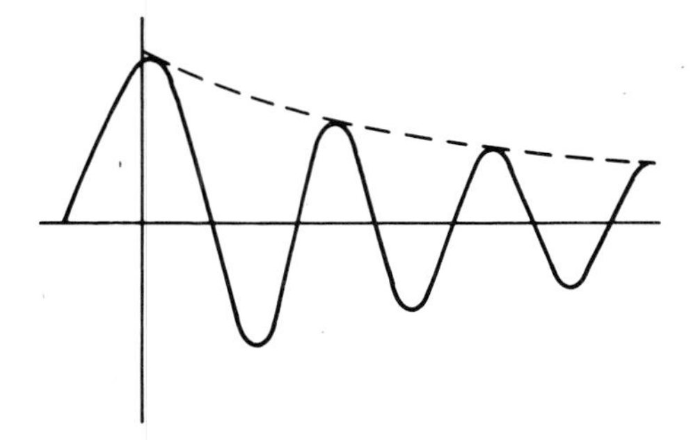

Figure 5.8

which can also be written as

$$x = Ce^{-kt}\sin(\gamma t+\alpha)$$

The graph of such a curve is shown in Figure 5.8. It is a damped sine curve, whose amplitude decreases as t increases due to the factor e^{-kt}. For any positive value of k, no matter how small, $\lim_{t\to\infty} e^{-kt} = 0$. The period of the motion is

$$\frac{2\pi}{\gamma} = \frac{2\pi}{\sqrt{1-k^2}}$$

which is always longer than 2π, the natural period of the undamped system, corresponding to $k = 0$. As k increases, getting closer to 1, the period becomes very long. The damping effect is also more pronounced for larger k. The graph of the motion for a value of k near 1 is shown in Figure 5.9.

Turn now to case (b). Here $k = 1$, and the differential equation becomes

$$(D^2+2D+1)\{x\} = 0 \tag{5.39}$$

Since $P(s)$ has -1 as a double root, the solution of (5.39) will be

$$\begin{aligned} x &= C_1 e^{-t} + C_2 te^{-t} \\ &= (C_1+C_2 t)e^{-t} \end{aligned}$$

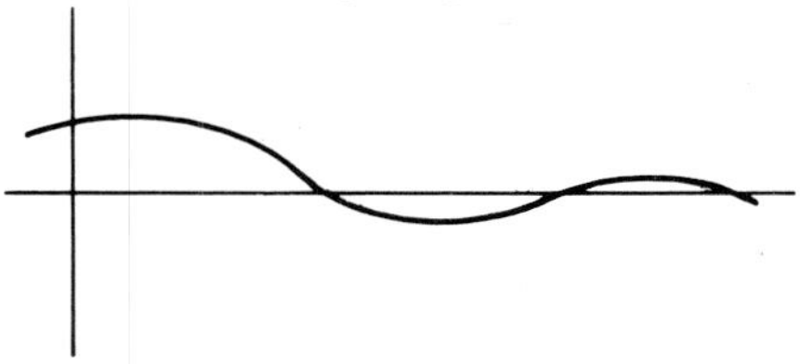

Figure 5.9

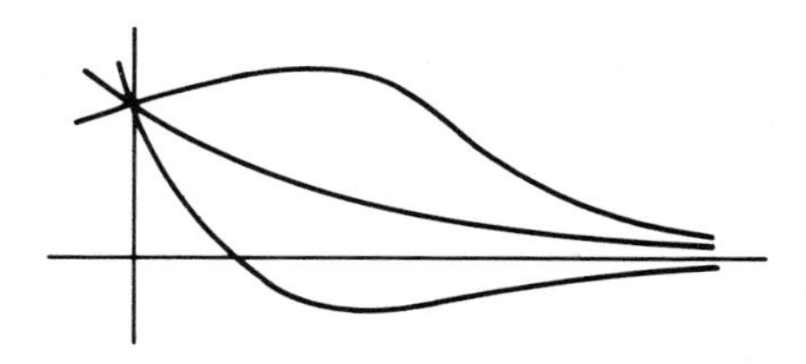

Figure 5.10

The shape of such a curve is quite different from the solution found in case (a). For $t \geq 0$, the graph will cross the t-axis at most once, and there is no periodic behavior. In Figure 5.10 we show several examples of possible shapes.

In case (c), $k > 1$, and the roots of $P(s)$ are real and distinct, $-k+\sqrt{k^2-1}$ and $-k-\sqrt{k^2-1}$. Since $\sqrt{k^2-1}$ is smaller than k, both of these roots are negative. For example, if $k = 3$, the roots are -0.17 and -5.83. The corresponding solutions of (5.38) have the form

$$x = C_1 e^{-at} + C_2 e^{-bt}$$

Again, there is no periodic behavior and the graph will cross the t-axis at most once. The shapes are similar to those in Figure 5.10.

How do these solutions translate into descriptions of the physical behavior of the mass-spring system? When the resistance supplied by the liquid is small, then the mass oscillates with a gradually decreasing amplitude and with a period somewhat longer than the natural period. As the resistance is increased, the damping effect is accentuated, and the period increases. However, at a certain point ($k = 1$ in our example) the behavior alters, and from this point on the behavior of the system is much simpler. If the mass is lifted and released, it will merely move downward toward the equilibrium position slowly and approach it asymptotically. If it is given a large enough vertical velocity, it might pass through the equilibrium position, but it will then move up toward it in the same asymptotic manner. No matter how it is started into motion, the mass will not pass through the equilibrium position twice.

Case (4). How does a damped mass-spring system behave under an external periodic driving force? The appropriate equation is now

$$\frac{d^2x}{dt^2} = -2k\frac{dx}{dt} - x + \sin(\beta t)$$

which we rewrite as

$$(D^2+2kD+1)\{x\} = \sin(\beta t) \tag{5.40}$$

The associated polynomial is $P(s) = s^2+2ks+1$. We have $P(\pm\beta i) = -\beta^2 \pm 2k\beta i+1$. Since $k > 0$ and β is real, $P(\pm\beta i)$ cannot vanish, and we can be sure that (5.40) always has a solution of the form $x = a\sin(\beta t)+b\cos(\beta t)$. Substituting this for x in (5.40) and solving, we find $a = (1-\beta^2)/\Delta$ and $b = -2k\beta/\Delta$, where $\Delta = (1-\beta^2)^2+(2k\beta)^2$. As we have seen above, the null space of the operator $D^2+2kD+1$ has for a basis functions of the form $e^{-kt}\sin(\gamma t+\alpha)$, or $(C_1+C_2 t)e^{-t}$, or e^{-at}, e^{-bt}, depending on the relative size of k. Each of these functions tends to zero as t becomes large. Thus, every solution of the equation (5.40) has the form

$$x = \left[\frac{1-\beta^2}{\Delta}\sin(\beta t) - \frac{2k\beta}{\Delta}\cos(\beta t)\right] + \text{something tending to zero}$$

The second term is called *transient*, because it is effective only for a short while, after which the motion of the moving mass is given essentially by the first term alone.

Let us therefore examine the behavior of the function

$$x_0 = x_0(t) = \frac{1-\beta^2}{\Delta}\sin(\beta t) - \frac{2k\beta}{\Delta}\cos(\beta t)$$

Such a linear combination of trigonometric functions can always be expressed in the form $C\sin(\beta t-\theta)$ for some choice of C and θ. Expanding this as

$$C\cos\theta\sin(\beta t) - C\sin\theta\cos(\beta t)$$

we see that we want to have

$$\begin{aligned} C\cos\theta &= \frac{1-\beta^2}{\Delta} \\ C\sin\theta &= \frac{2k\beta}{\Delta} \end{aligned} \tag{5.41}$$

From this we find the value of C by squaring both and adding. We obtain

$$C = \sqrt{\frac{(1-\beta^2)^2+(2k\beta)^2}{\Delta^2}} = \frac{1}{\sqrt{\Delta}}$$

Thus, the "steady state" motion of the oscillating mass has the simple form

$$x_0(t) = \frac{1}{\sqrt{\Delta}}\sin(\beta t-\theta) \tag{5.42}$$

It is again a sine curve of the same period as that of the external imposed motion of the supporting rod, but with a phase shift θ that could be explicitly determined from the equations (5.41). Its amplitude is $1/\sqrt{\Delta}$, which depends both upon the resistance factor k and upon the number β that determines the period of the external force, since $\Delta = (\beta^2 - 1)^2 + 4k^2\beta^2$.

What does this mathematical analysis of the equation (5.40) predict about the behavior of the physical system in case (4)? If we move the supporting rod up and down as specified by the function $f(t) = \sin(\beta t)$, the mass will start to move in a complicated manner. After some time determined by the strength of the damping terms e^{-kt}, and therefore a longer time if k is very small, the motion will settle down to one that is asymptotically a periodic harmonic motion given by (5.42). It will have the same frequency as the motion of the supporting rod, but it may be out of phase with it, depending upon the exact value of θ. The amplitude of the motion, which predicts the maximum displacement of the mass from equilibrium, will be $1/\sqrt{\Delta}$.

What is the effect of changing the frequency of the external driving force $f(t)$? The maximum amplitude will be greatest when Δ is least. If the resistance k is constant, we can easily find the β for which Δ is a minimum, namely,

$$\beta = \beta_0 = \begin{cases} 0 & \text{if } k \geq \dfrac{1}{\sqrt{2}} \\ \sqrt{1-2k^2} & \text{if } 0 < k < \dfrac{1}{\sqrt{2}} \end{cases} \tag{5.43}$$

We first consider the case when k is small. As β, the frequency of the driving force, is altered, the response of the moving mass will change. It will reach a peak if we choose $\beta = \beta_0 = \sqrt{1-2k^2}$, and for this choice, the maximum displacement $1/\sqrt{\Delta}$ turns out to be

$$\text{Maximum displacement} = \frac{1}{2k\sqrt{1-k^2}} \tag{5.44}$$

What we have described in case (4) is another instance of resonance. The response of the damped mass-spring system is greatest if the driving force is chosen to have β equal to the critical value $\beta_0 = \sqrt{1-2k^2}$. The frequency of the steady-state motion will be $\beta_0/2\pi$, which we note will be different from the natural frequency of the freely moving damped system, which was $(2\pi)^{-1}\sqrt{1-k^2}$. Formula (5.44) gives the value of this maximum displacement, and we note that because $k > 0$, this displacement is finite and does not grow as time increases. We do not have the catastrophic behavior that occurred in the undamped case (2). However, as we should have expected, the maximum displacement can be very large if k is taken quite small. A mass-spring system with very small resistance behaves much like the undamped case.

What is the behavior if k is large? Suppose that k exceeds the crucial value $1/\sqrt{2}$. By (5.43), the critical value of β is now $\beta_0 = 0$, and the corresponding minimum value for Δ is 1. In fact as β decreases toward zero, Δ decreases steadily toward 1, but it becomes arbitrarily large as β is increased. Since the amplitude of the motion (5.42) is $1/\sqrt{\Delta}$, this tells us that the amplitude is never larger than 1; that it is very small for large β; and that it steadily increases toward 1 as β decreases toward zero. In particular, there is no resonance phenomenon.

What we have just described is precisely the theory behind the shock absorber. A mass-spring system with a large resistance component is a physical system in which any periodic motion of the supporting rod R results in only a limited-amplitude motion of the suspended mass M, and there is no critical frequency which produces resonance.

Without going into the physics involved, we point out the fact that another important physical system leads to the same basic differential equation as does the mass-spring system. A simple LCR circuit consists of a resistor R, a capacitor C, and an inductor L, driven by an alternating voltage source $V(t)$ (see Figure 5.11). This system leads to the following equation for the current I:

$$L\frac{d^2I}{dt^2} + R\frac{dI}{dt} + \frac{1}{C}I = V(t)$$

Note that the inductance L is the analog of mass, the resistance R is akin to the damping resistance produced by the oil dashpot, and the reciprocal of the capacitance plays the role of the spring constant in the mechanical problem.

Since the equation is fundamentally the same, all the conclusions again apply. If $V(t)$ is identically zero and R is absent or very small, the system behaves like a simple oscillator; set into activity, it will produce a lightly damped oscillating current I. If the resistance is large, then it will not support

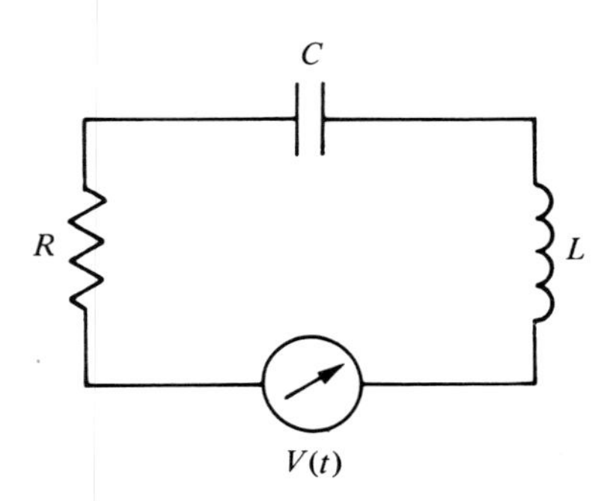

Figure 5.11

an alternating current. Much more interesting results happen if we choose $V(t)$ to be a simple alternating voltage like $\sin(\beta t)$. In the mechanical system, resonance could be pictured as resulting in violent unbounded oscillations; the classic story is of the corps of soldiers whose exactly timed marching, in resonance with the natural frequency of a bridge, led to its collapse. In the electrical system, resonance is what makes it possible to tune a receiver to match the frequency of a transmitter. Thus it has made possible our entire communications industry.

Exercises

1. Graph the curve given by (5.34) for $\omega = 3$, $\beta = 2$. Can you determine when $x = 0$?
2. Verify the shapes shown in Figure 5.10 by graphing the curves given by $x = (1+2t)e^{-t}$ and by $x = (1-2t)e^{-t}$.
3. Prove that $x = C_1 e^{-at} + C_2 e^{-bt}$ will cross $x = 0$ at most once for $t > 0$. ($a > 0$, $b > 0$.)
4. Verify that a solution of (5.40) is of the form $x = a\sin(\beta t) + b\cos(\beta t)$, where $a = (1-\beta^2)/\Delta$, $b = -2k\beta/\Delta$, and $\Delta = (\beta^2-1)^2 + 4k^2\beta^2$.
5. Discuss case (4) in detail for the special case $\beta = 1$.
6. Discuss case (4) in detail for the special case $k = 1$.
7. Show that (5.43) correctly gives the values of β for which Δ is minimum.
8. Discuss the behavior of case (4) when β is very small, (for example, $\beta \approx 0.01$) and (a) k is large, for example, $k \approx 500$; (b) k is small, for example, $k \approx 2$. Describe the behavior of the physical system in each case.

6

Systems of Differential Equations

Introduction

When a physical problem involves several interacting components or a number of interdependent quantities, the appropriate mathematical model is often taken to be a system of simultaneous differential equations. Recall, for example, the Two-Tank Problem discussed in Section 1.1, which led to the pair of first-order equations

$$(6.1) \qquad \begin{cases} \dfrac{dx}{dt} = 0.02y - 0.06x \\ \dfrac{dy}{dt} = 0.06x - 0.06y \end{cases}$$

with initial conditions $x(0) = 0$, $y(0) = 10$.

Similarly, the apparatus shown in Figure 6.1, consisting of two springs attached to two movable weights, leads to a pair of second-order differential equations. Typical values for the physical quantities involved lead to the following initial-value problem, assuming that we start from the equilibrium position by giving one mass an initial velocity.

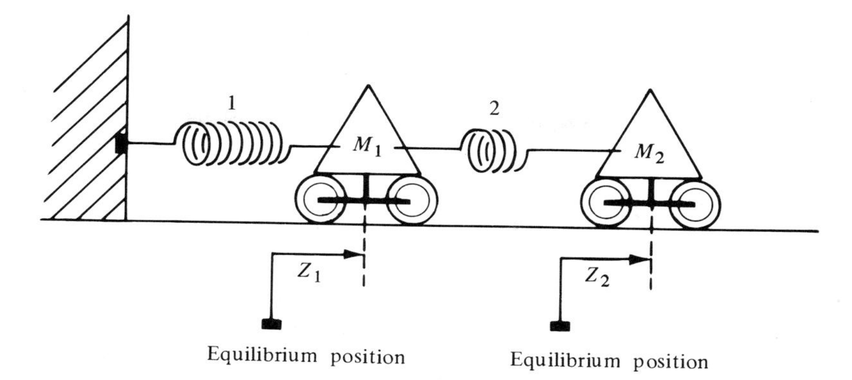

Figure 6.1 Two moving weights, with unequal springs

(6.2)
$$\begin{cases} \dfrac{d^2z_1}{dt^2} = -11z_1 + 6z_2 \\ \dfrac{d^2z_2}{dt^2} = -6z_2 + 6z_1 \end{cases}$$

$$z_1(0) = 0, \quad z_1'(0) = 0$$
$$z_2(0) = 0, \quad z_2'(0) = 1$$

The motion of a particle in orbit about a central attracting body can be described by a pair of equations of the form

(6.3)
$$\begin{cases} \dfrac{d^2x}{dt^2} = -C\dfrac{x}{r^3} \\ \dfrac{d^2y}{dt^2} = -C\dfrac{y}{r^3} \end{cases}$$

where $r = \sqrt{x^2+y^2}$.

In each case, a solution of such a system will be a pair of functions of t which turn the equations into identities when they are substituted for the variables. For example, the unique solutions of (6.2) are

(6.4)
$$z_1 = \frac{6}{13}\left\{\frac{1}{\sqrt{2}}\sin(\sqrt{2}\,t) - \frac{1}{\sqrt{15}}\sin(\sqrt{15}\,t)\right\}$$
$$z_2 = \frac{1}{13}\left\{\frac{9}{\sqrt{2}}\sin(\sqrt{2}\,t) + \frac{4}{\sqrt{15}}\sin(\sqrt{15}\,t)\right\}$$

which describe the complex behavior of the pair of weights.

The purpose of this chapter is to discuss various techniques for solving such systems, as well as to outline some of the general theory behind them. In Section 6.1, we show that the elementary methods developed in preceding sections for solving a single differential equation can also be used to solve certain systems. Sections 6.2, 6.4, and 6.5 deal with the use of vector and matrix methods which permit a more orderly approach to the solution of systems, an approach that is particularly useful in the important case of systems with constant coefficients. The discussion of these methods depends on the use of a few concepts from linear algebra, specifically eigenvalues and eigenvectors of a matrix. However, knowledge of linear algebra is not assumed and the small amount of needed information is developed here. This is not intended to provide any real depth in linear algebra, and the emphasis is upon providing an overview of the topic as it relates to the solution of differential equations, rather than to prove theorems or develop theory.

Finally, in Section 6.3, we offer a very brief discussion of some approximate methods for solving systems of differential equations, based on the earlier treatment of approximate solutions for single equations in Chapters 3 and 4. In particular, we discuss both the vector form of the Picard iteration procedure, and the use of numerical algorithms such as the Euler method, the Midpoint method, and others to solve systems. (We note that many computer-program packages available on a campus are designed to solve either a single differential equation or a system of equations. We give the usual warning: Without some knowledge of both numerical analysis and differential equations, it is easy to go wrong on a computer.)

6.1 Elementary Methods

The methods that have been explained in the preceding chapters are sufficient to solve many systems of differential equations, including some of the most common types. We start with a simple illustration. Suppose we are given the equations

$$
\begin{aligned}
\frac{dx}{dt} &= 3x - y \\
\frac{dy}{dt} &= x + y
\end{aligned}
\tag{6.5}
$$

with initial conditions $x(0) = 1$, $y(0) = 2$. We want to find two functions ϕ, ψ so that $x = \phi(t)$ and $y = \psi(t)$ satisfy (6.5) and also yield $x = 1$, $y = 2$ when $t = 0$. The procedure we will use is to eliminate either x or y from the equations, and then solve the new differential equation in the remaining variable. With the solution of this, we return to the original equations and solve for the other variable. The initial conditions will be used to pick out the unique solution of the system.

We start by rewriting (6.5) in operator form:

(6.6a) $$(D-3)\{x\} = -y$$

(6.6b) $$(D-1)\{y\} = \quad x$$

where, as before, $D = d/dt$. Let us apply the operator $D-3$ to both sides of the second equation to obtain

$$(D-3)(D-1)\{y\} = (D-3)\{x\}$$

Looking at (6.6a), we can simplify the right-hand side of this to

$$(D-3)(D-1)\{y\} = -y$$

Since we are dealing with operators with constant coefficients, we can multiply operators as in elementary algebra (Section 5.1), and have

$$(D^2-4D+3)\{y\} = -y$$

and finally, by collecting everything on the left-hand side, obtain the equation

(6.7) $$(D^2-4D+4)\{y\} = 0$$

This is a second-order linear differential equation of the sort we have solved in the previous chapter. The associated polynomial is $P(s) = s^2-4s+4 = (s-2)^2$. Thus, the general solution of (6.7) can at once be written down:

(6.8) $$y = C_1 e^{2t} + C_2 te^{2t}$$

To find a corresponding solution for x, we substitute (6.8) into (6.6b) as follows:

$$\begin{aligned}
x &= (D-1)\{y\} \\
&= (D-1)\{C_1 e^{2t}+C_2 te^{2t}\} \\
&= C_1(D-1)\{e^{2t}\} + C_2(D-1)\{te^{2t}\} \\
&= C_1(2e^{2t}-e^{2t}) + C_2(e^{2t}+2te^{2t}-te^{2t}) \\
&= (C_1+C_2)e^{2t} + C_2 te^{2t}
\end{aligned}$$

Next, we use the initial conditions to single out the unique solution desired. Set $t = 0$ and require that $x = 1$, $y = 2$. This gives a pair of algebraic linear equations for C_1 and C_2.

$$1 = (C_1 + C_2) + 0$$

$$2 = C_1 + 0$$

The solution of these is $C_1 = 2$, $C_2 = -1$, so that the desired solution of the system (6.5) is

$$\begin{aligned} x &= e^{2t} - te^{2t} = (1-t)e^{2t} \\ y &= (2-t)e^{2t} \end{aligned} \tag{6.9}$$

We remark that there was no particular reason for choosing to eliminate x first and solve for y. We could just as well have eliminated y and solved first for x. To do this, operate on both sides of (6.6a) with $D-1$.

$$\begin{aligned} (D-1)(D-3)\{x\} &= -(D-1)\{y\} \\ &= -x \end{aligned}$$

or

$$(D^2 - 4D + 4)\{x\} = 0$$

It is easy to check that this eventually leads to the particular solutions (6.9) as before. However, the form of the general solutions will seem different due to the different manner in which the constants C_1 and C_2 entered.

The method of elimination will work for any system of linear equations with constant coefficients. For example, it can be used to solve the Two-Tank Problem which we formulated as the equations (6.1). In operator form, these become

$$\begin{aligned} (D+0.06)\{x\} &= 0.02y \\ (D+0.06)\{y\} &= 0.06x \end{aligned} \tag{6.10}$$

Hence,

$$\begin{aligned} (D+0.06)(D+0.06)\{y\} &= (D+0.06)\{0.06x\} \\ &= (0.06)(D+0.06)\{x\} \\ &= (0.06)(0.02)y \end{aligned}$$

and calculating the product of the operators on the left, we have

$$(D^2 + 0.12D + 0.0024)\{y\} = 0 \tag{6.11}$$

The associated polynomial is $P(s) = s^2 + 0.12s + 0.0024$. (This is much more typical of realistic differential equations, since the coefficients are not integers and the quadratic polynomial does not factor neatly.) The roots of P are

$-0.06 \pm 2\sqrt{0.0003}$, or -0.02536 and -0.09464. Consequently, the general solution of (6.11) is

$$y = C_1 e^{-0.025t} + C_2 e^{-0.095t}$$

Substitute this into the second equation in (6.10) to obtain

$$\begin{aligned} 0.06x &= (D+0.06)\{y\} \\ &= C_1 (D+0.06)\{e^{-0.025t}\} + C_2 (D+0.06)\{e^{-0.095t}\} \\ &= C_1 [(-0.025)e^{-0.025t} + (0.06)e^{-0.025t}] \\ &\quad + C_2 [(-0.095)e^{-0.095t} + (0.06)e^{-0.095t}] \\ &= (0.035)C_1 e^{-0.025t} + (-0.035)C_2 e^{-0.095t} \end{aligned}$$

yielding the general solution for x as

$$x = 0.58C_1 e^{-0.025t} - 0.58C_2 e^{-0.095t}$$

In the Two-Tank Problem, the initial conditions were $x(0) = 0$, $y(0) = 10$. Putting $t = 0$ in our general solutions for x and y, we are led to the equations

$$\begin{aligned} 0 &= 0.58C_1 - 0.58C_2 \\ 10 &= C_1 + C_2 \end{aligned}$$

whose solutions are immediately $C_1 = C_2 = 5$. Thus, the specific solutions we are seeking are

$$\begin{aligned} x &= 2.9e^{-0.025t} - 2.9e^{-0.095t} \\ y &= 5.0e^{-0.025t} + 5.0e^{-0.095t} \end{aligned} \tag{6.12}$$

In the original statement of the Two-Tank Problem in Chapter 1, we raised the following question: "How long will it take for the toxic level in the outlet pipe to fall below 0.001 g/gal?" This translated into the mathematical question: "How large must t be for y to satisfy $y < 0.1$?" Using the formula given in (6.12), we would want to find a solution for the equation

$$5e^{-0.025t}[1+e^{-0.07t}] = 0.1 \tag{6.13}$$

Although it is not possible to find an exact solution of this, it is relatively easy to find approximate solutions to any desired accuracy. We first note that $e^{-0.07t}$ is considerably smaller than $e^{-0.025t}$; indeed, the cube of $e^{-0.025t}$ is $e^{-0.075t}$, and cubing a number smaller than 1 decreases it. Thus, as a first trial solution, we solve the easier equation

$$5e^{-0.025t} = 0.1$$

First, take natural logarithms of this to obtain

$$-0.025t = -3.912$$

and then

$$t = 156$$

If we try this out in (6.13), we get

$$(5)(0.02024)(1+0.00002) = 0.1012$$

The units of time were minutes. We can therefore answer the posed question as follows: The water flowing out of the outlet pipe from the second tank will be reasonably nontoxic two hours and forty minutes after the valves were opened.

A similar method will work for the equations (6.2) describing the motion of a pair of spring-driven masses. First rewrite these equations in operator form:

$$(6.14) \qquad \begin{aligned} (D^2+11)\{z_1\} &= 6z_2 \\ (D^2+6)\{z_2\} &= 6z_1 \end{aligned}$$

Then, operate on the first equation by D^2+6, obtaining

$$\begin{aligned} (D^2+6)(D^2+11)\{z_1\} &= (D^2+6)\{6z_2\} \\ &= (6)(D^2+6)\{z_2\} \end{aligned}$$

and use the second equation in (6.14) to simplify this.

$$\begin{aligned} (D^2+6)(D^2+11)\{z_1\} &= (6)(6z_1) \\ &= 36z_1 \end{aligned}$$

If we carry out the multiplication on the left and combine terms, we arrive at

$$(6.15) \qquad (D^4+17D^2+30)\{z_1\} = 0$$

The associated polynomial s^4+17s^2+30 factors neatly into $(s^2+15)(s^2+2)$, yielding the roots $\pm\sqrt{2}i$ and $\pm\sqrt{15}i$. The general solution for z_1 is then given by

$$(6.16) \qquad \begin{aligned} z_1 &= C_1 \sin(\sqrt{2}\,t) + C_2 \cos(\sqrt{2}\,t) \\ &\quad + C_3 \sin(\sqrt{15}\,t) + C_4 \cos(\sqrt{15}\,t) \end{aligned}$$

In order to find the formula for z_2 easily, we substitute this formula for z_1 into the first equation in (6.14) and have

$$\begin{aligned} 6z_2 &= (D^2+11)\{z_1\} \\ &= 9C_1 \sin(\sqrt{2}\,t) + 9C_2 \cos(\sqrt{2}\,t) - 4C_3 \sin(\sqrt{15}\,t) - 4C_4 \cos(\sqrt{15}\,t) \end{aligned}$$

which yields

$$z_2 = 1.5C_1 \sin(\sqrt{2}\,t) + 1.5C_2 \cos(\sqrt{2}\,t) - 0.66C_3 \sin(\sqrt{15}\,t) - 0.66C_4 \cos(\sqrt{15}\,t) \tag{6.17}$$

The initial conditions were $z_1(0) = z_1'(0) = 0$, $z_2(0) = 0$, $z_2'(0) = 1$, corresponding to the physical assumption that the first mass is centered and at rest, and the second mass is also centered but has been given a push to the right. To use these to find numerical values for C_1, C_2, C_3, and C_4, we need to differentiate (6.16) and (6.17) and then set $t = 0$. We arrive at the algebraic equations

$$\begin{aligned} 0 &= z_1(0) = C_2 + C_4 \\ 0 &= z_1'(0) = \sqrt{2}\,C_1 + \sqrt{15}\,C_3 \\ 0 &= z_2(0) = 1.5C_2 - 0.66C_4 \\ 1 &= z_2'(0) = 1.5\sqrt{2}\,C_1 - 0.66\sqrt{15}\,C_3 \end{aligned}$$

Solving these, we find $C_2 = C_4 = 0$, and

$$C_1 = \frac{6}{13\sqrt{2}} \approx 0.326, \qquad C_3 = -\frac{6}{13\sqrt{15}} \approx -0.119$$

Hence, the motion of the two weights is described by the functions

$$\begin{aligned} z_1 &= 0.326 \sin(\sqrt{2}\,t) - 0.119 \sin(\sqrt{15}\,t) \\ z_2 &= 0.489 \sin(\sqrt{2}\,t) + 0.079 \sin(\sqrt{15}\,t) \end{aligned} \tag{6.18}$$

which are the same as (6.4).

The motion described by (6.18) is complicated; it is a combination of two periodic motions with different periods, neither a rational multiple of the other. These motions sometimes reinforce each other and sometimes act in opposite directions. Each mass follows its own complicated oscillation. However, having the equations (6.18) makes it possible to answer many questions about this motion. For example, suppose we wanted to know if there were moments when both weights were as far to the right (see Figure 6.1) as it is possible for them to go. Clearly, since the sine and cosine functions have 1 as their maximum value and -1 as their minimum, the largest value which z_1 could have is $0.326 + 0.119 = 0.445$ and the largest value that z_2 could ever have is $0.489 + 0.079 = 0.568$. For this to occur simultaneously for both weights, we need to have a value of t such that

$$\begin{aligned} \sin(\sqrt{2}\,t) &= 1 \qquad & \sin(\sqrt{2}\,t) &= 1 \\ \sin(\sqrt{15}\,t) &= -1 \qquad & \sin(\sqrt{15}\,t) &= 1 \end{aligned}$$

Since we cannot have $\sin(\sqrt{15}\,t)$ both 1 and -1 at the same time t, this event cannot ever happen.

Finally, let us solve a system of three equations. Consider the following initial-value problem:

$$(6.19)\qquad \begin{aligned} \frac{dx}{dt} &= y + z & x(0) &= 0 \\ \frac{dy}{dt} &= 2x - z & y(0) &= 1 \\ \frac{dz}{dt} &= 4x + 4y & z(0) &= -1 \end{aligned}$$

Our objective is to eliminate all but one of the variables to obtain a single differential equation which can be solved. There are several ways to start. For example, if we differentiate the first equation and then substitute from the second and third, we obtain

$$\begin{aligned} D^2\{x\} &= D\{y\} + D\{z\} \\ &= (2x-z) + (4x+4y) \\ &= 6x + 4y - z \end{aligned}$$

or

$$(D^2-6)\{x\} = 4y - z$$

Adding this to the first equation, we eliminate z and have

$$(6.20)\qquad (D^2+D-6)\{x\} = 5y$$

Now, if we add the first two equations in (6.19), we obtain

$$\begin{aligned} D\{x\} + D\{y\} &= (y+z) + (2x-z) \\ &= y + 2x \end{aligned}$$

or

$$(6.21)\qquad (D-1)\{y\} = (-D+2)\{x\}$$

Apply the operator $D-1$ to both sides of (6.20):

$$\begin{aligned} (D-1)(D^2+D-6)\{x\} &= (D-1)\{5y\} \\ &= (5)(D-1)\{y\} \end{aligned}$$

If we now use (6.21) to replace the right-hand side, we can eliminate y entirely, and have

$$(D-1)(D^2+D-6)\{x\} = (5)(-D+2)\{x\}$$

or

$$(D^3 - 7D + 6)\{x\} = -5D\{x\} + 10x$$

This can be written as a third-order linear equation for x.

$$(D^3 - 2D - 4)\{x\} = 0$$

The associated polynomial $P(s) = s^3 - 2s - 4$ has 2 as a root and therefore $s - 2$ as a factor, and $P(s) = (s-2)(s^2 + 2s + 2)$. All the roots are then seen to be 2, $-1+i$, and $-1-i$. Hence, a general solution for x expressed in terms of real functions will be:

$$x = C_1 e^{2t} + C_2 e^{-t} \sin t + C_3 e^{-t} \cos t \tag{6.22}$$

To find a general solution for y, we can return to (6.20) and substitute (6.22) for x.

$$\begin{aligned} 5y &= (D^2 + D - 6)\{x\} \\ &= (D^2 + D - 6)\{C_1 e^{2t} + C_2 e^{-t} \sin t + C_3 e^{-t} \cos t\} \\ &= C_1 (D^2 + D - 6)\{e^{2t}\} + C_2 (D^2 + D - 6)\{e^{-t} \sin t\} \\ &\quad + C_3 (D^2 + D - 6)\{e^{-t} \cos t\} \\ &= C_1 (4e^{2t} + 2e^{2t} - 6e^{2t}) \\ &\quad + C_2(-2e^{-t} \cos t - e^{-t} \sin t + e^{-t} \cos t - 6e^{-t} \sin t) \\ &\quad + C_3 (2e^{-t} \sin t - e^{-t} \cos t - e^{-t} \sin t - 6e^{-t} \cos t) \\ &= C_2(-7e^{-t} \sin t - e^{-t} \cos t) + C_3 (e^{-t} \sin t - 7e^{-t} \cos t) \end{aligned}$$

Dividing by 5, we obtain the desired solution for y:

$$y = \tfrac{1}{5}(C_3 - 7C_2)\, e^{-t} \sin t - \tfrac{1}{5}(7C_3 + C_2)\, e^{-t} \cos t \tag{6.23}$$

To find z, we can return to either of the first two equations in (6.19), and substitute for x and y. Thus, we can write

$$\begin{aligned} z &= \frac{dx}{dt} - y \\ &= D\{C_1 e^{2t} + C_2 e^{-t} \sin t + C_3 e^{-t} \cos t\} \\ &\quad - \tfrac{1}{5}(C_3 - 7C_2)\, e^{-t} \sin t + \tfrac{1}{5}(C_2 + 7C_3)\, e^{-t} \cos t \\ &= 2C_1 e^{2t} + \tfrac{1}{5}(2C_2 - 6C_3)\, e^{-t} \sin t + \tfrac{1}{5}(6C_2 + 2C_3)\, e^{-t} \cos t \end{aligned}$$

Since we want the unique solution that obeys the assigned initial conditions, we set $t = 0$ and require that we have $x = 0$, $y = 1$, $z = -1$. This yields three equations for the coefficients C_i:

$$0 = C_1 + (0)\,C_2 + C_3$$

$$1 = 0 - \tfrac{1}{5}(7C_3 + C_2)$$

$$-1 = 2C_1 + 0 + \tfrac{1}{5}(6C_2 + 2C_3)$$

These can be rewritten as

$$0 = C_1 + C_3$$

$$5 = -C_2 - 7C_3$$

$$-5 = 10C_1 + 6C_2 + 2C_3$$

and the solutions turn out to be: $C_1 = \frac{1}{2}$, $C_2 = -\frac{3}{2}$, $C_3 = -\frac{1}{2}$. Putting these into the formulas for x, y, and z, we arrive finally at the desired solutions of the system (6.19).

$$x = (0.5)\,e^{2t} - (1.5)\,e^{-t}\sin t - (0.5)\,e^{-t}\cos t$$

$$y = 2e^{-t}\sin t + e^{-t}\cos t$$

$$z = e^{2t} - 2e^{-t}\cos t$$

(In Section 6.4, we will study a more systematic method for solving a system like this one, based on the use of matrices.)

Exercises

1. Solve the system $dx/dt = y,\ dy/dt = 4x;\quad x(0) = 2,\ y(0) = 4$
2. Solve: $du/dt = u + 2v,\ dv/dt = u;\quad u(0) = 1,\ v(0) = -1$
3. Solve: $du/dt = v + 2,\ dv/dt = u - 1;\quad u(0) = 3,\ v(0) = 4$
4. Solve: $dx_1/dt = 2x_1 - 3x_2 - 7,\ dx_2/dt = 5x_1 + 10x_2$
5. Solve: $dx/dt = x + 2y,\ dy/dt = 8x + y;\quad x(0) = 1,\ y(0) = 2$
6. Solve: $du/dt = 2u - v,\ dv/dt = 2v - u;\quad u(0) = 2,\ v(0) = 0$
7. Solve the system $dx/dt = y + t,\ dy/dt = x - t^2$, finding a solution that obeys $x(1) = 2,\ y(1) = 3$.
8. Rewrite the equation $y''' - 2ty'' + y' - e^t y = t^2$ as a first-order system.

PC9. (a) Find the solution of the Two-Tank Problem [equation (6.1)] with initial conditions $x(0) = 10,\ y(0) = 0$.

(b) What is the maximum concentration that is ever reached in tank II? When is it reached? (You may wish to use a pocket calculator.)

10. Solve the system of equations given in (6.19) by using the following steps.
    (a) Show that $(D^2+2)\{y\} = -4x+2z$.
    (b) Show that $(D^2+2D+2)\{y\} = 0$.
    (c) Find a general solution for y.
    (d) Show that $(D-2)\{x\}+(D-1)\{y\} = 0$.
    (e) Show that $(D-2)\{x\} = (2C_1+C_2)e^{-t}\sin t+(2C_2-C_1)e^{-t}\cos t$.
    (f) Find a solution for x.
    (g) Use one of the equations in (6.19) to find z.
    (h) Use the initial conditions to obtain expressions for x, y, and z and check with the result in the text.

11. Solve the system $dx/dt = z$, $dy/dt = -x$, $dz/dt = 3x+2y$, with initial conditions $x(0) = y(0) = z(0) = 1$.

12. Four tanks, each holding 100 gallons of liquid, are interconnected as shown in Figure 6.2. The flow between them is indicated by arrows and rates shown on the figure. At the start, tanks I and IV contain only pure

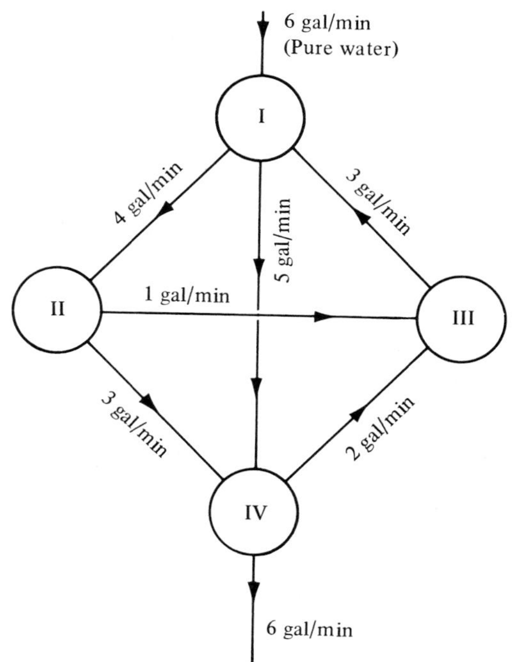

Figure 6.2

water, while tanks II and III each contain 10 grams of a toxic substance. Write a system of four differential equations to describe this situation. (Do not solve.)

6.2 Vector-Valued Functions

The key ideas that simplify a complicated mathematical problem are often very simple. The key idea that helps to reduce the difficulty in treating systems of differential equations is the concept of a *vector-valued function.* A numerical **vector** is simply an ordered list or array of numbers, written either horizontally or vertically, which we treat as though it were a single object. For example, we are using vector notation when we use coordinate form such as $(2,-1,7)$ for a point P in 3-space. Similarly, we might speak of $(2,0,1,-3,4)$ as a point (or as a vector) in 5-space. At times it is more convenient to display this vertically, writing

$$P=\begin{bmatrix}2\\0\\1\\-3\\4\end{bmatrix}$$

Vectors are added or subtracted entry by entry. Thus

$$\begin{bmatrix}2\\0\\1\\-3\\4\end{bmatrix}+\begin{bmatrix}7\\-2\\3\\3\\1\end{bmatrix}=\begin{bmatrix}9\\-2\\4\\0\\5\end{bmatrix}$$

We do not add or subtract vectors unless they are the same size—that is, have the same number of entries.

It is easy to see that the collection of numerical vectors of size n form a linear space, as described in Section 5.1. The algebraic operations on vectors are carried out componentwise, and the listed axioms hold. Thus, the sum of two vectors is defined by

$$\begin{bmatrix}a_1\\a_2\\a_3\\\vdots\\a_n\end{bmatrix}+\begin{bmatrix}b_1\\b_2\\b_3\\\vdots\\b_n\end{bmatrix}=\begin{bmatrix}a_1+b_1\\a_2+b_2\\a_3+b_3\\\vdots\\a_n+b_n\end{bmatrix}$$

and if β is a **scalar** (real or complex number),

$$\beta\begin{bmatrix} a_1 \\ a_2 \\ \vdots \\ a_n \end{bmatrix} = \begin{bmatrix} \beta a_1 \\ \beta a_2 \\ \vdots \\ \beta a_n \end{bmatrix}$$

The zero vector $\mathbf{0}$ is the vector all of whose entries are zero.

A **vector-valued function** of t is a function F whose values $F(t)$ are vectors of some specified size, for all choices of t. For example, the formula

$$F(t) = \begin{bmatrix} t^2+1 \\ 3t \\ t^3+2t-2 \end{bmatrix}$$

describes a function F whose values are 3-vectors. Thus, we have

$$F(0) = \begin{bmatrix} 1 \\ 0 \\ -2 \end{bmatrix} \qquad F(2) = \begin{bmatrix} 5 \\ 6 \\ 10 \end{bmatrix}$$

Vector-valued functions can be added and subtracted, and multiplied by scalars, carrying out the operation componentwise, if the various functions obey the same size restrictions. Thus, if we have a second function G whose values are 3-vectors, such as

$$G(t) = \begin{bmatrix} 2t-2 \\ 3 \\ t^2-3t \end{bmatrix}$$

then

$$F(t) + G(t) = \begin{bmatrix} t^2+2t-1 \\ 3t+3 \\ t^3+t^2-t-2 \end{bmatrix}$$

and

$$4F(t) - 3G(t) = \begin{bmatrix} 4t^2-6t+10 \\ 12t-9 \\ 4t^3-3t^2+17t-8 \end{bmatrix}$$

The variable t itself is a scalar, so it is also possible to rewrite this as

$$4F(t) - 3G(t) = \begin{bmatrix} 0 \\ 0 \\ 4 \end{bmatrix} t^3 + \begin{bmatrix} 4 \\ 0 \\ -3 \end{bmatrix} t^2 + \begin{bmatrix} -6 \\ 12 \\ 17 \end{bmatrix} t + \begin{bmatrix} 10 \\ -9 \\ -8 \end{bmatrix}$$

Such a function could be called a polynomial with vector coefficients.

The operations of elementary calculus, such as differentiation and integration, apply to vector-valued functions, again operating component by component. For example, using the same functions F and G, we have

$$F'(t) = \begin{bmatrix} 2t \\ 3 \\ 3t^2+2 \end{bmatrix} \qquad G'(t) = \begin{bmatrix} 2 \\ 0 \\ 2t \end{bmatrix}$$

and if $H(t)$ is the indefinite integral of G, written $H(t) = \int G(t)\,dt$ as usual, then

$$H(t) = \begin{bmatrix} t^2 - 2t + C_1 \\ 3t + C_2 \\ \frac{1}{3}t^3 - \frac{3}{2}t^2 + C_3 \end{bmatrix}$$

Note that the constants of integration are not all called C but rather C_1, C_2, C_3 since they are quite independent and need not have the same value. This is easier to appreciate if we first write G in polynomial form

$$G(t) = \begin{bmatrix} 0 \\ 0 \\ 1 \end{bmatrix} t^2 + \begin{bmatrix} 2 \\ 0 \\ -3 \end{bmatrix} t + \begin{bmatrix} -2 \\ 3 \\ 0 \end{bmatrix}$$

and then carry out the integration, obtaining

$$H(t) = \int G(t)\,dt = \begin{bmatrix} 0 \\ 0 \\ 1 \end{bmatrix} \frac{t^3}{3} + \begin{bmatrix} 2 \\ 0 \\ -3 \end{bmatrix} \frac{t^2}{2} + \begin{bmatrix} -2 \\ 3 \\ 0 \end{bmatrix} t + \begin{bmatrix} C_1 \\ C_2 \\ C_3 \end{bmatrix}$$

Here, we see that the "constant of integration" is merely an arbitrary vector with three constant entries.

How does the introduction of vector-valued functions simplify the treatment of systems of differential equations? Suppose we consider a general system of three equations:

$$\text{(6.24)} \qquad \begin{cases} \dfrac{du}{dt} = f(t, u, v, w) \\[2mm] \dfrac{dv}{dt} = g(t, u, v, w) \\[2mm] \dfrac{dw}{dt} = h(t, u, v, w) \end{cases}$$

Here, u, v, and w are each unknown scalar functions of t. Let us define a vector-valued function of t by

$$X = X(t) = \begin{bmatrix} u(t) \\ v(t) \\ w(t) \end{bmatrix}$$

Then, we can rewrite the system of equations (6.24) as

$$\frac{dX}{dt} = \begin{bmatrix} f(t,u,v,w) \\ g(t,u,v,w) \\ h(t,u,v,w) \end{bmatrix} \tag{6.25}$$

Using the fact that X is a vector whose components are u, v, and w, we can create a vector-valued function of the two variables t and X by the definition

$$F(t,X) = \begin{bmatrix} f(t,u,v,w) \\ g(t,u,v,w) \\ h(t,u,v,w) \end{bmatrix}$$

and then rewrite (6.25) in the simple form

$$\frac{dX}{dt} = F(t,X) \tag{6.26}$$

This single equation is equivalent to all three equations in (6.24), and it has exactly the same form and appearance as the general scalar differential equation of first order which we worked with in earlier chapters, except that now X is a vector-valued variable and F a vector-valued function.

As we will see shortly, this simple change in notation has many important consequences. One is in part psychological; since the final result (6.26) has the same appearance as the ordinary scalar first-order equation, we do not have to learn an entire new set of facts. For example, the existence and uniqueness theorems will look the same, and in fact their proofs are almost unchanged. Again, the computer algorithms for solving scalar equations turn out to have the same descriptions for the vector-valued equation (6.26), which again simplifies their understanding and even their use. Finally, we have an unexpected dividend; we no longer need to consider differential equations or differential systems of order higher than the first! The reason is that any such equation or system of equations can be replaced by a single vector-valued first-order equation like (6.26).

We illustrate this first with a simple example. Suppose we were given the second-order linear differential equation

$$\frac{d^2y}{dt^2} = t^2\frac{dy}{dt} - (t+1)\,y - e^{-t} \tag{6.27}$$

Introduce a vector variable

$$X = \begin{bmatrix} u \\ v \end{bmatrix}$$

where

$$u = y(t)$$
$$v = y'(t)$$

Then,

$$\frac{dX}{dt} = \begin{bmatrix} y'(t) \\ y''(t) \end{bmatrix} = \begin{bmatrix} v \\ t^2 v - (t+1)u - e^{-t} \end{bmatrix}$$

Thus, the second-order equation (6.27) can be replaced by the first-order vector equation

$$\frac{dX}{dt} = F(t, X)$$

where

$$F(t, X) = F\left(t, \begin{bmatrix} u \\ v \end{bmatrix}\right) = \begin{bmatrix} v \\ t^2 v - (t+1)u - e^{-t} \end{bmatrix}$$

The general case works in the same way. If X is a vector-valued variable of size n, then

$$X = \begin{bmatrix} x_1(t) \\ x_2(t) \\ \vdots \\ x_n(t) \end{bmatrix}$$

A vector-valued function $F(t, X)$ of the scalar variable t and the vector variable X will have the form

$$F(t, X) = \begin{bmatrix} f(t, x_1, x_2, \ldots, x_n) \\ g(t, x_1, x_2, \ldots, x_n) \\ \vdots \\ h(t, x_1, x_2, \ldots, x_n) \end{bmatrix} \begin{matrix} = \\ = \\ \\ = \end{matrix} \begin{bmatrix} f(t, X) \\ g(t, X) \\ \vdots \\ h(t, X) \end{bmatrix}$$

and the single differential equation $dX/dt = F(t, X)$ is the same as the system of n scalar equations

$$\begin{cases} \dfrac{dx_1}{dt} = f(t, x_1, x_2, \ldots, x_n) \\ \dfrac{dx_2}{dt} = g(t, x_1, x_2, \ldots, x_n) \\ \quad\vdots \qquad\qquad \vdots \\ \dfrac{dx_n}{dt} = h(t, x_1, x_2, \ldots, x_n) \end{cases}$$

Thus, any system of n first-order equations can be replaced by a single first-order vector differential equation.

The general form of an nth-order differential equation is

$$\frac{d^n y}{dt^n} = f(t, y, y', y'', \ldots, y^{(n-1)})$$

We introduce a special vector variable X, where

$$X = \begin{bmatrix} x_1 \\ x_2 \\ \vdots \\ x_n \end{bmatrix} \begin{matrix} = \\ = \\ \\ = \end{matrix} \begin{bmatrix} y(t) \\ y'(t) \\ \vdots \\ y^{(n-1)}(t) \end{bmatrix}$$

Then,

$$\frac{dX}{dt} = \begin{bmatrix} y'(t) \\ y''(t) \\ \vdots \\ y^{(n-1)}(t) \\ y^{(n)}(t) \end{bmatrix} = \begin{bmatrix} x_2 \\ x_3 \\ \vdots \\ x_n \\ f(t, x_1, x_2, \ldots, x_n) \end{bmatrix}$$
$$= F(t, X)$$

Thus, every nth-order scalar differential equation is equivalent to a special system of first-order equations, and therefore equivalent to a *single* first-order vector equation.

What can we do with a system of second-order equations, such as those for the two-mass problem given in (6.2)? The same device works. Each second-order equation is replaced by two first-order scalar equations, and we end up with a system of four first-order equations, which can again be replaced by a single first-order vector equation. We set

$$x_1 = z_1(t) \qquad x_2 = z_1'(t) \qquad x_3 = z_2(t) \qquad x_4 = z_2'(t)$$

and then it is easy to see that equations (6.2) can be replaced by the single vector equation

$$\frac{dX}{dt} = \begin{bmatrix} x_2 \\ -11x_1 + 6x_3 \\ x_4 \\ 6x_1 - 6x_3 \end{bmatrix} \tag{6.28}$$

In this way, the theory of systems of differential equations of order n can be reduced to the study of a single first-order vector equation of the standard form (6.26).

In this process, how do we translate the given set of initial conditions? In an initial-value problem such as the Two-Tank Problem, (6.1), or the problem arising from the motion of a pair of weights, (6.2), part of the information given was the value of certain of the measurable quantities at the initial moment $t = t_0$. When each of these systems of differential equations is translated into vector form, it is easily checked that these initial conditions translate into information about the value of the vector variable X at $t = t_0$. For example, in the problem involving the moving weights, we were given the values of z_1, z_2, z_1', and z_2' at $t = 0$, which corresponded to knowledge of the position and velocity of each weight at the starting time. In the vector form, this is the same as knowing that

$$X(0) = \begin{bmatrix} 0 \\ 0 \\ 0 \\ 1 \end{bmatrix}$$

Thus, any initial-value problem for a system of first-order equations translates into a vector problem of the same sort—namely, solve the vector equation $dX/dt = F(t, X)$ with an assigned value for the starting point $X(t_0)$. The same thing occurs when a higher-order equation is translated into vector form.

There are physical situations that do not give rise to initial-value problems. For example, in the moving-weights problem, we might be given information about the weights at *two different* moments of time. Such information is more difficult to translate appropriately into vector form; the resulting problems are called "boundary-value problems" rather than "initial-value problems," and will not be treated in any detail in this book. In the next chapter, we will discuss some examples of these to see how they arise and how to handle some of them.

There is a simple way to picture the solution of a system of two first-order equations, which can be adapted with some changes to the general case. Suppose we are considering the system

$$(6.29) \qquad \begin{aligned} \frac{du}{dt} &= f(t, u, v) \\ \frac{dv}{dt} &= g(t, u, v) \end{aligned}$$

with initial conditions $u(t_0) = a_0$, $v(t_0) = b_0$. The solution will be a pair of functions $\phi(t)$ and $\psi(t)$ such that $u = \phi(t)$ and $v = \psi(t)$ satisfy (6.29) identically and such that $\phi(t_0) = a_0$, $\psi(t_0) = b_0$. If we regard t as time, then

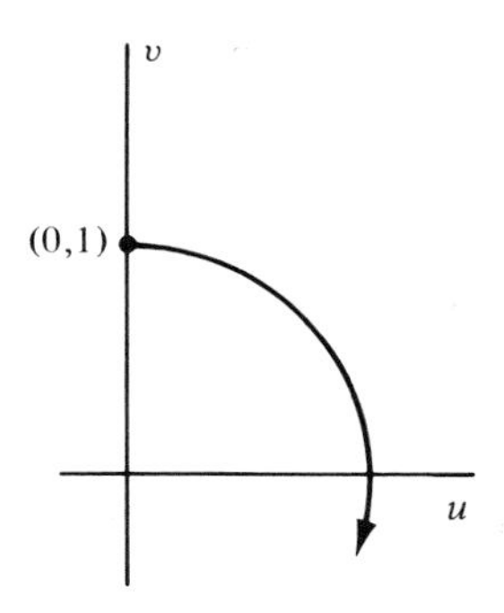

Figure 6.3

the equations

$$\begin{cases} u = \phi(t) \\ v = \psi(t) \end{cases} \tag{6.30}$$

describe a curve in the UV-plane, traced by a moving point. For example, the equations $u = \sin t$, $v = \cos t$ describe a circle, traced clockwise starting at the point $(0, 1)$ with $t = 0$. (See Figure 6.3.) The solution of (6.29) will be a curve in the UV-plane that starts at the point (a_0, b_0) at time $t = t_0$. The set of all solutions of (6.29) is a family of such curves, each starting at its own initial point.

We can now give geometric meaning to a very important mathematical result, namely, the uniqueness theorem for such a system of differential equations. This is the fundamental result that states that if the functions f and g are continuous and satisfy certain other requirements (for example, if they have continuous partial derivatives), then the system (6.29) will have exactly one solution for the pair of functions u and v that obey the given initial conditions $u(t_0) = a_0$, $v(t_0) = b$. This means that the solution curves in the UV-plane will never be such that two of them go through the same point. In particular, none of the curves will intersect each other. (We do not give a proof of this important theorem for systems; it would be very similar to the proof in Section 3.6 which we gave for a single equation.)

A simple illustration may be helpful. Consider the system

$$\begin{aligned} \frac{du}{dt} &= v \qquad u(0) = a_0 \\ \frac{dv}{dt} &= u \qquad v(0) = b_0 \end{aligned} \tag{6.31}$$

By the methods explained in Section 6.1, one finds that $(D^2 - 1)\{u\} = 0$, and from this, one obtains the general solution

$$u = C_1 e^t + C_2 e^{-t}$$
(6.32)
$$v = C_1 e^t - C_2 e^{-t}$$

In order for this curve to pass through the point (a_0, b_0) when $t = 0$, we must choose $C_1 = \frac{1}{2}(a_0 + b_0)$, $C_2 = \frac{1}{2}(a_0 - b_0)$. Making these replacements, we have the desired solution of (6.31):

$$u = a_0 \cosh t + b_0 \sinh t$$

$$v = a_0 \sinh t + b_0 \cosh t$$

It would be possible to sketch these curves in the UV-plane from these parametric equations. However, it is easier to eliminate t and obtain an equation in u and v directly. For example, if we square each of the equations (6.32), we obtain

$$u^2 = C_1^{\,2} e^{2t} + C_2^{\,2} e^{-2t} + 2C_1 C_2$$

$$v^2 = C_1^{\,2} e^{2t} + C_2^{\,2} e^{-2t} - 2C_1 C_2$$

so that

$$u^2 - v^2 = 4C_1 C_2 = a_0^{\,2} - b_0^{\,2}$$

Thus, if $C_1 C_2 \neq 0$, the curve given by (6.32) is an hyperbola, and the collection is the family of all hyperbolas of the form $u^2 - v^2 =$ constant. When $C_1 = 0$ and $C_2 \neq 0$, or $C_2 = 0$ and $C_1 \neq 0$, the curve is seen from (6.32) to be one of the rays that make up the lines $u = \pm v$. When $C_1 = C_2 = 0$, we have parametric equations that yield only a single point, the origin, corresponding to the solution $u(t) \equiv 0$, $v(t) \equiv 0$. Thus, we have obtained the complete picture for the solutions of the system (6.31): a family of hyperbolas, the rays that form the asymptotes, and the origin, together completely covering the UV-plane in such a fashion that exactly one of the curves goes through any selected point in the UV-plane (see Figure 6.4).

The general case is analogous. The vector equation $dX/dt = F(t, X)$ in n dimensions has a solution $X = \phi(t)$ which is a function taking values that are n-vectors, and obeying the initial condition $\phi(t_0) = a$, for any selected initial value a. The function ϕ can be regarded as describing the path of a moving point in n-space which starts at the point a. The complete family of solutions is, then, a family of curves that fills all or part of n-space in such a way that there is one and only one curve in the family through any initial point. In particular no two of these curves can intersect, for then there would be two solution curves through the point of intersection. [All of this presupposes that we are

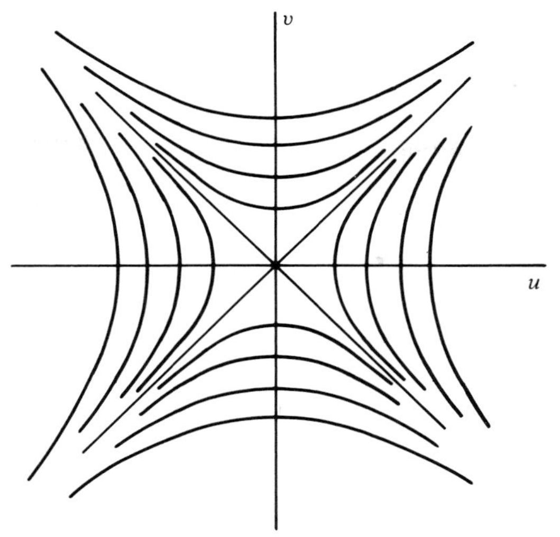

Figure 6.4

dealing with a system of differential equations that obey the restrictions on $F(t, X)$ needed for the uniqueness theorem. As we have seen in the case of scalar equations, there are differential equations which do not have unique solutions for special choices of the initial conditions. The same is true for systems.]

Exercises

1. Show that

$$3\begin{bmatrix} 2 \\ -1 \\ 4 \end{bmatrix} - 4\begin{bmatrix} 1 \\ 3 \\ -2 \end{bmatrix} + \begin{bmatrix} 5 \\ 0 \\ -11 \end{bmatrix} = \begin{bmatrix} 7 \\ -15 \\ 9 \end{bmatrix}$$

2. Rewrite the system of equations in Exercise 4 of Section 6.1 in vector form.

3. Show that the differential equation system in Exercise 11 of Section 6.1 is equivalent to

$$\frac{dX}{dt} = F(t, X), \qquad X(0) = \begin{bmatrix} 1 \\ 1 \\ 1 \end{bmatrix}$$

where

$$F(t, X) = \begin{bmatrix} x_3 \\ -x_1 \\ 3x_1 + 2x_2 \end{bmatrix}$$

4. Let

$$F(t) = \begin{bmatrix} 3t^2 \\ 5t-4 \\ t^2-3t \end{bmatrix} \quad \text{and} \quad G(t) = \begin{bmatrix} t \\ t^2-t \\ 2t-1 \end{bmatrix}$$

   (a) Write $2F(t)-3G(t)$ as a polynomial in t with vector coefficients.
   (b) Calculate $F'(t)$.
   (c) Calculate $\int_0^2 G(t)\,dt$.

5. Let

$$F(t, X) = \begin{bmatrix} x_1 - tx_2 \\ x_2 + tx_1 \end{bmatrix} \quad \text{where } X = \begin{bmatrix} x_1 \\ x_2 \end{bmatrix}$$

   (a) Restate the vector equation $dX/dt = F(t, X)$ as a system of scalar differential equations.
   (b) Rewrite the following second-order vector equation as a first-order vector equation:

$$\frac{d^2X}{dt^2} = 3\frac{dX}{dt} + F(t, X)$$

6. Rewrite the third-order scalar equation

$$\frac{d^3y}{dt^3} + 4\frac{d^2y}{dt^2} - 3\frac{dy}{dt} + 5y = t^2e^{-t}$$

   as a first-order vector equation.

7. Rewrite the differential system

$$\frac{d^2u}{dt^2} = -\frac{5}{u^2+v^2} \qquad \frac{d^2v}{dt^2} = -\frac{3}{u^2+v^2}$$

   as a first-order vector equation.

8. The general solution of a certain system of two equations is given by the formulas $u = 2C\sin(t+\beta)$, $v = 3C\cos(t+\beta)$. What curves in the UV-plane do these describe?

9. When a single second-order equation such as $y'' = -4y$ is converted into a first-order system and solved, the picture of the solution curves in the x_1x_2-plane is often called the *phase plane portrait* of y. Obtain these curves for this particular equation.

10. Why should you disbelieve someone who told you that the equations

$u = At^2 + B$, $v = At + B$, describe the solution of a pair of first-order differential equations? [*Hint:* By assigning values to the constants, sketch a few of the curves.]

6.3 Numerical Solution Methods

Most of the approximate methods for solving a single differential equation that we have discussed in Chapter 3 and Chapter 4 have analogs that work for systems of differential equations. This is true, for example, for the Picard method of successive approximations, discussed in Section 3.6. Given the vector equation

$$\frac{dX}{dt} = F(t, X), \qquad X(t_0) = X_0 \tag{6.33}$$

we can construct a sequence of vector-valued functions $\phi_0, \phi_1, \phi_2, \ldots$, each satisfying the initial conditions and converging to the exact solution $X = \phi(t)$. The method for constructing the sequence $\{\phi_n\}$ is the same as that in the scalar case. We start by choosing ϕ_0 to be the constant function whose value is X_0. Then, we define successive functions ϕ_n recursively by the formula

$$\begin{aligned} \phi'_{n+1}(t) &= F(t, \phi_n(t)) \\ \phi_{n+1}(t_0) &= X_0 \end{aligned} \tag{6.34}$$

Note that each new function is found by an integration step.

We illustrate this with the simple equations

$$\begin{aligned} \frac{du}{dt} &= v \qquad u(0) = 1 \\ \frac{dv}{dt} &= u \qquad v(0) = 0 \end{aligned} \tag{6.35}$$

which were discussed in Section 6.2 [equations (6.31)]. As before, we have

$$X = \begin{bmatrix} u \\ v \end{bmatrix}$$

and

$$F(t, X) = \begin{bmatrix} v \\ u \end{bmatrix}$$

and

$$X(0) = X_0 = \begin{bmatrix} 1 \\ 0 \end{bmatrix}$$

Accordingly, we start the Picard process by choosing ϕ_0 to be the constant function

$$\begin{bmatrix} 1 \\ 0 \end{bmatrix}$$

Following the instructions in (6.34), we next have

$$\begin{aligned} \phi_1'(t) &= F(t, \phi_0(t)) \\ &= \begin{bmatrix} 0 \\ 1 \end{bmatrix} \end{aligned}$$

Integrating, we have

$$\phi_1(t) = \begin{bmatrix} 0 \\ t \end{bmatrix} + \begin{bmatrix} C_1 \\ C_2 \end{bmatrix}$$

and choosing the constants of integration to have $\phi_1(0) = X_0$, we arrive at

$$\phi_1(t) = \begin{bmatrix} 1 \\ t \end{bmatrix}$$

Going on to the next step, we have

$$\begin{aligned} \phi_2'(t) &= F(t, \phi_1(t)) \\ &= \begin{bmatrix} t \\ 1 \end{bmatrix} \end{aligned}$$

so that, integrating, we have

$$\phi_2(t) = \begin{bmatrix} \frac{1}{2}t^2 \\ t \end{bmatrix} + \begin{bmatrix} C_1 \\ C_2 \end{bmatrix}$$

We again choose the constants so that $\phi_2(0) = X_0$, and have

$$\phi_2(t) = \begin{bmatrix} 1 + \frac{1}{2}t^2 \\ t \end{bmatrix}$$

If this process is continued, the following functions are obtained:

$$\phi_3(t) = \begin{bmatrix} 1 + \frac{1}{2}t^2 \\ t + \frac{1}{6}t^3 \end{bmatrix}$$

$$\phi_4(t) = \begin{bmatrix} 1 + \frac{1}{2}t^2 + \frac{1}{24}t^4 \\ t + \frac{1}{6}t^3 \end{bmatrix}$$

$$\phi_5(t) = \begin{bmatrix} 1 + \frac{1}{2}t^2 + \frac{1}{24}t^4 \\ t + \frac{1}{6}t^3 + \frac{1}{120}t^5 \end{bmatrix}$$

The functions $\{\phi_n\}$ converge to the vector-valued function

$$\phi(t) = \begin{bmatrix} 1 + \dfrac{t^2}{2!} + \dfrac{t^4}{4!} + \cdots \\ t + \dfrac{t^3}{3!} + \dfrac{t^5}{5!} + \cdots \end{bmatrix} = \begin{bmatrix} \cosh t \\ \sinh t \end{bmatrix}$$

which is easily recognized to be the exact solution of the system (6.35).

Perhaps this illustration is enough evidence to make it *plausible* that the Picard method works as well for systems as it did for single equations (Theorems 3.18 and 3.19), and therefore makes it possible to prove that a general vector-valued equation (6.33) with suitable restrictions on the function $F(t, X)$ always has a unique solution $X = \phi(t)$ obeying the given initial conditions. Since we have shown that a single nth-order differential equation is equivalent to a special type of first-order vector equation, this shows how it is possible to use the Picard method to prove existence and uniqueness theorems for these as well.

While it is very useful in discussing theoretical questions, the Picard method is not practical as a numerical method. For solving numerical problems, we use instead methods which produce a sequence of values that approximate the values of the true solution at a sequence of times $\{t_n\}$. The algorithms that are used are essentially the same as those discussed in Chapter 4 for scalar equations. We choose a step size $h = \Delta t$, and a sequence $\{t_n\}$ with $t_n = t_0 + nh$. We then construct a sequence $\{X_n\}$ in such a way that X_n is a good estimate for the exact value $X(t_n)$. The major difference now is that each term X_n is a vector rather than a number.

As before, the simple Euler method is a convenient illustration for the ideas involved. If we replace y by X, the statement of the Euler algorithm given in Section 4.1 is

$$\begin{aligned} X_0 &= X(t_0) \\ X_{n+1} &= X_n + hF(t_n, X_n) \end{aligned} \tag{6.36}$$

We do not need to make any changes in this in order to use it for vector-valued equations, and therefore for systems! Suppose we illustrate this for the same pair of equations we discussed earlier:

$$\frac{du}{dt} = v \qquad u(0) = 1$$

$$\frac{dv}{dt} = u \qquad v(0) = 0$$

As before, we set

$$X = \begin{bmatrix} u \\ v \end{bmatrix}$$

$$X(t_0) = \begin{bmatrix} 1 \\ 0 \end{bmatrix}$$

and

$$F(t, X) = \begin{bmatrix} v \\ u \end{bmatrix}$$

Let us write

$$X_n = \begin{bmatrix} u_n \\ v_n \end{bmatrix}$$

where u_n and v_n are numbers to be calculated. Then, following the instructions in (6.36), we have

$$X_0 = \begin{bmatrix} 1 \\ 0 \end{bmatrix}$$

$$X_{n+1} = X_n + h\begin{bmatrix} v_n \\ u_n \end{bmatrix}$$

Suppose we compute a few terms of the sequence $\{X_n\}$, with $h = 0.2$.

$$\begin{aligned}
X_0 &= \begin{bmatrix} 1 \\ 0 \end{bmatrix} \\
X_1 &= \begin{bmatrix} 1 \\ 0 \end{bmatrix} + (0.2)\begin{bmatrix} 0 \\ 1 \end{bmatrix} \\
&= \begin{bmatrix} 1 \\ 0.2 \end{bmatrix} \\
X_2 &= \begin{bmatrix} 1 \\ 0.2 \end{bmatrix} + (0.2)\begin{bmatrix} 0.2 \\ 1 \end{bmatrix} \\
&= \begin{bmatrix} 1.04 \\ 0.4 \end{bmatrix} \\
X_3 &= \begin{bmatrix} 1.04 \\ 0.4 \end{bmatrix} + (0.2)\begin{bmatrix} 0.4 \\ 1.04 \end{bmatrix} \\
&= \begin{bmatrix} 1.12 \\ 0.608 \end{bmatrix} \\
X_4 &= \begin{bmatrix} 1.12 \\ 0.608 \end{bmatrix} + (0.2)\begin{bmatrix} 0.608 \\ 1.12 \end{bmatrix} \\
&= \begin{bmatrix} 1.242 \\ 0.832 \end{bmatrix}
\end{aligned} \tag{6.37}$$

It is clear that this calculation is no more complicated in nature than in the case of a single scalar equation. Of course, the Euler method is still an unsatisfactory numerical method since it converges so slowly that h must be taken impractically small in order to obtain useful accuracy. However, we could instead use any of the methods discussed in Chapter 4 which have a smaller truncation error and which therefore are more accurate. To see how this is done, we choose one of the simpler methods, the Midpoint method given in formula (4.12). Repeating the description of this algorithm with X replacing y, we have

$$\begin{aligned} X_0 &= X(t_0) \\ X_n^* &= X_n + \tfrac{1}{2}hF(t_n, X_n) \\ m^* &= F(t_n + \tfrac{1}{2}h, X_n^*) \\ X_{n+1} &= X_n + hm^* \end{aligned} \tag{6.38}$$

We again illustrate this with the system (6.35), calculating the first few terms of the sequence $\{X_n\}$, and choosing $h = 0.2$.

$$\begin{aligned} X_0 &= \begin{bmatrix} 1 \\ 0 \end{bmatrix} \\ X_0^* &= \begin{bmatrix} 1 \\ 0 \end{bmatrix} + \tfrac{1}{2}(0.2)\begin{bmatrix} 0 \\ 1 \end{bmatrix} \\ &= \begin{bmatrix} 1 \\ 0.1 \end{bmatrix} \\ m^* &= F(0+0.1, X_0^*) \\ &= \begin{bmatrix} 0.1 \\ 1 \end{bmatrix} \\ X_1 &= \begin{bmatrix} 1 \\ 0 \end{bmatrix} + (0.2)\begin{bmatrix} 0.1 \\ 1 \end{bmatrix} \\ &= \begin{bmatrix} 1.02 \\ 0.2 \end{bmatrix} \\ X_1^* &= \begin{bmatrix} 1.02 \\ 0.2 \end{bmatrix} + \tfrac{1}{2}(0.2)\begin{bmatrix} 0.2 \\ 1.02 \end{bmatrix} \\ &= \begin{bmatrix} 1.04 \\ 0.302 \end{bmatrix} \end{aligned}$$

$$\begin{aligned} m^* &= F(0.2+0.1,\, X_1^*) \\ &= \begin{bmatrix} 0.302 \\ 1.04 \end{bmatrix} \\ X_2 &= \begin{bmatrix} 1.02 \\ 0.2 \end{bmatrix} + (0.2)\begin{bmatrix} 0.302 \\ 1.04 \end{bmatrix} \\ &= \begin{bmatrix} 1.080 \\ 0.408 \end{bmatrix} \end{aligned}$$

While computations of this sort can be done tediously by hand, they are ideally suited to the high-speed computer. The steps are routine and repetitive, and the use of vectors rather than scalars imposes only a relatively small amount of extra programming. Of course, more storage is needed, and more time is needed to carry out the intermediate calculations. However, the entire process is almost exactly the analog of the process used in the scalar case. Most computer-program libraries include a routine for solving differential equations, often the classical RK method (4.13). Since this routine is usually one that is designed to solve either a single scalar equation or a first-order vector equation, it can therefore be used to solve a system of first-order scalar equations, or a single differential equation of higher order, first converting it into a vector equation. The most complicated step for the user will be writing the program to calculate the function $F(t, X)$.

Much research is currently being done by mathematicians and computer scientists to learn more about the behavior of these numerical methods for solving differential equations, and better methods are constantly being devised to take care of special types of numerical instability that arise in certain cases. The same warning applies here that was issued at the start of Chapter 4: It is fatal to depend on the computer to supply numerical solutions of a system of differential equations if you do not know enough either about differential equations or about numerical analysis.

Exercises

1. Verify the formulas obtained for $\phi_3(t)$, $\phi_4(t)$, and $\phi_5(t)$ in the Picard solution (6.34) of equation (6.35).

2. Extend the Euler solution (6.37) of equation (6.35) to obtain X_5 and X_6, and verify that

$$X_7 = \begin{bmatrix} 1.896 \\ 1.687 \end{bmatrix}$$

3. Carry out two more steps in the Midpoint method solution (6.38) of equation (6.35) to obtain

$$X_3 = \begin{bmatrix} 1.184 \\ 0.632 \end{bmatrix} \qquad X_4 = \begin{bmatrix} 1.334 \\ 0.881 \end{bmatrix}$$

4. Compare X_2 and X_3 from Exercise 3 with the exact values of the solution of (6.35).

PC5. Apply the Midpoint method to the following system of equations, using $h = 0.2$, to obtain estimates for $u(0.4)$ and $v(0.4)$.

$$\frac{du}{dt} = uv \qquad u(0) = 1$$

$$\frac{dv}{dt} = u + v \qquad v(0) = -2$$

(You may wish to use a pocket calculator.)

6.4 Linear Systems and Matrix Methods

All of Chapter 5 was devoted to the study of the general linear differential equation

$$\frac{d^n y}{dt^n} = a_1(t)\frac{d^{n-1}y}{dt^{n-1}} + \cdots + a_{n-1}(t)\frac{dy}{dt} + a_n(t)\,y + g(t) \tag{6.39}$$

The rest of the present chapter will be devoted to a treatment of systems of linear equations. Most of the emphasis here will be upon the constant-coefficient case, since this is the most elementary type and also the one most frequently encountered. We can confine ourselves to first-order equations, because of the general reduction technique explained in Section 6.2. The form of the general linear system is the following:

$$\begin{cases} \dfrac{dx_1}{dt} = a_{11}(t)x_1 + a_{12}(t)x_2 + \cdots + a_{1n}(t)x_n + g_1(t) \\ \dfrac{dx_2}{dt} = a_{21}(t)x_1 + a_{22}(t)x_2 + \cdots + a_{2n}(t)x_n + g_2(t) \\ \vdots \\ \dfrac{dx_n}{dt} = a_{n1}(t)x_1 + a_{n2}(t)x_2 + \cdots + a_{nn}(t)x_n + g_n(t) \end{cases} \tag{6.40}$$

where the coefficient functions $a_{ij}(t)$ and the functions $g_i(t)$ are specified. In vector form, this has the standard form

$$\frac{dX}{dt} = F(t, X)$$

except that the function F is quite special in nature. This can be made more apparent, and the theory of linear systems simplified, by bringing in the use of matrices and matrix-valued functions.

A **matrix** is merely any rectangular array of quantities displayed in n rows and m columns, where n and m can be any positive integers. It is thus a generalization of the concept of vector. A typical matrix with numerical entries is

$$A = \begin{bmatrix} 2 & 3 & -1 \\ 4 & -7 & 2 \end{bmatrix} \tag{6.41}$$

While each entry represents a certain piece of information, derived perhaps from a measurement or a computation, we are interested in the totality of information displayed in the array; the key idea is therefore to treat the matrix A as a single object to which we can apply certain operations. However, at times it is convenient to think of a matrix as a list of vectors, written either horizontally or vertically. Thus, the 3-by-2 matrix A in (6.41) can also be regarded as a list of two 3-vectors (each written horizontally), or a list of three 2-vectors (each given vertically).

Two matrices of the same size (that is, having the same number of rows and of columns) are added or subtracted by carrying out this operation entry by entry. Thus, if we form the sum $A + A$, we have

$$\begin{bmatrix} 2 & 3 & -1 \\ 4 & -7 & 2 \end{bmatrix} + \begin{bmatrix} 2 & 3 & -1 \\ 4 & -7 & 2 \end{bmatrix} = \begin{bmatrix} 4 & 6 & -2 \\ 8 & -14 & 4 \end{bmatrix}$$

A matrix is multiplied by a scalar λ by multiplying each entry by the scalar. For example, we observe that $2A$ is the same matrix as $A + A$.

Multiplication of matrices is done according to a special row-by-column product, using the vectors that make up each matrix, and the multiplication can be carried out only if the sizes of the two matrices fit properly; we can calculate AB only when the length of the rows of A matches the length of the columns of B. (This implies that the number of columns in A must equal the number of rows in B.) We calculate the product of a row from A and a column from B by finding the sum of the products of corresponding entries. Thus

$$[2, \quad -3, \quad 5, \quad 1] \begin{bmatrix} 4 \\ 1 \\ -3 \\ 7 \end{bmatrix} = -3$$

since $(2)(4)+(-3)(1)+(5)(-3)+(1)(7) = -3$. The product of two matching matrices is computed by calculating the product of each row of the first by each column of the second, placing the results at appropriate positions to form the entries in the new matrix which is the product AB.

As an illustration,

$$\begin{bmatrix} 2 & 1 & -1 \\ 3 & -4 & 5 \end{bmatrix} \begin{bmatrix} 3 & 4 & 2 & 1 \\ 5 & -1 & 0 & 3 \\ 2 & 3 & 1 & 2 \end{bmatrix} = \begin{bmatrix} 9 & 4 & 3 & 3 \\ -1 & 31 & 11 & 1 \end{bmatrix}$$

In the product matrix on the right, the entry in row i and column j is always the result of calculating the product of the ith row of A and the jth column of B. Thus, the entry 11 came from the second row $[3 \quad -4 \quad 5]$ and the third column

$$\begin{bmatrix} 2 \\ 0 \\ 1 \end{bmatrix}$$

and we calculated $(3)(2)+(-4)(0)+(5)(1) = 11$.

If $A = [a_{ij}]$, with n rows and m columns, and $B = [b_{ij}]$, with m rows and p columns, then $AB = C = [c_{ij}]$, a matrix with n rows and p columns whose entries are given by the formula

$$c_{ij} = a_{i1}b_{1j} + a_{i2}b_{2j} + \cdots + a_{im}b_{mj} \tag{6.42}$$

Since vectors are merely special matrices, we can multiply matrices and vectors when they obey the size restriction. For example, we can multiply any square matrix with n rows and columns by any n-vector:

$$\begin{bmatrix} 2 & -1 & 3 \\ 4 & 5 & -2 \\ -1 & 0 & 4 \end{bmatrix} \begin{bmatrix} 5 \\ 2 \\ 1 \end{bmatrix} = \begin{bmatrix} 11 \\ 28 \\ -1 \end{bmatrix}$$

A **matrix-valued function** is a function F whose values are matrices of the same size. Such a function can be described by a matrix whose entries are ordinary scalar functions. For example, the matrix-valued function

$$F(t) = \begin{bmatrix} t^2 & 3t+1 & -2 \\ 1 & t^2+1 & 3t \\ t-1 & 3 & 2t \end{bmatrix}$$

has at $t = 1$ the value

$$F(1) = \begin{bmatrix} 1 & 4 & -2 \\ 1 & 2 & 3 \\ 0 & 3 & 2 \end{bmatrix}$$

Matrix-valued functions are differentiated and integrated entry by entry. Thus,

$$F'(t) = \begin{bmatrix} 2t & 3 & 0 \\ 0 & 2t & 3 \\ 1 & 0 & 2 \end{bmatrix}$$

$$\int_1^t F(t)\,dt = \begin{bmatrix} \frac{1}{3}t^3 & \frac{3}{2}t^2+t & -2t \\ t & \frac{1}{3}t^3+t & \frac{3}{2}t^2 \\ \frac{1}{2}t^2-t & 3t & t^2 \end{bmatrix} - \begin{bmatrix} \frac{1}{3} & \frac{5}{2} & -2 \\ 1 & \frac{4}{3} & \frac{3}{2} \\ -\frac{1}{2} & 3 & 1 \end{bmatrix}$$

$$= \begin{bmatrix} \frac{1}{3}(t^3-1) & \frac{3}{2}t^2+t-\frac{5}{2} & -2t+2 \\ t-1 & \frac{1}{3}t^3+t-\frac{4}{3} & \frac{3}{2}(t^2-1) \\ \frac{1}{2}(t-1)^2 & 3(t-1) & t^2-1 \end{bmatrix}$$

We can now return to equations (6.40) and use matrix-valued functions to find a simpler way to display the function $F(t, X)$ in such a way that its special structure becomes more apparent. We introduce the vector variable X, whose entries are the scalar variables x_i. We next introduce two functions $A(t)$ and $G(t)$, one matrix-valued and the other vector-valued.

$$A(t) = \begin{bmatrix} a_{11}(t) & a_{12}(t) & \cdots & a_{1n}(t) \\ a_{21}(t) & a_{22}(t) & \cdots & a_{2n}(t) \\ \vdots & & \vdots & \vdots \\ a_{n1}(t) & a_{n2}(t) & \cdots & a_{nn}(t) \end{bmatrix} \qquad G(t) = \begin{bmatrix} g_1(t) \\ g_2(t) \\ \vdots \\ g_n(t) \end{bmatrix}$$

Then, if we calculate the product of the matrix $A(t)$ by the column vector

$$X = \begin{bmatrix} x_1 \\ x_2 \\ \vdots \\ x_n \end{bmatrix}$$

we see that equations (6.40) are exactly equivalent to the single equation

$$\frac{dX}{dt} = A(t)X + G(t) \tag{6.43}$$

Let us illustrate this with several of the equations we have studied. Consider first the pair of equations (6.1) for the Two-Tank Problem.

$$\frac{dx}{dt} = 0.02y - 0.06x$$

$$\frac{dy}{dt} = 0.06x - 0.06y$$

We can now rewrite this in the matrix form

$$\frac{d}{dt}\begin{bmatrix} x \\ y \end{bmatrix} = \begin{bmatrix} -0.06 & 0.02 \\ 0.06 & -0.06 \end{bmatrix}\begin{bmatrix} x \\ y \end{bmatrix} \tag{6.44}$$

Similarly, the system of three equations (6.19)

$$\frac{dx}{dt} = y + z$$

$$\frac{dy}{dt} = 2x - z$$

$$\frac{dz}{dt} = 4x + 4y$$

in matrix form becomes

$$\frac{d}{dt}\begin{bmatrix} x \\ y \\ z \end{bmatrix} = \begin{bmatrix} 0 & 1 & 1 \\ 2 & 0 & -1 \\ 4 & 4 & 0 \end{bmatrix}\begin{bmatrix} x \\ y \\ z \end{bmatrix} \tag{6.45}$$

In both of these examples, the coefficient matrix is constant, and the function $G(t)$ is missing. This is such an important type that we will devote the next section to a discussion of general methods for solving such systems, based on the use of matrices.

To see another type of linear system, suppose we consider a single linear equation of third order:

$$\frac{d^3y}{dt^3} = (t^2+1)\frac{d^2y}{dt^2} - 2\frac{dy}{dt} + 5ty - e^{-2t} \tag{6.46}$$

Following the usual procedure of setting $x_1 = y$, $x_2 = dy/dt$, and $x_3 = d^2y/dt^2$, we can first rewrite (6.46) as

$$\frac{dX}{dt} = \frac{d}{dt}\begin{bmatrix} x_1 \\ x_2 \\ x_3 \end{bmatrix} = \begin{bmatrix} x_2 \\ x_3 \\ (t^2+1)x_3 - 2x_2 + 5tx_1 - e^{-2t} \end{bmatrix}$$

Then, introducing the matrix form, we can rewrite this as in (6.43):

$$\begin{aligned} \frac{dX}{dt} &= \begin{bmatrix} 0 & 1 & 0 \\ 0 & 0 & 1 \\ 5t & -2 & t^2+1 \end{bmatrix}\begin{bmatrix} x_1 \\ x_2 \\ x_3 \end{bmatrix} + \begin{bmatrix} 0 \\ 0 \\ -e^{-2t} \end{bmatrix} \\ &= A(t)X + G(t) \end{aligned}$$

We therefore see that the general nth-order linear equation, discussed in

Chapter 5, can always be reduced to a first-order vector linear equation in matrix form as in (6.43). The formal appearance of this equation is the same as that of the scalar first-order linear equation, discussed first in Section 2.2, namely,

$$\frac{dy}{dt} = a(t)\,y + g(t) \tag{6.47}$$

where $a(t)$ and $g(t)$ were scalar functions. Because of this similarity, much of the theory of matrix equations follows closely the pattern given earlier. In particular, we can use the theoretical developments in Chapter 5 to describe the general structure of the solutions to such a linear system. For example, we can introduce the differential operator $T = D - A(t) = (d/dt) - A(t)$, now operating on vector-valued functions of t of size n. T is linear, and we again use $\mathcal{N}_T$ to denote its null space, the collection of all functions X such that $T\{X\} = 0$. This will be a linear space of vector-valued functions, and it will have dimension n. This means that there are n vector-valued functions $\phi_1(t), \phi_2(t), \ldots, \phi_n(t)$ which are linearly independent and are such that every function $\psi(t)$ in $\mathcal{N}_T$ can be expressed (uniquely) in the form

$$\psi(t) = c_1\phi_1 + c_2\phi_2 + \cdots + c_n\phi_n \tag{6.48}$$

where c_i is a scalar. In particular, each of the functions ϕ_k satisfies the homogeneous equation

$$\frac{dX}{dt} = A(t)X \tag{6.49}$$

Finally, every solution of the original nonhomogeneous equation (6.43) can be obtained by combining the null space $\mathcal{N}_T$ with any one particular solution of the equation; that is, if the function $\phi_0(t)$ is one solution of (6.43), then the most general solution of (6.43) has the form

$$X(t) = \phi_0(t) + \{c_1\phi_1(t) + c_2\phi_2(t) + \cdots + c_n\phi_n(t)\}$$

where the c_i are scalar coefficients.

The proof of all this follows the pattern of the proof of Theorem 5.7 generalized to vector-valued functions. We first choose n special solutions $\phi_1, \phi_2, \ldots, \phi_n$ of equation (6.49), obeying a selected set of initial conditions, namely,

$$\phi_1(t_0) = \begin{bmatrix} 1 \\ 0 \\ 0 \\ \vdots \\ 0 \end{bmatrix} \qquad \phi_2(t_0) = \begin{bmatrix} 0 \\ 1 \\ 0 \\ \vdots \\ 0 \end{bmatrix} \qquad \cdots \qquad \phi_n(t_0) = \begin{bmatrix} 0 \\ 0 \\ 0 \\ \vdots \\ 1 \end{bmatrix}$$

These special functions form a generating set for the null space $\mathcal{N}_T$. For, let ψ be any member of $\mathcal{N}_T$, calculate the vector

$$\psi(t_0) = \begin{bmatrix} b_1 \\ b_2 \\ \vdots \\ b_n \end{bmatrix}$$

and form the linear combination $\psi_0 = \sum_1^n b_i \phi_i$. If we compute $\psi_0(t_0)$, we have

$$\psi_0(t_0) = b_1 \begin{bmatrix} 1 \\ 0 \\ 0 \\ \vdots \\ 0 \end{bmatrix} + b_2 \begin{bmatrix} 0 \\ 1 \\ 0 \\ \vdots \\ 0 \end{bmatrix} + \cdots + b_n \begin{bmatrix} 0 \\ 0 \\ 0 \\ \vdots \\ 1 \end{bmatrix} = \begin{bmatrix} b_1 \\ b_2 \\ \vdots \\ b_n \end{bmatrix}$$

Thus, ψ and ψ_0 are both members of $\mathcal{N}_T$, and $\psi(t_0) = \psi_0(t_0)$. By the uniqueness theorem for differential equations, $\psi = \psi_0$. Thus, every member of $\mathcal{N}_T$ can be expressed as a linear combination of the special functions $\{\phi_i\}$, so that these generate $\mathcal{N}_T$.

A similar argument shows that the $\{\phi_i\}$ are linearly independent and hence a basis. Indeed, suppose that $\sum_1^n c_i \phi_i$ were the zero function, for some choice of coefficients c_i. Then, as above, we compute the value which this sum takes for $t = t_0$, and arrive at a vector whose entries are the c_i, so that $c_1 = c_2 = \cdots = c_n = 0$, proving independence.

We can take one final step, and combine all these functions ϕ_k into one matrix-valued function

$$\Phi(t) = [\phi_1(t), \phi_2(t), \ldots, \phi_n(t)] \tag{6.50}$$

which will be itself a solution of the same equation (6.49), so that

$$\frac{d}{dt}\Phi(t) = A(t)\Phi(t)$$

Moreover, because the vector-valued functions ϕ_k that make up the columns of the matrix-valued function $\Phi(t)$ are linearly independent solutions of the differential equation (6.49), $\Phi(t)$ must be such that its determinant does not vanish; thus, for each t, $\Phi(t)$ is a *nonsingular* matrix.

A simple example will undoubtedly help to clarify all this. Suppose we consider the pair of equations

$$\begin{aligned} \frac{dx_1}{dt} &= 3x_1 \\ \frac{dx_2}{dt} &= 2x_2 \end{aligned} \tag{6.51}$$

Writing this in matrix form, we have

$$\frac{dX}{dt} = \begin{bmatrix} 3 & 0 \\ 0 & 2 \end{bmatrix} \begin{bmatrix} x_1 \\ x_2 \end{bmatrix} = AX$$

[Since the coefficient matrix is constant, we write A rather than $A(t)$.]

The fact that the matrix A is diagonal makes the solution of this system extremely simple. Indeed, the two equations in (6.51) are entirely independent of each other, and may be solved separately. (In such cases, one says that the variables x_1 and x_2 are "uncoupled.") Solving them, we have

$$x_1 = C_1 e^{3t}$$

$$x_2 = C_2 e^{2t}$$

The vector solutions ϕ_k are seen to be

$$\phi_1(t) = \begin{bmatrix} e^{3t} \\ 0 \end{bmatrix} \qquad \phi_2(t) = \begin{bmatrix} 0 \\ e^{2t} \end{bmatrix}$$

and the most general solution of (6.51) is given by $X = C_1 \phi_1 + C_2 \phi_2$.

The matrix-valued solution Φ of equation (6.51) will, according to (6.50), be

$$\Phi(t) = \begin{bmatrix} e^{3t} & 0 \\ 0 & e^{2t} \end{bmatrix}$$

and the determinant of this is e^{5t}, which is never zero.

Such a matrix function $\Phi(t)$ which is a solution of the homogeneous equation $dX/dt = A(t)X$ on an interval I, and which has a nonvanishing determinant there, is sometimes called a *fundamental solution*. It can be used to give a simple description of the entire null space $\mathcal{N}_T$. Indeed, the most general member of $\mathcal{N}_T$ has the form $\psi(t) = \Phi(t)C$, where

$$C = \begin{bmatrix} c_1 \\ c_2 \\ \vdots \\ c_n \end{bmatrix}$$

is an arbitrary constant vector. This fact is nothing more than equation (6.48) in disguise, since $\Phi(t) = [\phi_1(t), \phi_2(t), \ldots, \phi_n(t)]$, and since the product of this and the vector C is $\sum_1^n c_i \phi_i(t)$. What is not so obvious is that knowledge of such a matrix solution $\Phi(t)$ makes it possible (in theory) to solve the nonhomogeneous equation

$$\frac{dX}{dt} = A(t)X + G(t)$$

for any choice of the vector-valued function $G(t)$. The method is analogous to the process for solving the ordinary scalar linear equation (2.15) outlined in Section 2.2, and is a vector/matrix version of the so-called "variation of parameters" method that was explained in Section 5.4. It is quite useful for developing a general understanding of linear systems, but does not seem to be very practical as a solution method. Since it involves some additional matrix theory, we have left this to a starred exercise. (See Exercise 11.)

In the next section, we outline methods by which one finds these solutions ϕ_k and Φ, in the case of a matrix equation of the form (6.49) where the coefficient function $A(t)$ is *constant*, leaving the more general case to be treated in advanced treatises on differential equations. (See reference 7 in the Bibliography.)

Exercises

1. Write the following system of equations in the matrix form $AX = C$.

$$\begin{aligned} 2x_1 + x_2 + x_3 + x_4 &= 3 \\ x_1 - 2x_2 - 3x_3 - 3x_4 &= -4 \\ x_2 + 2x_3 + 5x_4 &= -2 \\ x_2 + x_3 - 4x_4 &= 11 \\ x_1 + x_2 + 2x_3 + 4x_4 &= 1 \end{aligned}$$

2. Verify the following matrix operations.

$$\begin{bmatrix} 2 & -1 \\ 5 & 3 \\ -2 & 0 \end{bmatrix} \begin{bmatrix} 3 & -2 \\ 1 & 4 \end{bmatrix} = \begin{bmatrix} 5 & -8 \\ 18 & 2 \\ -6 & 4 \end{bmatrix}$$

$$\begin{bmatrix} 2 & 0 & 1 \\ -1 & 3 & 0 \\ 3 & 1 & 0 \\ 1 & -1 & 4 \end{bmatrix} \begin{bmatrix} 5 & -2 & 0 & -1 \\ 0 & 1 & 3 & 4 \\ 1 & 0 & -2 & 1 \end{bmatrix} = \begin{bmatrix} 11 & -4 & -2 & -1 \\ -5 & 5 & 9 & 13 \\ 15 & -5 & 3 & 1 \\ 9 & -3 & -11 & -1 \end{bmatrix}$$

3. If

$$A(t) = \begin{bmatrix} 2t^3e^{-t} & 3t+t^2 \\ 2t & e^t \end{bmatrix}$$

find (a) $A'(t)$; (b) $\det\{A(t)\}$; (c) $A(t)^{-1}$. *Note:* If B is a 2-by-2 matrix

$$\begin{bmatrix} a & b \\ c & d \end{bmatrix}$$

and $\Delta = \det\{B\} = ad - bc \neq 0$, then the matrix inverse is

$$B^{-1} = \begin{bmatrix} \dfrac{d}{\Delta} & \dfrac{-b}{\Delta} \\ \dfrac{-c}{\Delta} & \dfrac{a}{\Delta} \end{bmatrix} \qquad BB^{-1} = I = \begin{bmatrix} 1 & 0 \\ 0 & 1 \end{bmatrix}$$

4. Let

$$F(t) = \begin{bmatrix} t^2+1 & 3t-2 \\ e^{-t} & e^{2t} \end{bmatrix}$$

Find $\int_0^1 F(t)\,dt$.

5. If $A(t)$ and $B(t)$ are 2-by-2 matrix functions, prove that

$$\frac{d}{dt}\{A(t)\,B(t)\} = A'(t)\,B(t) + A(t)\,B'(t)$$

6. Show that

$$\begin{bmatrix} 3 \\ 1 \end{bmatrix} e^{3t}$$

and

$$\begin{bmatrix} -1 \\ 1 \end{bmatrix} e^{-t}$$

are solutions of the homogeneous equation

$$\frac{dX}{dt} = \begin{bmatrix} 0 & 1 \\ 3 & 2 \end{bmatrix} X$$

7. (a) Show that

$$\Phi(t) = \begin{bmatrix} e^t & te^t \\ e^t & (1+t)e^t \end{bmatrix}$$

is a matrix solution of the homogeneous equation

$$\frac{dX}{dt} = \begin{bmatrix} 0 & 1 \\ -1 & 2 \end{bmatrix} X$$

(b) Find a vector solution of this equation satisfying the initial condition

$$X(0) = \begin{bmatrix} 2 \\ 3 \end{bmatrix}$$

8. (a) Verify that

$$\Phi(t) = \begin{bmatrix} t^3 & t^2 \\ t & 0 \end{bmatrix}$$

is a matrix solution for the system of equations

$$\frac{dx_1}{dt} = \frac{2}{t}x_1 + tx_2$$

$$\frac{dx_2}{dt} = \frac{1}{t}x_2$$

(b) Find a solution of this system such that $x_1(1) = 3$, $x_2(1) = -1$.

9. (a) Write the following system in matrix form.

$$\frac{dx}{dt} = y \qquad \frac{dy}{dt} = 4x$$

(b) Show that

$$\Phi(t) = \begin{bmatrix} e^{2t} & e^{-2t} \\ 2e^{2t} & -2e^{-2t} \end{bmatrix}$$

is a matrix solution of the resulting equation.

(c) Use this to find a solution of part (a) satisfying the conditions $x(0) = 2$, $y(0) = 4$.

10. Consider the system

$$\frac{du}{dt} = 3e^{-2t}v$$

$$\frac{dv}{dt} = e^{2t}u$$

(a) Change this to a homogeneous matrix equation $dX/dt = A(t)X$.

(b) Show that the following vector functions are solutions:

$$\phi_1(t) = \begin{bmatrix} e^{-3t} \\ -e^{-t} \end{bmatrix} \qquad \phi_2(t) = \begin{bmatrix} 3e^{t} \\ e^{3t} \end{bmatrix}$$

(c) Use these to find a solution of the system satisfying the initial conditions $u(0) = 4$, $v(0) = -3$.

*11. Let $\Phi(t)$ be a nonsingular matrix solution of the homogeneous equation $dX/dt = A(t)X$. Prove that a solution of the inhomogeneous equation $dX/dt = A(t)X + G(t)$ is given by $\phi_0(t) = \Phi(t)u(t)$, where $u(t) =$

$\int \Phi(t)^{-1}G(t)\,dt$ and $\Phi(t)^{-1}$ is the matrix inverse of $\Phi(t)$. [See note in Exercise 3.]

12. Use the method of Exercise 11 to find a solution of

$$\frac{du}{dt} = 3e^{-2t}v + e^{2t}$$

$$\frac{dv}{dt} = e^{2t}u + e^{t}$$

[Use Exercise 10 to find $\Phi(t)$.]

13. Solve the equations

$$\frac{dx_1}{dt} = \frac{2}{t}x_1 + tx_2 + t^4$$

$$\frac{dx_2}{dt} = \frac{1}{t}x_2 + 2t^2$$

by applying Exercise 8 and the method of Exercise 11.

6.5 Constant-Coefficient Systems

The purpose of this section is to develop some systematic procedures for solving homogeneous systems of linear differential equations with constant coefficients, using some simple concepts from linear algebra. The standard form of such a system is therefore

$$\frac{dX}{dt} = AX \tag{6.52}$$

where A is an n-by-n matrix with numerical entries. Restated in different words, the object of this section is to find ways to determine the null space $\mathcal{N}_T$ of the vector differential operator $T = (d/dt) - A$, where A is a given n-by-n matrix with numerical entries.

We have already seen a number of examples of systems such as (6.52), including (6.44) and (6.45) in the previous section. Any scalar nth-order linear homogeneous equation with constant coefficients is equivalent to such a system, so that the work of this section will give another way to cover the contents of part of Chapter 5. Thus, the third-order equation

$$\frac{d^3y}{dt^3} = 3\frac{d^2y}{dt^2} + \frac{dy}{dt} - 3y \tag{6.53}$$

becomes the system

$$\frac{dX}{dt} = \begin{bmatrix} 0 & 1 & 0 \\ 0 & 0 & 1 \\ -3 & 1 & 3 \end{bmatrix} X \qquad \text{where } X = \begin{bmatrix} y \\ y' \\ y'' \end{bmatrix} \tag{6.54}$$

Although the methods to be discussed in this section work in any case, we will assume that the entries of A are real numbers; even so, it will still be necessary to use complex numbers at times to solve the problems.

As a starting point, take $n = 1$ and consider the solution of the simple scalar differential equation $dx/dt = ax$, where a is merely a real number (that is, a square matrix with one row and one column). We expect the null space of the operator to be one-dimensional, and in fact it is generated by the single function $x = e^{at}$. Thus, in this simple case, all the solutions of the differential equation are constant multiples of a single exponential function. Our main result is a simple generalization of this fact. Before we come to it, we first need to introduce an important idea from linear algebra and the theory of matrices.

DEFINITION **6.1** A real or complex number λ is called an **eigenvalue** (or characteristic value) for the square matrix A if there is a nonzero vector V such that $AV = \lambda V$. Such a vector is called an **eigenvector** (or characteristic vector) associated with the specific eigenvalue λ, and any such associated pair consisting of a number λ and a vector V is called an **eiv pair**. Any nonzero multiple of an eigenvector is also an eigenvector, associated with the same eigenvalue.

We shall take up the study of methods for finding eiv pairs shortly. For the present, we observe that it is easy to check directly that the matrix

$$A = \begin{bmatrix} 0 & 1 & 0 \\ 0 & 0 & 1 \\ -3 & 1 & 3 \end{bmatrix}$$

has the following three eiv pairs of associated eigenvalues and eigenvectors:

$$\lambda_1 = 3,\ V_1 = \begin{bmatrix} 1 \\ 3 \\ 9 \end{bmatrix} \qquad \lambda_2 = 1,\ V_2 = \begin{bmatrix} 1 \\ 1 \\ 1 \end{bmatrix} \qquad \lambda_3 = -1,\ V_3 = \begin{bmatrix} 1 \\ -1 \\ 1 \end{bmatrix}$$

The connection between eigenvalue and eigenvector pairs is seen in the following important result.

THEOREM 6.1 *If λ is an eigenvalue for the matrix A, and V is an associated eigenvector, then the vector function $X = Ve^{\lambda t}$ is a solution of the equation $dX/dt = AX$.*

Proof Since V is constant, $dX/dt = V\lambda e^{\lambda t}$. Because of the properties of λ and V, $AV = \lambda V$; hence, $AX = AVe^{\lambda t} = \lambda V e^{\lambda t} = dX/dt$, verifying that X is a solution.

COROLLARY *If A is an n-by-n matrix having n distinct eigenvalues $\lambda_1, \lambda_2, \ldots, \lambda_n$, and if V_i is an eigenvector associated with λ_i, then the general solution of $dX/dt = AX$ is given by*

$$X = \psi(t) = c_1 V_1 e^{\lambda_1 t} + c_2 V_2 e^{\lambda_2 t} + \cdots + c_n V_n e^{\lambda_n t}$$

where the c_i are arbitrary scalars.

Proof This follows at once from the theorem since the functions $V_i e^{\lambda_i t}$ are each solutions and are linearly independent (see Exercise 23), and the null space $\mathcal{N}_T$ has dimension n; the $V_i e^{\lambda_i t}$ then must form a basis for $\mathcal{N}_T$.

We can illustrate this immediately with equation (6.54). Using the information given about eigenvalues for the matrix A, we can conclude at once that the general solution of (6.54) is

$$X = c_1 \begin{bmatrix} 1 \\ 3 \\ 9 \end{bmatrix} e^{3t} + c_2 \begin{bmatrix} 1 \\ 1 \\ 1 \end{bmatrix} e^{t} + c_3 \begin{bmatrix} 1 \\ -1 \\ 1 \end{bmatrix} e^{-t}$$

Since we recall that

$$X = \begin{bmatrix} y \\ y' \\ y'' \end{bmatrix}$$

we discover that the general solution of (6.53) is

$$y = c_1 e^{3t} + c_2 e^{t} + c_3 e^{-t}$$

(This could have been discovered more easily by the methods explained in Chapter 5; however, this example is used here to show the connections between the new methods in the present section and the work discussed earlier.)

This key result reduces the solution of a large class of linear systems to the algebraic problem of finding the eigenvalues and eigenvectors of a given matrix. A first step in this direction is the following.

THEOREM 6.2 *The eigenvalues of a matrix A are the real or complex numbers* *that are the roots of the polynomial* $d(\lambda)$, *where*

$$d(\lambda) = \det\{\lambda I - A\}$$

In this, the matrix I is the identity matrix

$$\begin{bmatrix} 1 & 0 & 0 & \cdots & 0 \\ 0 & 1 & 0 & \cdots & 0 \\ \vdots & & & & \vdots \\ 0 & 0 & 0 & \cdots & 1 \end{bmatrix}$$

so that if $A = [a_{ij}]$,

$$(6.55) \qquad d(\lambda) = \det\begin{bmatrix} \lambda - a_{11} & -a_{12} & -a_{13} & \cdots & -a_{1n} \\ -a_{21} & \lambda - a_{22} & -a_{23} & \cdots & -a_{2n} \\ \vdots & & & & \vdots \\ -a_{n1} & -a_{n2} & & \cdots & \lambda - a_{nn} \end{bmatrix}$$

To prove Theorem 6.2, it is only necessary to return to Definition 6.1 and rewrite the characteristic property of λ and V in the form $(\lambda I - A)V = 0$, using the fact that $IV = V$ for any vector V. Regarding the entries of the vector V for the moment as variables to be determined, the equation $(\lambda I - A)V = 0$ becomes a system of n algebraic linear equations with right side zero, and such an algebraic system has a solution which is not all zero (ruled out since the eigenvector V is required to be nonzero!) only when the determinant of the coefficient matrix $\lambda I - A$ is zero.

Since $d(\lambda)$ is a polynomial of degree n, and therefore has at most n distinct roots, we also find the following to be true.

COROLLARY *An n-by-n matrix has at most n distinct eigenvalues.*

These two results give us a method for finding all the eigenvalues of a matrix; unfortunately, this method has practical limitations. We must first compute an n-by-n determinant like (6.55) and then find the roots, both real and complex, of the resulting nth-degree polynomial. This is easy when $n = 2$, but becomes much more difficult as n increases. For this reason, most of our illustrations and exercises are confined to the case where $n \leq 4$. There are special numerical methods for handling larger values of n, but these are best left to a more advanced course in linear algebra or numerical analysis.

We also note that Theorem 6.2 does not explain how to find an eigenvector associated with an eigenvalue λ that you have found. The procedure for determining an associated eigenvector is easiest to explain by means of examples.

There are special matrices for which one can determine both the eigenvalues and eigenvectors at sight. Consider the diagonal matrix

$$A = \begin{bmatrix} 5 & 0 & 0 & 0 \\ 0 & -3 & 0 & 0 \\ 0 & 0 & 2 & 0 \\ 0 & 0 & 0 & 7 \end{bmatrix}$$

This matrix has four eigenvalues, and they are 5, -3, 2, and 7. This could have been found by using Theorem 6.2 but is much more easily seen by using Definition 6.1 and solving the equation $AV = \lambda V$. It is also easy in this case to find an eigenvector for each eigenvalue.

Thus, with 5 we associate $V = \begin{bmatrix} 1 \\ 0 \\ 0 \\ 0 \end{bmatrix}$

with -3 the vector $\begin{bmatrix} 0 \\ 1 \\ 0 \\ 0 \end{bmatrix}$

and so on. Of course, any nonzero multiple of these would also do. For example, we could also associate

$$\begin{bmatrix} -9 \\ 0 \\ 0 \\ 0 \end{bmatrix}$$

with 5, since $A\begin{bmatrix} -9 \\ 0 \\ 0 \\ 0 \end{bmatrix} = 5\begin{bmatrix} -9 \\ 0 \\ 0 \\ 0 \end{bmatrix}$

Another class of matrices for which it is easy to determine the eigenvalues are triangular matrices. Consider the matrix

$$A = \begin{bmatrix} 2 & 6 & -1 \\ 0 & -3 & 5 \\ 0 & 0 & 4 \end{bmatrix}$$

We have

$$d(\lambda) = \det\begin{bmatrix} \lambda-2 & -6 & 1 \\ 0 & \lambda+3 & -5 \\ 0 & 0 & \lambda-4 \end{bmatrix}$$
$$= (\lambda-2)(\lambda+3)(\lambda-4)$$

so that we conclude that the eigenvalues are 2, -3, and 4, the diagonal entries in the matrix A. This time, it is a little harder to find the corresponding eigenvectors. If we write

$$V = \begin{bmatrix} a \\ b \\ c \end{bmatrix}$$

and the equation $AV = 2V$, we have

$$\begin{aligned} 2a + 6b - c &= 2a \\ -3b + 5c &= 2b \\ 4c &= 2c \end{aligned}$$

Solving these, starting with the last, we have $c = 0$, $b = 0$, and $2a = 2a$. In order not to have $V = 0$, we take $a = 1$, and we have found the vector

$$\begin{bmatrix} 1 \\ 0 \\ 0 \end{bmatrix}$$

to associate with the eigenvalue 2.

Following the same procedure with the eigenvalue -3, the equation $AV = -3V$ leads to the equations

$$\begin{aligned} 2a + 6b - c &= -3a \\ -3b + 5c &= -3b \\ 4c &= -3c \end{aligned}$$

We find $c = 0$ and then $5a + 6b = 0$. We could choose $a = 1$ and $b = -\frac{5}{6}$, but an integral choice is $b = -5$, $a = 6$. Thus,

$$\begin{bmatrix} 6 \\ -5 \\ 0 \end{bmatrix}$$

is an eigenvector to associate with $\lambda = -3$.

Finally, for $\lambda = 4$, we want to solve $AV = 4V$, which leads to

$$\begin{aligned} 2a + 6b - c &= 4a \\ -3b + 5c &= 4b \\ 4c &= 4c \end{aligned}$$

If we set $c = 1$, we find $b = \frac{5}{7}$ and $a = \frac{23}{14}$. Multiplying the resulting vector by 14 gives integer entries, so we can use

$$\begin{bmatrix} 23 \\ 10 \\ 14 \end{bmatrix}$$

as the eigenvector to associate with $\lambda = 4$.

Thus, for the matrix A we have found the following three eiv pairs:

$$\lambda_1 = 2,\ V_1 = \begin{bmatrix} 1 \\ 0 \\ 0 \end{bmatrix} \qquad \lambda_2 = -3,\ V_2 = \begin{bmatrix} 6 \\ -5 \\ 0 \end{bmatrix} \qquad \lambda_3 = 4,\ V_3 = \begin{bmatrix} 23 \\ 10 \\ 14 \end{bmatrix}$$

To illustrate again the connection all this has with the solution of differential equations, consider the system

$$\begin{aligned} \frac{dx}{dt} &= 2x + 6y - z \\ \frac{dy}{dt} &= \quad -3y + 5z \\ \frac{dz}{dt} &= \qquad 4z \end{aligned} \tag{6.56}$$

The matrix for this system is the one we have just worked with, so that we can now use our results to give the general solution of (6.56), namely,

$$\begin{bmatrix} x \\ y \\ z \end{bmatrix} = c_1 \begin{bmatrix} 1 \\ 0 \\ 0 \end{bmatrix} e^{2t} + c_2 \begin{bmatrix} 6 \\ -5 \\ 0 \end{bmatrix} e^{-3t} + c_3 \begin{bmatrix} 23 \\ 10 \\ 14 \end{bmatrix} e^{4t}$$

or, in component form,

$$\begin{aligned} x &= c_1 e^{2t} + 6c_2 e^{-3t} + 23c_3 e^{4t} \\ y &= \qquad -5c_2 e^{-3t} + 10c_3 e^{4t} \\ z &= \qquad\qquad 14c_3 e^{4t} \end{aligned}$$

Even though a matrix A has only real entries, some of its eigenvalues and eigenvectors can be complex. Consider the matrix

$$A = \begin{bmatrix} 3 & -5 \\ 2 & 1 \end{bmatrix}$$

Since

$$d(\lambda) = \det\begin{bmatrix} \lambda-3 & 5 \\ -2 & \lambda-1 \end{bmatrix} = (\lambda-3)(\lambda-1) + 10$$

$$= \lambda^2 - 4\lambda + 13$$

we solve the quadratic equation $d(\lambda) = 0$ and find that the eigenvalues of A are $2+3i$ and $2-3i$. To find the eigenvector corresponding to the first of these, we write

$$AV = \begin{bmatrix} 3 & -5 \\ 2 & 1 \end{bmatrix}\begin{bmatrix} a \\ b \end{bmatrix} = \lambda V = (2+3i)\begin{bmatrix} a \\ b \end{bmatrix}$$

which leads at once to the pair of algebraic equations

$$3a - 5b = (2+3i)a$$

$$2a + b = (2+3i)b$$

Either of these leads to the same equation, namely,

$$(1-3i)a = 5b$$

so that a solution for V is given by $a = 5$, $b = 1-3i$. Hence, one eiv pair is seen to be

$$\lambda_1 = 2 + 3i, \qquad V_1 = \begin{bmatrix} 5 \\ 1-3i \end{bmatrix}$$

If the same process is followed with the second eigenvalue, $2-3i$, one arrives at the eiv pair

$$\lambda_2 = 2 - 3i, \qquad V_2 = \begin{bmatrix} 5 \\ 1+3i \end{bmatrix}$$

We observe at once that this pair could have been obtained from the first pair by taking the complex conjugate of all the numbers involved. This is a general property of real matrices, as described in Exercise 8.

We can use complex eiv pairs in the solution of differential equations in the same way as we use real eiv pairs. Thus, if we were solving the vector differential equation

$$\frac{dX}{dt} = \begin{bmatrix} 3 & -5 \\ 2 & 1 \end{bmatrix} X \tag{6.57}$$

(whose matrix is the matrix A we have just studied), the computations above tell us that the general solution of (6.57) is

$$X = c_1 \begin{bmatrix} 5 \\ 1-3i \end{bmatrix} e^{(2+3i)t} + c_2 \begin{bmatrix} 5 \\ 1+3i \end{bmatrix} e^{(2-3i)t}$$

This answer, which involves complex-valued functions, can be replaced by one which involves only real functions. The key is a simple observation, which we record here but whose proof we leave to the exercises; see Exercise 27.

LEMMA *Let $dX/dt = AX$ be a vector equation, whose matrix A has only real entries. Then, the real or imaginary part of a solution is also a solution. Explicitly, if $U(t)+iV(t)$ is a solution, so are the vector functions $U(t)$ and $V(t)$.*

To apply this to the example we are studying, we take $c_1 = 1$ and $c_2 = 0$, and thus obtain the complex solution

$$X = \begin{bmatrix} 5 \\ 1-3i \end{bmatrix} e^{(2+3i)t}$$

In order to find the real and imaginary parts of this, we must also use the Cotes-Euler relation

$$e^{\alpha+i\beta} = e^{\alpha}(\cos\beta + i\sin\beta)$$

[see (3.17) and (3.20)]. Using this, we write our solution as

$$X = \left(\begin{bmatrix} 5 \\ 1 \end{bmatrix} + i \begin{bmatrix} 0 \\ -3 \end{bmatrix} \right) (e^{2t}\cos(3t) + ie^{2t}\sin(3t))$$

The real part of this is

$$\phi_1(t) = \begin{bmatrix} 5 \\ 1 \end{bmatrix} e^{2t}\cos(3t) - \begin{bmatrix} 0 \\ -3 \end{bmatrix} e^{2t}\sin(3t)$$

and the imaginary part is

$$\phi_2(t) = \begin{bmatrix} 5 \\ 1 \end{bmatrix} e^{2t}\sin(3t) + \begin{bmatrix} 0 \\ -3 \end{bmatrix} e^{2t}\cos(3t)$$

and ϕ_1 and ϕ_2 are two real solutions of (6.57) that are linearly independent, and thus form a basis for the space of all solutions of (6.57). We could also have started from the conjugate-complex solution

$$X = \begin{bmatrix} 5 \\ 1+3i \end{bmatrix} e^{(2-3i)t}$$

$$= \left(\begin{bmatrix} 5 \\ 1 \end{bmatrix} + i \begin{bmatrix} 0 \\ 3 \end{bmatrix} \right) (e^{2t}\cos(-3t) + ie^{2t}\sin(-3t))$$

and taking real and imaginary parts, obtain the pair of real solutions

$$\psi_1(t) = \begin{bmatrix} 5 \\ 1 \end{bmatrix} e^{2t}\cos(-3t) - \begin{bmatrix} 0 \\ 3 \end{bmatrix} e^{2t}\sin(-3t)$$

$$\psi_2(t) = \begin{bmatrix} 5 \\ 1 \end{bmatrix} e^{2t}\sin(-3t) + \begin{bmatrix} 0 \\ 3 \end{bmatrix} e^{2t}\cos(-3t)$$

However, it is at once evident that $\psi_1 = \phi_1$ and $\psi_2 = -\phi_2$ so that we have obtained nothing new. Thus, each *complex* pair of solutions of a homogeneous system with a real-coefficient matrix yields two independent *real* solutions, which can be obtained from either solution of the complex pair.

What happens if the polynomial $d(\lambda)$ has multiple roots? In this case, the n-by-n matrix A will have fewer than n *distinct* eigenvalues; two cases arise, depending upon the total number of independent eigenvectors associated with the multiple eigenvalue.

The first case is the simplest, and is illustrated by the diagonal matrix

$$A = \begin{bmatrix} 2 & 0 & 0 \\ 0 & 3 & 0 \\ 0 & 0 & 3 \end{bmatrix}$$

Here, $d(\lambda) = (\lambda-2)(\lambda-3)^2$ so the eigenvalues are 2 and 3. If we ask for eigenvectors associated with $\lambda = 3$, and solve for vectors

$$V = \begin{bmatrix} a \\ b \\ c \end{bmatrix}$$

such that $AV = 3V$, we find that any vector of the form

$$V = \begin{bmatrix} 0 \\ b \\ c \end{bmatrix}$$

will do. This is a two-dimensional space of vectors, and has a basis consisting of

$$\begin{bmatrix} 0 \\ 1 \\ 0 \end{bmatrix} \quad \text{and} \quad \begin{bmatrix} 0 \\ 0 \\ 1 \end{bmatrix}$$

Thus, in this case, we would speak of $\lambda = 3$ as an eigenvalue of *multiplicity two*, associated with *two* independent eigenvectors. A is a 3-by-3 matrix having only two distinct eigenvalues but a total of three independent eigenvectors:

$$\begin{bmatrix} 1 \\ 0 \\ 0 \end{bmatrix}$$

associated with $\lambda = 2$, and

$$\begin{bmatrix} 0 \\ 1 \\ 0 \end{bmatrix} \quad \text{and} \quad \begin{bmatrix} 0 \\ 0 \\ 1 \end{bmatrix}$$

associated with $\lambda = 3$. We would treat these in the same way we did when a matrix had all of its eigenvalues distinct. For example, if A were the matrix for the differential equation $dX/dt = AX$, we would write down its solutions as

$$X = c_1 \begin{bmatrix} 1 \\ 0 \\ 0 \end{bmatrix} e^{2t} + c_2 \begin{bmatrix} 0 \\ 1 \\ 0 \end{bmatrix} e^{3t} + c_3 \begin{bmatrix} 0 \\ 0 \\ 1 \end{bmatrix} e^{3t}$$

The second case is illustrated by the matrix

$$A = \begin{bmatrix} 2 & 0 & 0 \\ 0 & 3 & 1 \\ 0 & 0 & 3 \end{bmatrix}$$

Again, $d(\lambda) = (\lambda - 2)(\lambda - 3)^2$, and the eigenvalues are again 2 and 3. However, when we attempt to find eigenvectors associated with $\lambda = 3$, only one is found; if we ask for

$$V = \begin{bmatrix} a \\ b \\ c \end{bmatrix}$$

with $AV = 3V$, we have

$$\begin{aligned} 2a &= 3a \\ 3b + c &= 3b \\ 3c &= 3c \end{aligned}$$

and this has only a one-dimensional space of solutions with $a = 0$, $b =$ anything, $c = 0$, and thus having a basis such as

$$\begin{bmatrix} 0 \\ 1 \\ 0 \end{bmatrix}$$

In this case, a 3-by-3 matrix A having only two distinct eigenvalues yields a total of only two independent eigenvectors, instead of three.

What happens in this case should not be too surprising, in the light of the treatment of linear equations in Chapter 5. When this second case occurs, the differential equation dX/dt has solutions that are not of the simple form $X = Ve^{\lambda t}$, but instead of the form $X = P(t)e^{\lambda t}$, where $P(t)$ is a vector polynomial in t.

We can illustrate this with the example we have just treated. The corresponding differential system is

$$\frac{dx}{dt} = 2x$$

$$(6.58) \qquad \frac{dy}{dt} = 3y + z$$

$$\frac{dz}{dt} = 3z$$

If we start to solve this at the bottom, we have $z = C_1 e^{3t}$. Putting this in the second equation, we have

$$\frac{dy}{dt} = 3y + C_1 e^{3t}$$

and the general solution of this (using either the methods of Chapter 5 or those of Chapter 2) is

$$y = C_2 e^{3t} + C_1 te^{3t}$$

and finally, $x = C_3 e^{2t}$. Putting these in vector form, we have a basis for $\mathcal{N}_T$:

$$(6.59) \qquad X = \begin{bmatrix} x \\ y \\ z \end{bmatrix} = C_3 \begin{bmatrix} 1 \\ 0 \\ 0 \end{bmatrix} e^{2t} + C_2 \begin{bmatrix} 0 \\ 1 \\ 0 \end{bmatrix} e^{3t} + C_1 \begin{bmatrix} 0 \\ t \\ 1 \end{bmatrix} e^{3t}$$

The two eigenvectors are visible in this solution in the usual way, but the last term in (6.59) brings in a new type of solution, $P(t)e^{\lambda t}$.

We now examine this with the aim of finding a more systematic procedure which can be used in general to solve this class of differential equations. We note that in the example just completed, the polynomial function $P(t)$ turned out to be

$$P(t) = \begin{bmatrix} 0 \\ t \\ 1 \end{bmatrix} = \begin{bmatrix} 0 \\ 1 \\ 0 \end{bmatrix} t + \begin{bmatrix} 0 \\ 0 \\ 1 \end{bmatrix}$$

and we also note that the coefficient of t is the eigenvector

$$\begin{bmatrix} 0 \\ 1 \\ 0 \end{bmatrix}$$

corresponding to $\lambda = 3$. Another way of stating this is to write

$$P'(t) = \begin{bmatrix} 0 \\ 1 \\ 0 \end{bmatrix}$$

Let us try something like this in the general case. Suppose we want to solve the vector equation

$$\frac{dX}{dt} = AX \tag{6.60}$$

and λ is an eigenvalue of the matrix A which is associated with an eigenvector V. We already know that $X = Ve^{\lambda t}$ is a solution of (6.60). Can we find another solution of (6.60) of the form $X = P(t)e^{\lambda t}$, where $P(t)$ is a vector-valued polynomial of degree 1, with $P'(t) = V$? If so, we can write $P(t) = Vt + U$, where U is a constant vector to be determined. Trying this, we have

$$\begin{aligned} X &= (Vt + U)e^{\lambda t} \\ \frac{dX}{dt} &= Ve^{\lambda t} + (Vt + U)\lambda e^{\lambda t} \\ AX &= A(Vt + U)e^{\lambda t} \\ &= (AVt + AU)e^{\lambda t} \end{aligned} \tag{6.61}$$

Since V is an eigenvector for A, $AV = \lambda V$, so that using this and requiring that $dX/dt = AX$, we need to have

$$Ve^{\lambda t} + (Vt + U)\lambda e^{\lambda t} = (\lambda Vt + AU)e^{\lambda t}$$

If we cancel the common terms on each side and divide by $e^{\lambda t}$, we end up with the equation

$$V + \lambda U = AU$$

which we prefer to write in the form

$$(A - \lambda I)\, U = V \tag{6.62}$$

All the steps above are reversible, so that if we can find a vector U satisfying the equation (6.62), then the vector function given by (6.61) is a solution of (6.60).

Let us test this against the example considered earlier, namely, the differential system (6.58). The matrix A is

$$A = \begin{bmatrix} 2 & 0 & 0 \\ 0 & 3 & 1 \\ 0 & 0 & 3 \end{bmatrix}$$

and we have $\lambda = 3$,

$$V = \begin{bmatrix} 0 \\ 1 \\ 0 \end{bmatrix}$$

Writing

$$U = \begin{bmatrix} a \\ b \\ c \end{bmatrix}$$

we want to solve

$$(A - \lambda I)\, U = \begin{bmatrix} -1 & 0 & 0 \\ 0 & 0 & 1 \\ 0 & 0 & 0 \end{bmatrix} \begin{bmatrix} a \\ b \\ c \end{bmatrix} = V = \begin{bmatrix} 0 \\ 1 \\ 0 \end{bmatrix}$$

This is equivalent to

$$-a = 0$$

$$c = 1$$

$$0 = 0$$

so that a solution is

$$U = \begin{bmatrix} 0 \\ 0 \\ 1 \end{bmatrix}$$

According to (6.61), another solution to the equation $dX/dt = AX$, and therefore to the system (6.58), is given by

$$X = (Vt+U)e^{3t} = \left(\begin{bmatrix} 0 \\ 1 \\ 0 \end{bmatrix} t + \begin{bmatrix} 0 \\ 0 \\ 1 \end{bmatrix}\right) e^{3t}$$

$$= \begin{bmatrix} 0 \\ t \\ 1 \end{bmatrix} e^{3t}$$

in agreement with (6.59).

Let us illustrate this with another example. Consider the equation $dX/dt = AX$, where

$$A = \begin{bmatrix} 1 & 1 \\ -4 & -3 \end{bmatrix}$$

It is easily seen that $d(\lambda) = \lambda^2 + 2\lambda + 1$, so that the only eigenvalue is $\lambda = -1$, which turns out to be associated with only one eigenvector,

$$V = \begin{bmatrix} 1 \\ -2 \end{bmatrix}$$

This gives us one solution of the differential equation, namely,

$$X = \begin{bmatrix} 1 \\ -2 \end{bmatrix} e^{-t}$$

Since A is a 2-by-2 matrix, we need another independent solution, and we find it by solving the matrix equation (6.62). We have

$$(A-\lambda I)U = (A+I)U$$

$$= \begin{bmatrix} 2 & 1 \\ -4 & -2 \end{bmatrix} \begin{bmatrix} a \\ b \end{bmatrix} = V = \begin{bmatrix} 1 \\ -2 \end{bmatrix}$$

or

$$2a + b = 1$$

$$-4a - 2b = -2$$

and a solution of this is $a = 0$, $b = 1$. Accordingly, we conclude that a second and independent solution of our differential equation is given by (6.61) and is

$$X = (Vt+U)e^{t}$$

$$= \left(\begin{bmatrix} 1 \\ -2 \end{bmatrix} t + \begin{bmatrix} 0 \\ 1 \end{bmatrix}\right) e^{-t} = \begin{bmatrix} t \\ -2t+1 \end{bmatrix} e^{-t}$$

When we are missing additional solutions due to a further deficiency of eigenvectors, the process can be extended.

Consider the equation $dX/dt = AX$, where

$$A = \begin{bmatrix} 1 & 2 & 0 \\ 0 & 1 & -3 \\ 0 & 0 & 1 \end{bmatrix}$$

This has only one eigenvalue, $\lambda = 1$, and it is associated with only *one* eigenvector,

$$V = \begin{bmatrix} 1 \\ 0 \\ 0 \end{bmatrix}$$

The usual procedure gives us only one of the solutions for the equation, namely,

$$X = \begin{bmatrix} 1 \\ 0 \\ 0 \end{bmatrix} e^t \tag{6.63}$$

The new procedure gives us another solution

$$X = \left(\begin{bmatrix} 1 \\ 0 \\ 0 \end{bmatrix} t + \begin{bmatrix} 0 \\ \frac{1}{2} \\ 0 \end{bmatrix} \right) e^t = \begin{bmatrix} t \\ \frac{1}{2} \\ 0 \end{bmatrix} e^t \tag{6.64}$$

We have

$$(A - \lambda I)\, U = \begin{bmatrix} 0 & 2 & 0 \\ 0 & 0 & -3 \\ 0 & 0 & 0 \end{bmatrix} \begin{bmatrix} a \\ b \\ c \end{bmatrix} = V = \begin{bmatrix} 1 \\ 0 \\ 0 \end{bmatrix}$$

which yields $a =$ anything, $b = \frac{1}{2}$, $c = 0$, and we can take

$$U = \begin{bmatrix} 0 \\ \frac{1}{2} \\ 0 \end{bmatrix}$$

To find a third solution, independent of the first two, we look for a solution $X = P(t)e^t$, where $P(t)$ is a vector polynomial of degree 2 such that $P'(t)$ is the polynomial of degree 1 that appears in (6.61). The algebraic analysis is very similar, and we are led to ask for a vector W such that

$$(A - I)\, W = U$$

where U is the vector we have just found. Accordingly, we solve

$$\begin{bmatrix} 0 & 2 & 0 \\ 0 & 0 & -3 \\ 0 & 0 & 0 \end{bmatrix} W = \begin{bmatrix} 0 \\ \frac{1}{2} \\ 0 \end{bmatrix}$$

and find

$$W = \begin{bmatrix} 0 \\ 0 \\ -\frac{1}{6} \end{bmatrix}$$

The desired solution is then

$$X = \left(\begin{bmatrix} 1 \\ 0 \\ 0 \end{bmatrix} t^2/2 + \begin{bmatrix} 0 \\ \frac{1}{2} \\ 0 \end{bmatrix} t + \begin{bmatrix} 0 \\ 0 \\ -\frac{1}{6} \end{bmatrix} \right) e^t$$

$$= \begin{bmatrix} \frac{1}{2}t^2 \\ \frac{1}{2}t \\ -\frac{1}{6} \end{bmatrix} e^t$$

The detailed treatment of the general case of an n-by-n matrix A is quite simple when one can use certain facts about matrices, usually presented in a course in linear algebra. Since we are not attempting to treat this topic in any depth, we stop here and merely describe the facts that can be shown to hold. (See reference 7 in the Bibliography.)

Suppose that λ_0 is an eigenvalue of the square matrix A which has multi plicity r, meaning that it is an r-fold root of the characteristic polynomial $d(\lambda)$. With λ_0 there may be associated from one to r independent eigenvectors, depending upon the algebraic nature of the matrix A. If there are r eigenvectors, then, as we have seen, we can produce r independent solutions of the differential equation $dX/dt = AX$, each of the form $X = Ve^{\lambda_0 t}$. If there are only k independent eigenvectors associated with λ_0, then in addition to the k solutions formed from these in the usual way, there will be $r-k$ additional solutions, each of the form $X = P(t)e^{\lambda_0 t}$, where $P(t)$ is an appropriate vector polynomial. The case when there is only one eigenvector associated with λ_0 is the simplest, for then one has a set of r solutions of the simple form

$$X_1 = Ve^{\lambda_0 t}$$

$$X_2 = (Vt + U_1)e^{\lambda_0 t}$$

$$X_3 = (\tfrac{1}{2}Vt^2 + U_1 t + U_2)e^{\lambda_0 t}$$

$$\vdots$$

where

$$\begin{aligned}(A-\lambda_0 I)\,U_1 &= V\\ (A-\lambda_0 I)\,U_2 &= U_1\\ (A-\lambda_0 I)\,U_3 &= U_2\\ \text{etc.}\end{aligned}$$

There is an entirely different approach to the study of solutions of a homogeneous linear equation $dX/dt = AX$ which is based on further developments in the theory of matrices. The corresponding scalar equation $dx/dt = ax$ has the solution $x = e^{at}$. It is natural to wonder if it is possible to solve the vector equation in an analogous fashion, writing

$$X = e^{At}$$

where we define the exponential symbol as

$$e^{At} = I + At + A^2\frac{t^2}{2!} + A^3\frac{t^3}{3!} + \cdots$$

and where $A^2, A^3, \ldots$ are the various powers of the matrix A, namely, $A^2 = AA$, $A^3 = AA^2$, etc. This can in fact be done, and it leads to an extremely elegant treatment of linear systems.

We choose not to discuss this, but refer the reader to the more advanced texts listed in the Bibliography. Since a random matrix A will almost certainly have distinct eigenvalues (see Exercise 25), the simple treatment explained in this section is usually sufficient; as a practical matter, one does not often encounter situations requiring the more elaborate theory. Other computational questions, such as the difficulty in finding the eigenvalues of a large matrix, take precedence, and we therefore go no further into this topic.

Exercises

Note: The abbreviation "eiv pair" means an eigenvalue and associated eigenvector for a given matrix A.

1. (a) Show that $\lambda_1 = 4$, $\quad v_1 = \begin{bmatrix} 2 \\ -1 \end{bmatrix}$,

   and $\lambda_2 = -1$, $\quad v_2 = \begin{bmatrix} 4 \\ 3 \end{bmatrix}$,

   are eiv pairs for the matrix $\begin{bmatrix} 2 & -4 \\ -\frac{3}{2} & 1 \end{bmatrix}$.

(b) Show that $\lambda_1 = 3+5i$, $v_1 = \begin{bmatrix} 1 \\ i \end{bmatrix}$,

and $\lambda_2 = 3-5i$, $v_2 = \begin{bmatrix} 1 \\ -i \end{bmatrix}$,

are eiv pairs for the matrix $\begin{bmatrix} 3 & 5 \\ -5 & 3 \end{bmatrix}$.

2. (a) Show that $\lambda_1 = 1$, $v_1 = \begin{bmatrix} 1 \\ -1 \end{bmatrix}$,

and $\lambda_2 = 3$, $v_2 = \begin{bmatrix} 3 \\ -1 \end{bmatrix}$,

are eiv pairs for the matrix $\begin{bmatrix} 4 & 3 \\ -1 & 0 \end{bmatrix}$.

(b) Show that the matrix $\begin{bmatrix} 2 & 1 \\ -1 & 4 \end{bmatrix}$ has an eigenvalue $\lambda = 3$ of multiplicity 2 but only one eigenvector $v = \begin{bmatrix} 1 \\ 1 \end{bmatrix}$.

3. Show that $\lambda_1 = 0$, $v_1 = \begin{bmatrix} 1 \\ 2 \\ -3 \end{bmatrix}$

$\lambda_2 = 1$, $v_2 = \begin{bmatrix} 1 \\ 1 \\ -2 \end{bmatrix}$

and $\lambda_3 = 2$, $v_3 = \begin{bmatrix} 1 \\ 0 \\ -3 \end{bmatrix}$

are eiv pairs for the matrix $\begin{bmatrix} 2 & -1 & 0 \\ 3 & 0 & 1 \\ -3 & 3 & 1 \end{bmatrix}$

4. Find the eiv pairs for the matrix $\begin{bmatrix} 2 & -1 \\ 0 & 3 \end{bmatrix}$.

5. Find the eiv pairs for

(a) $\begin{bmatrix} 7 & -1 \\ 3 & 3 \end{bmatrix}$ (b) $\begin{bmatrix} 5 & 6 \\ 2 & 1 \end{bmatrix}$

6. Find the eiv pairs for the matrix

$$\begin{bmatrix} 2 & -1 & 4 \\ 0 & 3 & 1 \\ 0 & 0 & -1 \end{bmatrix}$$

7. Check that the eigenvalues of the matrix

$$A = \begin{bmatrix} 0 & 1 & 1 \\ 2 & 0 & -1 \\ 4 & 4 & 0 \end{bmatrix}$$

are 2, $-1+i$, and $-1-i$, and find the associated eigenvectors.

8. If A is a square matrix with real entries, and λ and V are an associated eiv pair, show that $\bar{\lambda}$ (the complex conjugate of λ) and $\bar{V}$ are also an associated eiv pair, where $\bar{V}$ is the vector whose respective entries are the complex conjugates of the entries of V.

9. Solve the differential equation $dX/dt = AX$ if

$$A = \begin{bmatrix} 2 & -1 & 4 \\ 0 & 3 & 1 \\ 0 & 0 & -1 \end{bmatrix}$$

10. Solve the system of differential equations

$$\frac{dx}{dt} = 2x - y$$

$$\frac{dy}{dt} = 3x + z$$

$$\frac{dz}{dt} = -3x + 3y + z$$

11. Solve the system

$$\frac{dx}{dt} = -2x + 3y$$

$$\frac{dy}{dt} = -2y + z$$

$$\frac{dz}{dt} = z$$

12. For

$$A = \begin{bmatrix} 2 & 0 & 1 \\ 0 & 1 & 0 \\ 1 & 0 & 2 \end{bmatrix}$$

find three independent eiv pairs.

13. Solve $dX/dt = AX$ with

$$A = \begin{bmatrix} 0 & 1 & -1 \\ 2 & 0 & 0 \\ -3 & -\frac{5}{2} & 1 \end{bmatrix}$$

14. (a) If

$$A = \begin{bmatrix} 3 & -4 \\ 1 & -2 \end{bmatrix}$$

show that

$$\frac{dX}{dt} = AX - \begin{bmatrix} 6t^2+1 \\ 2t^2+3 \end{bmatrix}$$

has a solution of the form $X = b_0 + b_1 t + b_2 t^2$, where the b_i are constant vectors.

(b) Find a solution obeying the initial condition $X(0) = \begin{bmatrix} 0 \\ 0 \end{bmatrix}$.

15. Find a solution obeying $X(0) = \begin{bmatrix} 1 \\ 1 \end{bmatrix}$

for the equation

$$\frac{dX}{dt} = \begin{bmatrix} 2 & 3 \\ 4 & 1 \end{bmatrix} X - \begin{bmatrix} 3 \\ 1 \end{bmatrix} e^{-t}$$

16. Knowing that $\lambda_1 = -3$, solve the system

$$\frac{du}{dt} = -9u + 2v$$

$$\frac{dv}{dt} = -2v - 2w$$

$$\frac{dw}{dt} = 9u - 9w$$

with initial conditions $u(0) = 1$, $v(0) = 0$, $w(0) = 2$.

17. Solve the equation

$$\frac{dX}{dt} = \begin{bmatrix} 5 & 2 & 1 \\ 4 & 7 & -4 \\ 0 & 0 & 9 \end{bmatrix} X$$

18. If

$$A = \begin{bmatrix} 2 & 0 \\ 0 & 3 \end{bmatrix}$$

(a) Show that

$$A^n = \begin{bmatrix} 2^n & 0 \\ 0 & 3^n \end{bmatrix}$$

(b) Verify that

$$e^{At} = \begin{bmatrix} e^{2t} & 0 \\ 0 & e^{3t} \end{bmatrix}$$

19. If

$$A = \begin{bmatrix} 2 & 1 \\ 0 & 3 \end{bmatrix}$$

(a) Verify by induction that

$$A^n = \begin{bmatrix} 2^n & 3^n - 2^n \\ 0 & 3^n \end{bmatrix}$$

(b) Show that

$$e^{At} = \begin{bmatrix} e^{2t} & e^{3t} - e^{2t} \\ 0 & e^{3t} \end{bmatrix}$$

20. If

$$A = \begin{bmatrix} 2 & 1 \\ 0 & 2 \end{bmatrix}$$

verify that

(a) $A^n = \begin{bmatrix} 2^n & n2^{n-1} \\ 0 & 2^n \end{bmatrix}$ (b) $e^{At} = \begin{bmatrix} e^{2t} & te^{2t} \\ 0 & e^{2t} \end{bmatrix}$

21. If

$$A = \begin{bmatrix} 2 & -3 \\ 0 & 0 \end{bmatrix}$$

find A^n and compute e^{At}.

22. Let

$$A = \begin{bmatrix} 1 & 1 & 0 \\ 0 & 1 & -1 \\ 0 & 0 & -1 \end{bmatrix}$$

Find a formula for A^n, depending on whether n is even or odd.

*23. If A is an n-by-n matrix that has n distinct eigenvalues, show that the associated eigenvectors are linearly independent.

*24. Let A be an n-by-n matrix with n distinct eigenvalues. Prove that there is a nonsingular matrix P such that $P^{-1}AP$ is diagonal.

25. (a) Would you expect a randomly chosen 10-by-10 matrix A to be nonsingular, that is, $\det(A) \neq 0$?
    (b) Would you expect it to have ten distinct eigenvalues?

26. What is the relationship between the eigenvalues of a matrix A and the eigenvalues of A^2?

27. Show that if A is a matrix with only real entries and if $U(t)+iV(t)$ is a solution of the vector equation $dX/dt = AX$, then so also are the vector functions $U(t)$ and $V(t)$.

28. If

$$\frac{dX}{dt} = \begin{bmatrix} 0 & 1 & 1 \\ 2 & 0 & -1 \\ 4 & 4 & 0 \end{bmatrix} X$$

    (a) Find $X = \phi(t)$.
    (b) Find the particular solution so that

$$\phi(0) = \begin{bmatrix} 0 \\ 1 \\ -1 \end{bmatrix}$$

29. Solve the Two-Tank Problem (6.1) using the method of this section.

30. The Two-Mass Problem (6.2) can be expressed as

$$\begin{cases} \dfrac{dx_1}{dt} = x_2 \\ \dfrac{dx_2}{dt} = -11x_1 + 6x_3 \\ \dfrac{dx_3}{dt} = x_4 \\ \dfrac{dx_4}{dt} = -6x_3 + 6x_1 \end{cases}$$

    Use the method of this section to solve for $x_1(t)$ and $x_3(t)$.

31. Three tanks, each holding 100 gallons of liquid, are interconnected as shown in Figure 6.5. The flow is indicated by arrows and rates shown in

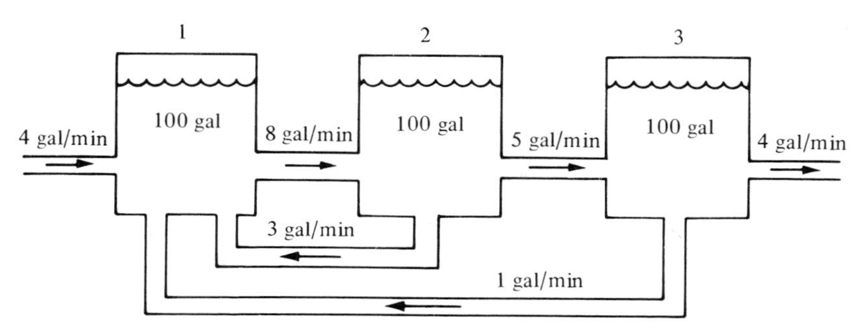

Figure 6.5

the figure. Initially, tank 1 is filled with pure water, tank 2 has 10 grams of red dye in its 100 gallons of water, and tank 3 has 20 grams of green dye in its 100 gallons. If x_i is the amount of red dye in tank i, and y_i is the amount of green dye in tank i, write the equation governing the flow. (Do not solve.)

7

Models Again

Summary

This chapter is a return in part to the spirit of Chapter 1, discussing the way in which differential equations arise out of real problems; it is also a look toward further developments in mathematics, particularly those dealing with partial differential equations and orthogonal (Fourier) series. Our purpose in introducing these topics is not to develop them systematically but rather to motivate them and to show how the tools of earlier chapters can be used to answer interesting and important questions.

Thus, Section 7.1 contains a treatment of Kepler's laws and their relationship to Newton's gravitational theory; Section 7.2 deals with waves, and specifically with the motion of a freely vibrating string and a freely vibrating drumhead. This section provides a very brief introduction to some aspects of partial differential equations with emphasis upon separation of variables. (The standard abbreviation for partial differential equation is PDE). In Section 7.3, we treat three very specific applied problems which illustrate the use of differential equations to optimize a design. In the first, we want a water clock with uniform hour markings; in the second, we want a mirror for a solar furnace; in the third, we want the fastest ski slide. (The last two conceal classical illustrations, the focal property of a parabola and the brachistochrone property of the cyloid.) The last section contains the details of an analytic treatment of one of the differential equations which has been reoccurring as an example throughout the book, namely, $dy/dt = t^2 + y^2$. The results obtained in this section finally explain the peculiar behavior that we have noted earlier in the course of trying to solve this equation by other methods.

We call attention to the collection of word problems that follows this chapter. These problems vary in difficulty and in the background they demand of the student. All can be done by the methods explained in the text, and many call for skill in building good models.

This list, and the chapter itself, can be used in a variety of ways. Since the sections are independent of one another, they can serve as a basis for individual study or group projects. Each section contains material which can be extended further, using for example the books in the appended list of references. (For example, Section 7.3 contains enough about the calculus of variations to permit the reader to attack a number of simple problems on his own.)

7.1 Kepler and Newton

Johannes Kepler (1571–1630) succeeded Tycho Brahe in Prague as astronomer to Emperor Rudolph and, using the excellent observational data that had been obtained at the observatory over many years, made a study of the orbits of the planets. Kepler believed that there must be other mathematical and geometric regularities beyond those already discovered by Copernicus. In 1609, he announced that he had discovered three empirical laws. The first asserted that the path of a planet is not to be looked at as a combination of various circular motions, but is in fact a geometric ellipse with the sun at one focus; the second assertion was that the speed of a moving planet is not constant, but that the line from the sun to the planet sweeps out equal areas in equal times; the third assertion was that there is a systematic relationship between the period of each of the planets and its distance from the sun, so that (measured in the right units) the cube of the average distance is the square of the period.

Kepler's Laws, as they came to be called, arose out of Tycho Brahe's empirical observations and described striking mathematical patterns in the motions of the planets. However, they left unanswered many questions. What caused the motion of the planets? What lay behind and explained these mathematical relationships? Isaac Newton was born in 1642 and entered Trinity College, Cambridge, in 1661. Already, by 1650, various ideas were current about the forces that drove the planets about the sun, and the concept of "action at a distance" had been proposed. Light, of course, was the most familiar example of such a phenomenon, and the fact that the intensity of light seemed to vary inversely with the square of the distance was also appreciated. Cambridge University closed because of the plague, and Newton spent 1665–1666 on his mother's farm, studying on his own. These were vintage years, producing the kernel of his work in mechanics, optics, and mathematics. During this period, he worked out most of the details of his theory of gravitation, showing that an inverse square law of attraction explained Kepler's Laws. Using modern methods, we will now retrace this discovery.

We first examine the meaning of Kepler's Second Law. We suppose that the motion of a planet P is given by some formula which, in polar coordinates, takes the form

$$r = f(\theta)$$

(See Figure 7.1.) As the angle increases from zero to θ, the line joining the origin to P sweeps out an area that is given by the integral

$$A(\theta) = \int_0^\theta \tfrac{1}{2}r^2\,d\theta = \tfrac{1}{2}\int_0^\theta f^2$$

Kepler's Second Law states that the rate of change of $A(\theta)$ with respect to time is constant—that is, $dA/dt = \text{const}$. Since $dA/d\theta = \frac{1}{2}f(\theta)^2 = \frac{1}{2}r^2$, we have

$$\text{Const} = \frac{dA}{dt} = \frac{dA}{d\theta}\frac{d\theta}{dt} = \tfrac{1}{2}r^2\frac{d\theta}{dt}$$

This argument is reversible, so we arrive at a simple mathematical formulation for the Second Law:

$$r^2\frac{d\theta}{dt} = C \tag{7.1}$$

We can obtain a second equivalent form that is more enlightening by recasting these relations in Cartesian coordinate form. We start from the formula for polar coordinates:

$$\begin{aligned} x &= r\cos\theta \\ y &= r\sin\theta \end{aligned} \tag{7.2}$$

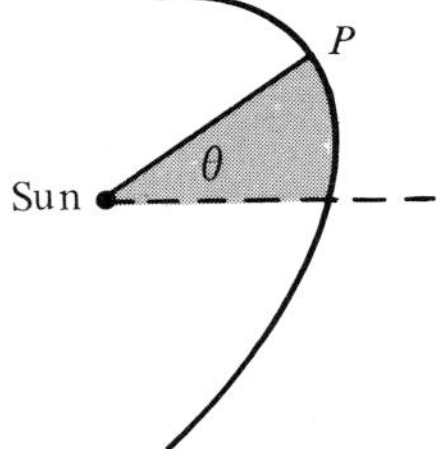

Figure 7.1

and differentiate these with respect to t, obtaining

$$\frac{dx}{dt} = \cos\theta \frac{dr}{dt} - r\sin\theta \frac{d\theta}{dt}$$

$$\frac{dy}{dt} = \sin\theta \frac{dr}{dt} + r\cos\theta \frac{d\theta}{dt}$$

Then we find by combining these with (7.2) and (7.1) that

$$x\frac{dy}{dt} - y\frac{dx}{dt} = r^2\frac{d\theta}{dt} = C \tag{7.3}$$

If we differentiate this, remembering that C is constant, we obtain another version of the Second Law:

$$\frac{d}{dt}\left\{x\frac{dy}{dt} - y\frac{dx}{dt}\right\} = 0$$

Carrying out the differentiation, this becomes

$$x\frac{d^2y}{dt^2} - y\frac{d^2x}{dt^2} = 0 \tag{7.4}$$

Now, set $\rho = (1/x)\, d^2x/dt^2$; this will be a (scalar) function of t. Using (7.4), we have a pair of differential equations

$$\begin{aligned} \frac{d^2x}{dt^2} &= \rho(t)x \\ \frac{d^2y}{dt^2} &= \rho(t)y \end{aligned} \tag{7.5}$$

If we set

$$X = \begin{bmatrix} x \\ y \end{bmatrix}$$

then this system becomes $d^2X/dt^2 = \rho(t)X$, which merely states that the acceleration vector for the moving planet is always a scalar multiple of the position vector. If now we make the Newtonian assumption that force is proportional to acceleration, we see that Kepler's Second Law has been shown to be equivalent to the following statement:

Each planet moves as though it were under the action of a force of attraction which always acts along the line joining the planet to the sun.

This statement is usually called the "hypothesis of a central force," and it is thus equivalent to "equal areas in equal times," and has nothing to do with the precise form of the central force.

We next show that Kepler's First Law—that the orbits are ellipses—is equivalent to the assumption that the central force is one which depends only upon the distance r from the planet to the sun, and that it varies inversely with the square of the distance. Let us first see what this assumption says about the nature of the function $\rho(t)$. If the force of attraction at P has magnitude k/r^2 and acts along the line from P toward the sun (see Figure 7.2), then its x and y components are

$$F_x = -\frac{k}{r^2}\cos\theta$$

$$F_y = -\frac{k}{r^2}\sin\theta$$

This would lead to

$$\text{Mass}\,\frac{d^2x}{dt^2} = -\frac{k}{r^2}\cos\theta = -\frac{k}{r^3}x$$

$$\text{Mass}\,\frac{d^2y}{dt^2} = -\frac{k}{r^2}\sin\theta = -\frac{k}{r^3}y$$

Comparing this with equations (7.5), we see that the inverse square law for gravitation is equivalent, mathematically, to the assertion that $\rho(t) = -K/r^3$, where K is a constant, $K = k/\text{mass}$.

Without making this assumption yet, let us obtain another deduction from the simple equation (7.3). If we differentiate the equation $r^2 = x^2 + y^2$ with respect to t, we obtain

$$r\frac{dr}{dt} = x\frac{dx}{dt} + y\frac{dy}{dt} \tag{7.6}$$

Recall equation (7.3):

$$C = x\frac{dy}{dt} - y\frac{dx}{dt}$$

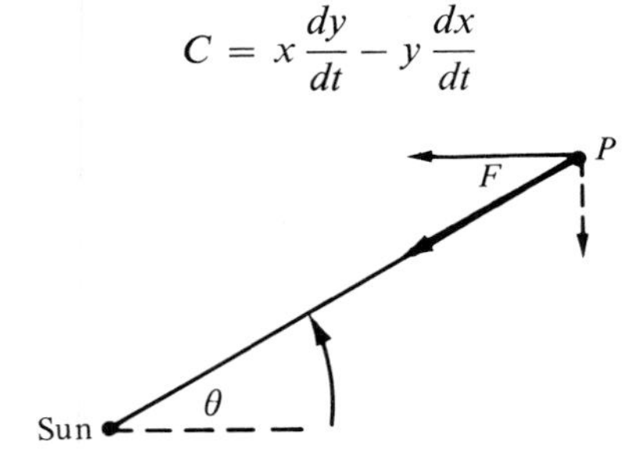

Figure 7.2

Squaring these and adding, we have

$$r^2\left(\frac{dr}{dt}\right)^2 + C^2 = (x^2+y^2)\left[\left(\frac{dx}{dt}\right)^2 + \left(\frac{dy}{dt}\right)^2\right]$$

or

$$\left(\frac{dr}{dt}\right)^2 + \frac{C^2}{r^2} = \left(\frac{dx}{dt}\right)^2 + \left(\frac{dy}{dt}\right)^2 \tag{7.7}$$

which is a statement about the speed of the moving planet. Now return to (7.6) and differentiate again, obtaining

$$\left(\frac{dr}{dt}\right)^2 + r\frac{d^2r}{dt^2} = \left(\frac{dx}{dt}\right)^2 + x\frac{d^2x}{dt^2} + \left(\frac{dy}{dt}\right)^2 + y\frac{d^2y}{dt^2}$$

and use (7.7) and (7.5) to simplify this, to obtain the simple-appearing equation

$$\frac{d^2r}{dt^2} = \frac{C^2}{r^3} + r\rho(t) \tag{7.8}$$

This is the general differential equation for the motion of a particle in any central force field, and its solution will give the distance r as a function of time. If we want information about the shape of the orbit of the particle, we must proceed further.

Suppose we set $r = 1/u$. (An unmotivated trick!) Then

$$\begin{aligned}\frac{dr}{dt} &= -\frac{1}{u^2}\frac{du}{dt} = -\frac{1}{u^2}\frac{du}{d\theta}\frac{d\theta}{dt} = -r^2\frac{du}{d\theta}\frac{d\theta}{dt}\\ &= -C\frac{du}{d\theta}\end{aligned}$$

using (7.1). Now, differentiate again:

$$\begin{aligned}\frac{d^2r}{dt^2} &= -C\frac{d}{dt}\left\{\frac{du}{d\theta}\right\} = -C\frac{d}{d\theta}\left\{\frac{du}{d\theta}\right\}\frac{d\theta}{dt}\\ &= -C\frac{d^2u}{d\theta^2}\frac{d\theta}{dt} = -C^2u^2\frac{d^2u}{d\theta^2}\end{aligned}$$

[Here again we have used (7.1).] If we substitute these expressions into (7.8), we obtain the differential equation

$$\frac{d^2u}{d\theta^2} + u = -\frac{\rho(t)}{C^2u^3} \tag{7.9}$$

Note that if the central force depends only on the distance r, ρ can be expressed in terms of u, and (7.9) becomes a second-order differential equation in u, with θ as the independent variable. Its solution will give an expression for u in terms of θ, and thus for r in terms of θ. This will be the desired formula for the path $r = f(\theta)$ of the moving planet.

Let us see how this works in the case of an inverse square law for gravitation. As we have seen, this is equivalent to setting $\rho(t) = -K/r^3$. The equation (7.9) then becomes

$$\frac{d^2u}{d\theta^2} + u = \frac{K}{C^2}$$

Rewriting the equation as $(D^2+1)\{u\} = K/C^2$, we see that the null space consists of the functions $B\cos(\theta-\beta)$, where B and β are arbitrary constants, and a particular solution is $u = K/C^2$. Thus

$$u = \frac{K}{C^2} + B\cos(\theta-\beta)$$

Replacing u by $1/r$, we obtain the equation of an orbit as

$$r = \frac{1}{a+B\cos(\theta-\beta)}$$

which is recognizable as the polar coordinate equation for a conic section. (Although the orbits of the planets are ellipses, there is nothing in theory that prevents them from being parabolas or hyperbolas; however, a satellite on a hyperbolic or parabolic path could make only one trip past the sun.)

Exercises

1. Explain why Kepler's Laws require that a planet move more slowly when it is far from the sun than when it is close to the sun.
2. Check the calculations leading to (7.8).
3. Show that the assumption that the path of a planet is the ellipse $r = 1/(2-\cos\theta)$, combined with equation (7.9), means that $\rho(t) = -2C^2/r^3$.
4. Show that if a central force of *repulsion* varies inversely with the square of the distance, then the orbits are hyperbolas. Can you think of a physical example of this type?

7.2 Waves

If a long rope is stretched between two poles and you stand near one end and give the rope a flip, you produce a local distortion of the rope which travels rapidly along the rope to the other end. We are interested in understanding this mathematically, and more generally in studying the general behavior of vibrations of a stretched string or a stretched membrane such as a drumhead. This study will lead us to the solution of certain partial differential equations.

Many physical phenomena lead to mathematical models which involve rates of change of a quantity in several different directions, or with respect both to time and distance. Such rates of change must be viewed as partial derivatives, since there is more than one independent variable in such problems. Thus, the resultant models are **partial differential equations** rather than the ordinary differential equations that have been solved in preceding chapters. Certain standard equations occur in many different models and have been studied extensively. The equation

$$\frac{\partial^2 u}{\partial x^2}+\frac{\partial^2 u}{\partial y^2}+\frac{\partial^2 u}{\partial z^2}=\frac{\partial u}{\partial t} \tag{7.10}$$

is used to describe the flow of heat in a solid object, and the equation

$$\frac{\partial^2 u}{\partial x^2}+\frac{\partial^2 u}{\partial y^2}+\frac{\partial^2 u}{\partial z^2}=\frac{1}{c^2}\frac{\partial^2 u}{\partial t^2} \tag{7.11}$$

is used for a simplified model of a sound wave in air. We will look at certain special ways to solve such equations. Our investigation will lead to the study of ordinary differential equations of the sort that are called two-point boundary-value problems, and these in turn will lead in a natural way to Fourier series. The purpose is not to present a thorough treatment of any of these concepts but merely to show how they arose out of an effort to answer certain interesting questions.

Let us now return to our rope and find a mathematical description of the phenomenon we see. We think of the undisturbed rope as the x-axis, and use u to measure the vertical displacement of the rope at any point. If the rope is given a flip, its instantaneous shape could be described by writing $u = \phi(x)$, where ϕ is some specific function. We observe that this shape moves down the rope in the manner shown in Figure 7.3. This can be described mathematically by writing

$$u = u(x,t) = \phi(x-ct)$$

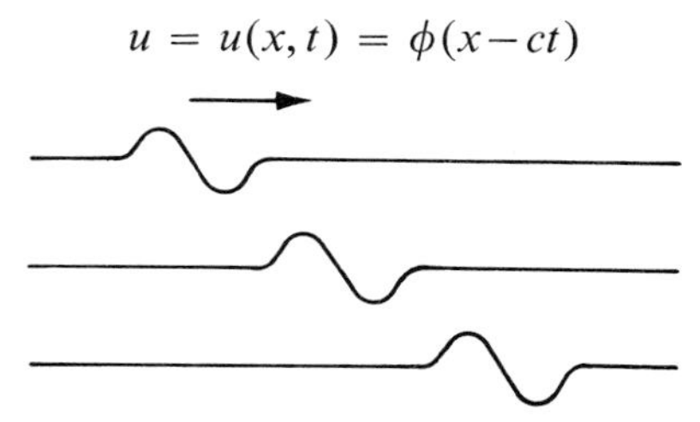

Figure 7.3

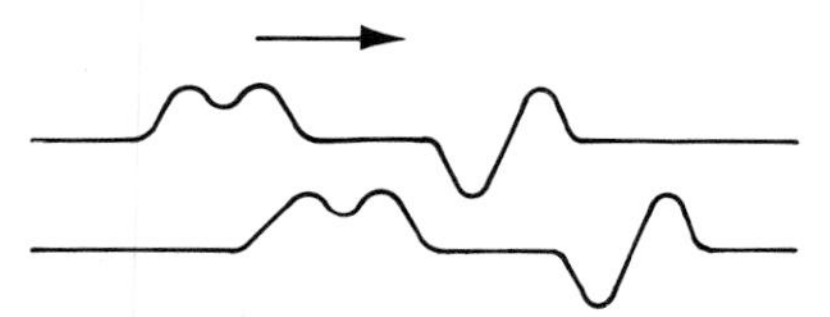

Figure 7.4

This tells us what the vertical displacement of the rope is at each position x on the rope and at each moment t. The number c determines the velocity with which the traveling wave moves down the rope. For example, if the original shape of the wave at time $t = 0$ had a maximum at the point x_0, so that this is the highest point on the graph of ϕ, then $u(x, t)$ will achieve this same height when $x - ct = x_0$, that is, at the point $x = x_0 + ct$. Clearly, as t increases, this point x moves to the right at velocity c.

Two different waves can be combined. For example, if we take two functions ϕ and ψ, and write

$$u = \phi(x-ct) + \psi(x-ct)$$

then we have combined two waves, both moving toward the right. If $F = \phi + \psi$, then this equation can be written merely as $u = F(x-ct)$, and we see that the result can be treated as a single complicated wave traveling toward the right along the rope. (See Figure 7.4.)

The same analysis shows that the formula

$$u = \psi(x+ct)$$

describes a traveling wave that moves along the rope toward the *left*. What will the combined equation

$$u = \phi(x-ct) + \psi(x+ct) \tag{7.12}$$

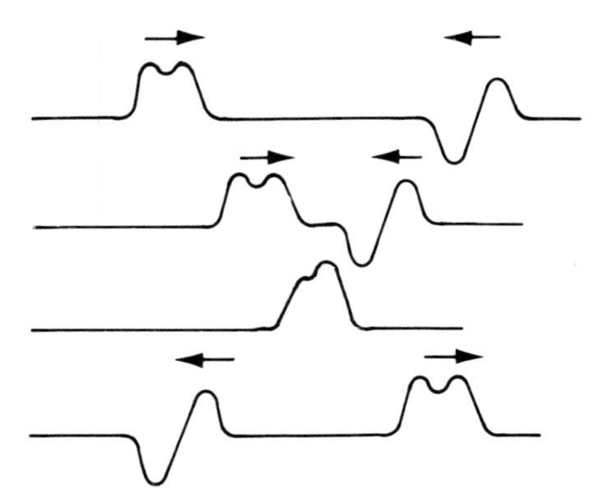

Figure 7.5

describe? It should be two superimposed traveling waves, one moving along the rope toward the right and one toward the left. In Figure 7.5, we show successive pictures of the result.

THEOREM 7.1 *Assuming that* ϕ *and* ψ *are of class* $\mathcal{C}^2$, *the function* u *given in* (7.12) *must obey the partial differential equation*

$$\frac{\partial^2 u}{\partial x^2} - \frac{1}{c^2}\frac{\partial^2 u}{\partial t^2} = 0 \tag{7.13}$$

Conversely, any solution of this equation for all x and t must have the form (7.12).

The first half of this is a routine exercise in differentiation. From (7.12) we have

$$\frac{\partial u}{\partial x} = \phi'(x-ct) + \psi'(x+ct)$$

$$\frac{\partial^2 u}{\partial x^2} = \phi''(x-ct) + \psi''(x+ct)$$

and

$$\frac{\partial u}{\partial t} = (-c)\phi'(x-ct) + (c)\psi'(x+ct)$$

$$\frac{\partial^2 u}{\partial t^2} = (-c)^2\phi''(x-ct) + (c)^2\psi''(x+ct)$$

Substituting these into (7.13), we see that the equation is satisfied no matter what the functions ϕ and ψ are. (One important difference between partial differential equations and ordinary differential equations is that the solution of a partial differential equation often involves arbitrary functions, whereas the solution of an ordinary differential equation involves only constants.)

The proof of the second half of the theorem is more complicated, and involves a special lemma whose proof illustrates the remark just made. We introduce two new variables, $y = x-ct$, $z = x+ct$, and set

$$u = u(x,t) = u\left(\frac{y+z}{2}, \frac{z-y}{2c}\right) = V(y,z)$$

Then, using the chain rule of differentiation,

$$\frac{\partial u}{\partial x} = \frac{\partial V}{\partial y}\frac{\partial y}{\partial x} + \frac{\partial V}{\partial z}\frac{\partial z}{\partial x} = (1)\frac{\partial V}{\partial y} + (1)\frac{\partial V}{\partial z}$$

and

$$\frac{\partial^2 u}{\partial x^2} = \frac{\partial^2 V}{\partial y^2}\frac{\partial y}{\partial x} + \frac{\partial}{\partial z}\left(\frac{\partial V}{\partial y}\right)\frac{\partial z}{\partial x} + \frac{\partial}{\partial y}\left(\frac{\partial V}{\partial z}\right)\frac{\partial y}{\partial x} + \frac{\partial}{\partial z}\left(\frac{\partial V}{\partial z}\right)\frac{\partial z}{\partial x}$$

$$= \frac{\partial^2 V}{\partial y^2} + 2\frac{\partial^2 V}{\partial y\,\partial z} + \frac{\partial^2 V}{\partial z^2}$$

Likewise,

$$\frac{\partial u}{\partial t} = \frac{\partial V}{\partial y}\frac{\partial y}{\partial t} + \frac{\partial V}{\partial z}\frac{\partial z}{\partial t} = (-c)\frac{\partial V}{\partial y} + (c)\frac{\partial V}{\partial z}$$

and

$$\frac{\partial^2 u}{\partial t^2} = (-c)^2\frac{\partial^2 V}{\partial y^2} - 2c^2\frac{\partial^2 V}{\partial y\,\partial z} + (c)^2\frac{\partial^2 V}{\partial z^2}$$

Substituting these into the partial differential equation (7.13), we obtain

$$0 = \frac{\partial^2 u}{\partial x^2} - \frac{1}{c^2}\frac{\partial^2 u}{\partial t^2}$$

$$= 4\frac{\partial^2 V}{\partial y\,\partial z}$$

Thus, the function V must satisfy the simpler partial differential equation $\partial^2 V/\partial y\,\partial z = 0$. We now invoke the following simple result.

LEMMA **7.1** *Any function V that satisfies the partial differential equations $\partial V/\partial y\,\partial z = 0$ for all y and z must have the form $V = \phi(y) + \psi(z)$, where ϕ and ψ are arbitrary functions.*

Proof Since

$$\left(\frac{\partial}{\partial y}\right)\left(\frac{\partial V}{\partial z}\right) = 0$$

$\partial V/\partial z$ must be a function of y and z that is independent of y. Thus, $\partial V/\partial z = b(z)$. Let $\psi(z)$ be the indefinite integral of $b(z)$. Then,

$$\frac{\partial}{\partial z}\{V - \psi(z)\} = b(z) - b(z) = 0$$

Hence, $V - \psi(z)$ is independent of z, and must have the form $\phi(y)$. Hence, $V = \phi(y) + \psi(z)$.

Since $u(x, t) = V(y, z)$, we have shown that the most general solution of (7.13) is $u = \phi(y) + \psi(z) = \phi(x - ct) + \psi(x + ct)$.

We now digress to show that a physical analysis of the motion of a stretched rope also leads, with some simplifying assumptions, to the *same* partial differential equation (7.13). This means that *every* motion of the rope is described by (7.12)!

Consider a thin, flexible rope stretched between two points and slightly displaced from the equilibrium position, as shown in Figure 7.6. We ignore the effects of gravity, and assume that the motion of the rope is produced solely by the tension force in the rope. We assume that the tension T is constant in magnitude but changes in direction and at any point acts along the tangent to the rope. At the moment t_0 we are examining the rope and the shape is given by $u = g(x, t_0) = h(x)$. We take two nearby positions corresponding to x_1 and x_2, and examine the vertical forces acting on the segment $P_1 P_2$. The slope at P_1 is $h'(x_1)$, and at P_2 is $h'(x_2)$. Thus, the vertical component of the tension is roughly $-Th'(x_1)$ at P_1 and $Th'(x_2)$ at P_2. The resultant force on the segment $P_1 P_2$ is

$$Th'(x_2) - Th'(x_1) = T(h'(x_2) - h'(x_1))$$

If we use the mean-value theorem, we can rewrite the right-hand side as $Th''(\bar{x})\,\Delta x$, where $\Delta x = x_2 - x_1$, and $\bar{x}$ is a value between x_1 and x_2. Restated, the vertical force on the segment $P_1 P_2$ is now given by

$$Th''(\bar{x})\,\Delta x, \quad \text{which can also be written as} \quad T\frac{\partial^2 u}{\partial x^2}\,\Delta x$$

If the density of the rope is k, the mass of the segment $P_1 P_2$ is essentially $k\,\Delta x$. The vertical acceleration of the segment is $\partial^2 u/\partial t^2$, computed for example at the point $\bar{x}$. Thus, the standard relation *force = mass times acceleration* leads to

$$T\frac{\partial^2 u}{\partial x^2}\,\Delta x = k\,\Delta x \quad \frac{\partial^2 u}{\partial t^2}$$

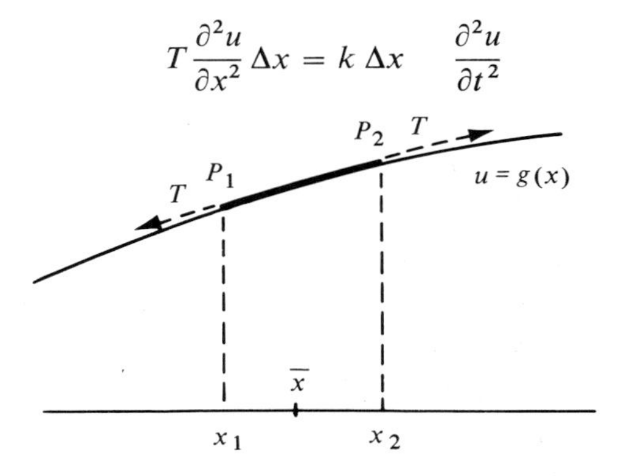

Figure 7.6

and canceling Δx, we arrive at the desired differential equation

$$\frac{\partial^2 u}{\partial x^2} - \frac{k}{T}\frac{\partial^2 u}{\partial t^2} = 0 \tag{7.14}$$

The argument given above is not a mathematical argument, at least not a rigorous one; we have ignored many aspects of the situation and made many crude approximations. It should be regarded merely as plausible or suggestive, something that may justify taking (7.14) as a model for the violin string. [More rigorous reasoning leads to a much more complicated differential equation, taking into account the change in tension as the string stretches and changes length, the fact that the motion of a point on the string may not be exactly vertical, etc. However, equation (7.14) remains a usable approximation to this more accurate equation.]

Suppose we now attempt to answer a specific question about the vibrating string. At $t = 0$, let us pluck the string and try to predict the results. More specifically, let us assume that at $t = 0$ the string is motionless but distorted into a particular shape given by a function f as shown in Figure 7.7. We also assume that the length of the string is L, and that the two ends are kept fixed.

All of these statements can be translated into conditions on the unknown function $u(x, t)$. Table 7.1 gives the translations.

Table 7.1

At $t = 0$, $u = f(x)$.	The initial shape is f.
At $t = 0$, $u_t = 0$.	Initially, the string is motionless.
For all $t \geq 0$, $u(0, t) = 0$.	The left end of the string is fixed.
For $t \geq 0$, $u(L, t) = 0$.	The right end of the string is fixed.
$u_{xx} - (1/c^2)\, u_{tt} = 0$.	The string vibrates freely.

Using Theorem 7.1, we know that a general solution of the differential equation is given by

$$u(x, t) = \phi(x - ct) + \psi(x + ct) \tag{7.15}$$

By adding a constant to ϕ and subtracting the same constant from ψ, we can assume that the functions ϕ and ψ take the same value at some point x_0 between zero and L. We now want to choose ϕ and ψ so that each of the required conditions given in Table 7.1 is satisfied.

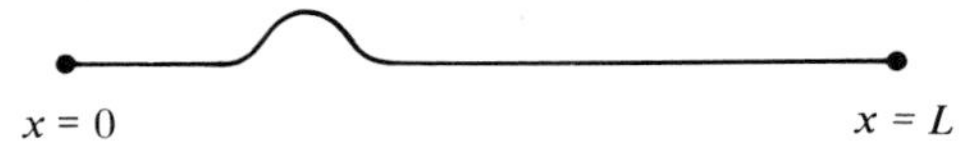

Figure 7.7

The first two conditions require that

$$\text{(7.16)} \qquad \phi(x) + \psi(x) = f(x) \qquad 0 \leq x \leq L$$

$$\text{(7.17)} \qquad -c\phi'(x) + c\psi'(x) = 0 \qquad 0 \leq x \leq L$$

The second of these yields $\phi' = \psi'$, and since $\phi(x_0) = \psi(x_0)$, we have $\phi = \psi$. Using this in (7.16), we obtain

$$\phi(x) = \tfrac{1}{2}f(x) \qquad 0 \leq x \leq L$$

and thus arrive at the desired solution

$$\text{(7.18)} \qquad u(x,t) = \tfrac{1}{2}[f(x-ct) + f(x+ct)]$$

However, this solution is as yet incomplete, since the function f was defined only on the interval $[0, L]$ and (7.18) would require that f be defined for all real numbers if we are to use formula (7.18) to calculate the values of u for large values of t. We also note that we as yet have done nothing about satisfying the remaining two conditions given in the table.

The left endpoint condition was that $u(0,t) = 0$ for all $t \geq 0$. If we set $x = 0$ in (7.18) we see that we must satisfy the functional equation

$$\text{(7.19)} \qquad 0 = f(-ct) + f(ct) \qquad t \geq 0$$

The right endpoint condition was that $u(L,t) = 0$ for $t \geq 0$. This becomes the requirement that

$$\text{(7.20)} \qquad 0 = f(L-ct) + f(L+ct) \qquad t \geq 0$$

We now see that equations (7.19) and (7.20) are exactly what we need in order to extend the definition of f in the proper way from the interval $[0, L]$ to the entire set of real numbers. For (7.19) permits us to determine the value of f at any negative number, once we have determined f at all the positive numbers, while (7.20) enables us to find the values of f on each of the intervals $[L, 2L]$, $[2L, 3L]$, $[3L, 4L]$, etc., in each case using the values of f determined on the previous interval.

A simple illustration may clarify this. Suppose that the initial shape was given to be that in Figure 7.7. Then, the extended definition of f for all real numbers, according to equations (7.19) and (7.20), will be as shown in Figure 7.8. (See Exercise 2.)

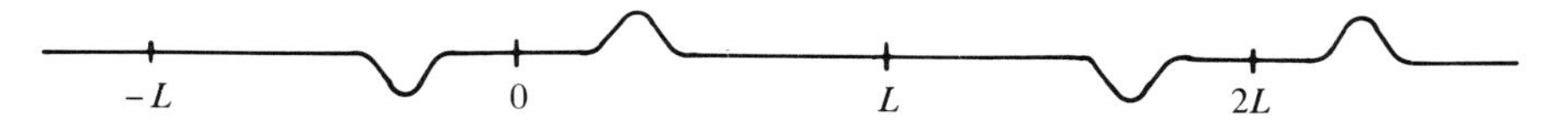

Figure 7.8

What would be the behavior of the corresponding wave as predicted by (7.18)? From this formula and the discussion we have given earlier, we see that u is made up of two traveling waves, one going to the right and the other to the left, each of the same shape as the initial displacement, but half the height. The picture of the extended f in Figure 7.8 predicts what happens to each of the traveling waves when it reaches the fixed end of the string toward which it is going. Each is "reflected," and reverses direction. In Figure 7.9 we have shown successive pictures of the vibrating string.

The method we have just developed is a rather special one. It works for the differential equation (7.14) because there is a general formula for its solution, as explained in Theorem 7.1. The differential equations that occur in practice seldom if ever have general solutions that can be written down easily. In such cases, other methods must be used. We will now take up one of these methods, called *separation of variables*, and will apply it first to the same vibrating string problem. The key idea is to find certain special solutions of the partial differential equation, and then combine these special solutions in order to find more general solutions. For simplicity, we assume that units have now been chosen so that the partial differential equation we want to

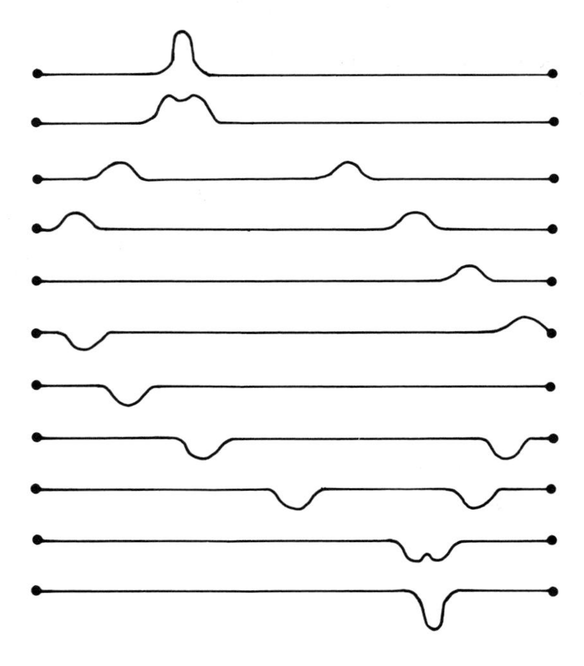

Figure 7.9

solve has the form

$$u_{xx} - u_{tt} = 0 \tag{7.21}$$

and that we want to find a solution which obeys the following boundary conditions.

$$u(0,t) = 0 \qquad \text{all } t \ge 0 \tag{7.22}$$

$$u(L,t) = 0 \qquad \text{all } t \ge 0 \tag{7.23}$$

$$u(x,0) = f(x) \qquad 0 \le x \le L \tag{7.24}$$

$$u_t(x,0) = 0 \qquad 0 \le x \le L \tag{7.25}$$

The key idea is to look for solutions of (7.21) that have the form $u(x,t) = A(x)\,B(t)$, and that satisfy the simpler of the boundary conditions. If we substitute this formula for u into (7.21), we obtain

$$A''(x)\,B(t) - A(x)\,B''(t) = 0$$

which we choose to rewrite as

$$\frac{A''(x)}{A(x)} = \frac{B''(t)}{B(t)} \tag{7.26}$$

This is supposed to hold for all choices of x and of t. However, x and t are independent variables, and the only way (7.26) can hold is for each side of the equation to be constant. If we call this constant β, as yet knowing nothing about its possible value except that it is real, we first replace (7.26) by

$$\frac{A''(x)}{A(x)} = \beta = \frac{B''(t)}{B(t)}$$

Looking at each half separately, we obtain two ordinary differential equations.

$$A''(x) - \beta A(x) = 0$$

$$B''(t) - \beta B(t) = 0$$

There are three cases to be considered, depending on the sign of the auxiliary number β; we could have $\beta < 0$, or $\beta = 0$, or $\beta > 0$. The last two lead to situations in which the boundary conditions (7.22) and (7.23) cannot be satisfied. (See Exercise 10.) We therefore turn to the first case, and since $\beta < 0$, we set $\beta = -\omega^2$, where ω is real and $\omega > 0$. Making this change in the differential equations above, we now have the pair

$$A''(x) + \omega^2 A(x) = 0 \tag{7.27}$$

$$B''(t) + \omega^2 B(t) = 0 \tag{7.28}$$

Each is easily solved, and we have

$$A(x) = C_1 \sin(\omega x) + C_2 \cos(\omega x) \tag{7.29}$$
$$B(t) = C_3 \sin(\omega t) + C_4 \cos(\omega t)$$

We now return to the boundary conditions and try to satisfy them. We start with (7.22); since $u = A(x)B(t)$, $u(0,t) = A(0)B(t)$. Since this is supposed to be zero for all $t \geq 0$, we conclude that we want $A(0) = 0$. Using (7.23) in the same way, we conclude that we want $A(L) = 0$. If we impose these on the solution for A in (7.29), we discover first that $C_2 = 0$, and then that we need to have

$$\sin(L\omega) = 0 \tag{7.30}$$

This puts a limitation on the possible values of ω, and we see that we should choose $\omega = \pi/L, 2\pi/L, 3\pi/L, \ldots$.

Returning to condition (7.25), we see that $u_t = A(x)B'(t)$, so that $u_t(x,0) = A(x)B'(0)$. In order to have this zero for all x in $[0, L]$, we require that $B'(0) = 0$. From (7.29), we see that this implies that $C_3 = 0$.

Putting all this together, we emerge with a collection of special solutions of (7.21) which obey all of the boundary conditions except (7.24) and have the form

$$u(x,t) = C\cos\left(\frac{n\pi}{L}t\right)\sin\left(\frac{n\pi}{L}x\right) \tag{7.31}$$

It is easy to describe these waves. The initial shape, at $t = 0$, is a sine curve with n extrema; the cosine term supplies an oscillating amplitude factor. Successive pictures of the wave with $n = 2$ are shown in Figure 7.10. It can also be represented in the traveling wave form (7.12) by making use of trigonometric identities. (See Exercise 6.)

It is clear that in general, no single solution u of the form (7.31) will satisfy the boundary condition (7.24). However, the situation is better if we try a linear combination of the solutions (7.31) such as

$$u(x,t) = \sum_1^\infty C_n \cos\left(\frac{n\pi}{L}t\right)\sin\left(\frac{n\pi}{L}x\right) \tag{7.32}$$

In order to satisfy (7.24), we need to have $u(x,0) = f(x)$, and according to (7.32), this requires that we have

$$f(x) = \sum_1^\infty C_n \sin\left(\frac{n\pi}{L}x\right) \tag{7.33}$$

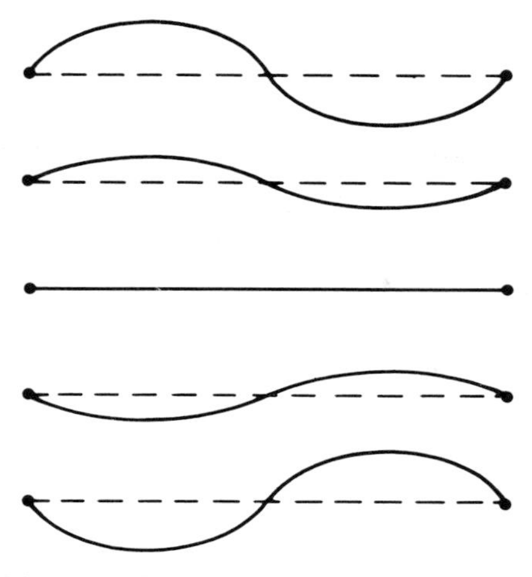

Figure 7.10

Such expansions of a function into a series of sines (or cosines) had been studied by Euler and others about 1750; but they were investigated in much more detail by Fourier at the start of the nineteenth century, with the result that we now call them Fourier series. The key to such expansions is the following special integration identity (see Exercise 7).

$$\int_0^L \sin\left(\frac{n\pi}{L}x\right)\sin\left(\frac{k\pi}{L}x\right)dx = \begin{cases} 0 & \text{if } n \neq k \\ \tfrac{1}{2}L & \text{if } n = k \end{cases} \tag{7.34}$$

As a result, we can find the correct values of the coefficients C_n to use in (7.33) by calculating the integrals

$$C_n = \frac{2}{L}\int_0^L f(x)\sin\left(\frac{n\pi}{L}x\right)dx \tag{7.35}$$

Having found these, we can use (7.32) to obtain the desired solution $u(x, t)$.

In order to be completely correct, the statements above need a certain number of qualifications. Not every function $f(x)$ can be expressed as a Fourier series of the form (7.33) for all points in the interval $[0, L]$, and not every series whose coefficients C_n are found by the formula (7.35) is convergent. However, if $f(x)$ is a function in $\mathcal{C}^2$ that obeys the endpoint conditions $f(0) = f(L) = 0$, then (7.35) is valid.

An illustration of what sometimes happens is seen in the following example. Take $f(x) = x - 1$ on the interval $[0, 2]$. Then, the Fourier process yields the series expansion

$$x - 1 = -\frac{2}{\pi}\sin(\pi x) - \frac{2}{2\pi}\sin(2\pi x) - \frac{2}{3\pi}\sin(3\pi x) - \cdots \tag{7.36}$$

This series does in fact converge as indicated for $0 < x < 2$. However, it is clearly incorrect for $x = 0$ or $x = 2$, since the left and right sides of (7.36) do not agree there.

Let us now leave the vibrating string and instead consider the behavior of a circular drumhead. This time, the vertical displacement of the vibrating membrane is given by a function $u = u(x, y, t)$, and the partial differential equation that governs the motion in the simplest model is

$$\frac{\partial^2 u}{\partial x^2} + \frac{\partial^2 u}{\partial y^2} = \frac{1}{c^2}\frac{\partial^2 u}{\partial t^2} \tag{7.37}$$

The rim of the drumhead is fixed, and we again consider the problem of predicting the movement of the membrane caused by an initial displacement. The corresponding boundary conditions for this problem are as follows

(7.38) $u(x, y, 0) = f(x, y)$ [Initial shape]

(7.39) $u_t(x, y, 0) = 0$ [Motionless at $t = 0$]

(7.40) $u(x, y, t) = 0$ if $t \geq 0$ and $x^2 + y^2 = R^2$ [Rim fixed]

For simplicity, we take $R = 1$ and $c = 1$, and we will consider only those motions of the drumhead that are circularly symmetric. This means that the initial shape f is to be a function of $r = (x^2 + y^2)^{1/2}$ alone, and we will look for solutions u which depend only on r and t.

If $u = g(r, t)$, then

$$u_x = \frac{\partial u}{\partial r}\frac{\partial r}{\partial x} = \frac{\partial u}{\partial r}\frac{x}{r}$$

and

$$\begin{aligned} u_{xx} &= \frac{\partial^2 u}{\partial r^2}\left(\frac{x}{r}\right)\left(\frac{x}{r}\right) + \frac{\partial u}{\partial r}\frac{1}{r} + \frac{\partial u}{\partial r}\left(-\frac{x}{r^2}\right)\frac{\partial r}{\partial x} \\ &= \frac{\partial^2 u}{\partial r^2}\frac{x^2}{r^2} + \frac{\partial u}{\partial r}\frac{1}{r} - \frac{x^2}{r^3}\frac{\partial u}{\partial r} \end{aligned}$$

In the same way, we have

$$u_{yy} = \frac{\partial^2 u}{\partial r^2}\frac{y^2}{r^2} + \frac{\partial u}{\partial r}\frac{1}{r} - \frac{y^2}{r^3}\frac{\partial u}{\partial r}$$

Hence,

$$\begin{aligned} u_{xx} + u_{yy} &= \frac{\partial^2 u}{\partial r^2}\left(\frac{x^2 + y^2}{r^2}\right) + 2\frac{\partial u}{\partial r}\frac{1}{r} - \frac{(x^2 + y^2)}{r^3}\frac{\partial u}{\partial r} \\ &= \frac{\partial^2 u}{\partial r^2} + \frac{1}{r}\frac{\partial u}{\partial r} \end{aligned}$$

Accordingly, for functions of the special form $g(r,t)$, the equation (7.37) becomes

$$\frac{\partial^2 u}{\partial r^2} + \frac{1}{r}\frac{\partial u}{\partial r} = \frac{\partial^2 u}{\partial t^2} \tag{7.41}$$

The boundary conditions are now as follows:

(7.42) If $t = 0$, $u = f(r)$ $\quad 0 \le r \le 1$

(7.43) If $t = 0$, $u_t = 0$ $\quad 0 \le r \le 1$

(7.44) If $r = 1$, $u = 0$ $\quad$ for all $t \ge 0$

We again try the method of separation of variables, by setting $u = A(r)B(t)$. Putting this into (7.41), we obtain

$$\frac{A'' + (1/r)A'}{A} = \frac{B''}{B} \tag{7.45}$$

Again, the left-hand side is a function of r alone and the right-hand side is a function of t alone. Thus, each must be constant. Calling the constant $-\omega^2$ as before, we arrive at two ordinary differential equations:

$$A'' + \frac{1}{r}A' + \omega^2 A = 0 \tag{7.46}$$

$$B'' + \omega^2 B = 0 \tag{7.47}$$

Solving (7.47) and applying boundary condition (7.43) gives us the simple function $B(t) = C\cos\omega t$, where there are as yet no restrictions on ω. However, things are more complicated when it comes to equation (7.46). This is a special case of the Bessel differential equation, and it does not have any simple solutions that can be expressed in terms of elementary functions such as exponential and trigonometric. However, we can obtain a power-series solution for (7.46). We assume that there is a solution of the form

$$A(r) = a_0 + a_1 r + a_2 r^2 + a_3 r^3 + \cdots \tag{7.48}$$

Then the standard method, as explained in Chapter 3, results in the following *single* solution. [*Note:* This is at first sight strange. Since we are solving a second-order equation, one would normally expect two independent solutions; in fact there are two, but the other one cannot be expressed as a power series of the form (7.48).]

$$A(r) = a_0\left(1 - \frac{\omega^2}{4}r^2 + \frac{\omega^4}{(4)(16)}r^4 - \frac{\omega^6}{(4)(16)(36)}r^6 + \frac{\omega^8}{[(2)(4)(6)(8)]^2}r^8 - \cdots\right) \tag{7.49}$$

Series of this form arise in a number of different applications so that it has been useful to give names to some of these. The Bessel function of index zero, $J_0(x)$, is the function defined by the series

$$J_0(x) = \sum_{k=0}^{\infty} \frac{(-1)^k x^{2k}}{4^k (k!)^2}$$

(7.50)

$$= 1 - \frac{x^2}{(2)^2} + \frac{x^4}{[(2)(4)]^2} - \frac{x^6}{[(2)(4)(6)]^2} + \cdots$$

The series for the Bessel function converges for all x, and defines a very well-behaved function having a graph similar to that shown in Figure 7.11. Numerical values of $J_0(x)$ have been tabulated, and computer programs exist which generate these values. Thus, in many ways, $J_0(x)$ is as useful and available a function as $\cos x$.

Returning to equation (7.49), we see that the desired solution to the differential equation (7.46) can be expressed in terms of the Bessel function J_0 in the form

(7.51) $$A(r) = a_0 J_0(\omega r)$$

Putting this together with the solution for $B(t)$ gives us the desired solutions of (7.41), namely,

(7.52) $$u(r, t) = C \cos(\omega t) J_0(\omega r)$$

We next impose boundary condition (7.44), setting $r = 1$ in (7.52) to obtain

$$u(1, t) = C \cos(\omega t) J_0(\omega)$$
$$= 0$$

Since this is to hold for all $t \geq 0$, we are led to require that the number ω be chosen so that $J_0(\omega) = 0$. This limits the acceptable values of ω to be the

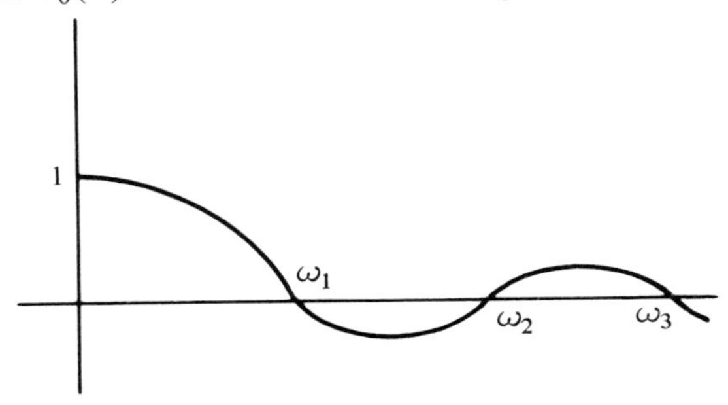

Figure 7.11

zeros of the Bessel function J_0. These form an infinite set, corresponding to the points in Figure 7.11 where the graph of J_0 crosses the horizontal axis. Let us label these values $\omega_1, \omega_2, \omega_3, \ldots$; many of these have been tabulated, and much is known about their size and distribution. However, for $J_0(x)$ there is no simple formula for the zeros such as exists for the zeros of the function $\cos x$.

Returning to (7.52), we have now found an infinite sequence of functions of the form

$$u(r,t) = C\cos(\omega_k t) J_0(\omega_k r) \tag{7.53}$$

each of which satisfies the partial differential equation (7.41) and also satisfies boundary conditions (7.43) and (7.44). In general, none of them will satisfy boundary condition (7.42).

There is a lot of similarity between what we have done so far, and the corresponding treatment of the vibrating string. For example, it is instructive to compare (7.31) with (7.53). In order to obtain a solution of (7.41) which also satisfies the boundary condition (7.42), we take a linear combination of all the special solutions (7.53), writing

$$u(r,t) = \sum_{1}^{\infty} C_k \cos(\omega_k t) J_0(\omega_k r) \tag{7.54}$$

Then, setting $t = 0$, we need to have

$$f(r) = \sum_{0}^{\infty} C_k J_0(\omega_k r) \tag{7.55}$$

Comparing this with (7.33), we see that we are again dealing with a Fourier-type series expansion of the function $f(r)$, this time as a series of Bessel functions rather than sine functions.

We are content to leave the problem here; the purpose of this section is to give an overview of the subject, not to develop a complete theory. It may be shown that there is an integral identity similar to (7.34) which makes it possible to determine the coefficients C_n, and if the function f is sufficiently smooth and obeys the obvious requirement $f(1) = 0$, then the series in (7.55) does converge to the function $f(r)$, and the function $u(r,t)$ in (7.54) is the desired solution of (7.41), obeying all the given boundary conditions. Further details of all this can be found in references 13, 14, and 16 of the Bibliography.

One last comment of a physical nature is worth making. The special functions $u(r,t)$ given in (7.53) describe simple modes of vibration of the drumhead. The shape of several of these is suggested by Figure 7.12. The period of each is determined by the cosine factor in (7.53), and these periods

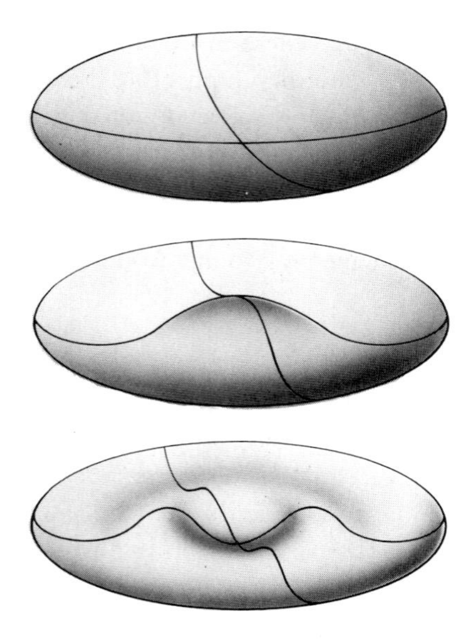

Figure 7.12

turn out to be $2\pi/\omega_1$, $2\pi/\omega_2$, $2\pi/\omega_3$, etc. The rate of growth of the numbers ω_k is sufficiently irregular that there is no simple arithmetic relationship between them. Thus, the natural periods of a vibrating drumhead do not have the same simple properties that hold for the natural periods of a vibrating string. This explains why the sound of a drum is far different from the sound of a violin.

Exercises

1. Verify that functions of the form $u = u(x,t) = \phi(x^2+t)+\psi(x^2-t)$ satisfy the partial differential equation

$$\frac{\partial^2 u}{\partial x^2} - \frac{1}{x}\frac{\partial u}{\partial x} = 4x^2 \frac{\partial^2 u}{\partial t^2}$$

2. (a) Using (7.20), show that for all x, $f(x)+f(2L-x) = 0$.
   (b) Using (7.19), show that for all x, $f(x+2L) = f(x)$.
   (c) Use these to justify Figure 7.8.
3. What kind of "traveling wave" is described by the equation $u = \phi(x^2 - ct)$?
4. Verify the details of Figure 7.5.

5. Verify the details of Figure 7.9.

6. (a) Show that the special solution given in (7.31) can be written in the form

$$u(x,t) = \frac{C}{2}\left[\sin\left(\frac{n\pi}{L}x + \frac{n\pi}{L}t\right) + \sin\left(\frac{n\pi}{L}x - \frac{n\pi}{L}t\right)\right]$$

   (b) Give a verbal description of this in terms of two traveling waves.

7. Verify the identity (7.34).

8. Verify (7.36) by using (7.35) to obtain the Fourier coefficients C_n.

9. Carry out the power-series solution of (7.46) to obtain the result given in (7.49).

10. Show why the choices $\beta = 0$ or $\beta > 0$ do not lead to functions that can satisfy conditions (7.22) and (7.23).

7.3 Three Problems in Design

In this section, we discuss three specific applied problems, each dealing with the determination of a curve that is required to achieve a desired geometric or physical objective. Each is solved by means of differential equations, but the last one, while similar in statement to the others, turns out to involve much greater mathematical depth.

An Ideal Water Clock

One of the simplest instruments for measuring time without reference to the sun or stars was the water clock, or *clepsydra* (supposedly invented by Plato). If a large glass cylinder is filled with water and a small hole is opened in the bottom, the water level will decrease smoothly; if suitable marks are painted on the side, the elapsed time can be measured. The velocity with which the water leaks out of the hole depends at each moment upon the height of the water in the vessel above the opening. Without going into the physical analysis behind it, we take for granted that the correct formula is

$$\text{Velocity} = b\sqrt{h} \tag{7.56}$$

where b is a constant and h is the height of the water.

Since h is decreasing, the exit velocity will decrease, and the rate at which the water level falls will not be constant. In particular, this means that the hour

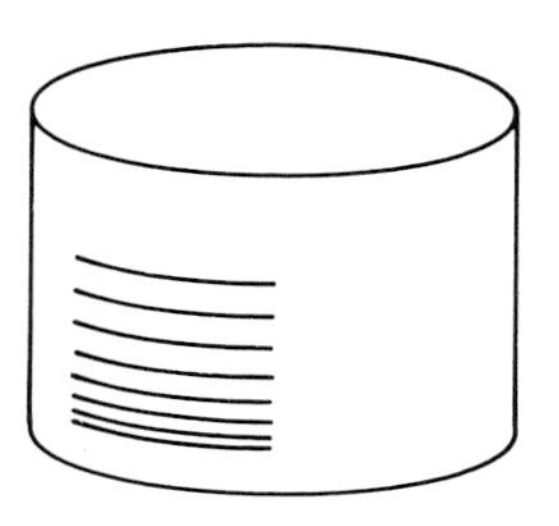

Figure 7.13

marks on the side of the jar should not be spaced uniformly, but must be increasingly closer together toward the bottom of the jar. (See Figure 7.13.)

We also observe that the rate at which the water level falls is affected by the shape of the jar. If it is conical, then with the same size hole in the bottom, the water level will fall more rapidly than if it is a cylinder, since the area of cross section will decrease as the water drains out.

We propose to answer the following question: What should be the shape of the jar in order that the hour marks can be uniformly spaced?

Let us first analyze the cylindrical water clock. If the height of the water at time t is h, and c is the effective area of the hole, then the volume of water leaving the jar between t and $t+\Delta t$ is, according to (7.56),

$$\Delta V = cb\sqrt{h}\,\Delta t \tag{7.57}$$

If the radius of the jar is R and the resulting change in the height of the water is Δh, a negative number, then we also have

$$\Delta V = -\pi R^2\,\Delta h \tag{7.58}$$

If we equate (7.57) and (7.58), we arrive at a differential equation

$$\frac{dh}{dt} = -K\sqrt{h} \tag{7.59}$$

where $K = cb/\pi R^2$. If H is the water level at the start, this gives us the initial conditions $t = 0$, $h = H$.

The equation (7.59) is separable and easily solved. We obtain

$$h = H - K\sqrt{H}\,t + \tfrac{1}{4}K^2t^2 \tag{7.60}$$

This formula clearly shows the nonuniformity of the hour markings for a cylindrical water clock, since the formula for h is nonlinear.

Suppose now that the jar is not cylindrical but is in the shape of a symmetric bowl with circular cross sections, whose radius at height h is given as $r = \phi(h)$. Repeating the analysis given above, (7.57) is unchanged but (7.58) is replaced by

$$\Delta V = -\pi[\phi(h)]^2 \, \Delta h$$

As before, this leads to a differential equation for h, namely,

$$\frac{dh}{dt} = -k\frac{\sqrt{h}}{[\phi(h)]^2} \tag{7.61}$$

where $k = cb/\pi$. If we want to determine ϕ so that the hour markings of the water clock will be uniformly spaced, then the equation (7.61) must have a solution of the form

$$h = H - \beta t$$

for some constant β. In particular, it must be true that $dh/dt = -\beta$, a constant. Examining (7.61), we see that this implies that we must have

$$-k\frac{\sqrt{h}}{[\phi(h)]^2} = -\beta$$

Solving for ϕ, we discover that the correct choice for the shape of a water clock is to take

$$\phi(h) = Ch^{1/4}$$

A Mirror for a Solar Furnace

It is easy to see that it would be wrong to use a large spherical mirror for a high-intensity solar furnace, since parallel light rays striking a spherical surface do not all reflect as rays that pass through the same point. (See Figure 7.14.) We propose to answer the question: What should be the shape of a

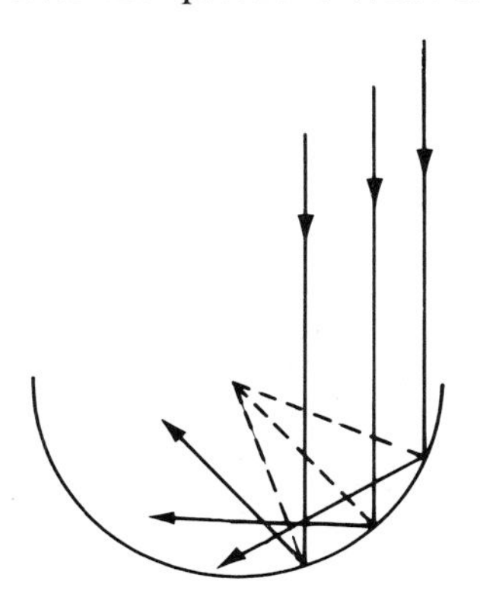

Figure 7.14

mirror if every vertical ray is to reflect to a single point? In this case, we all know the answer in advance, since our question describes one of the celebrated geometric properties of the parabola. However, the point of this treatment is to see that this answer is forced upon us, and we do not need to know the answer in advance.

We suppose that the shape of the mirror is given by a function $y = \phi(x)$, and that we want every ray of light coming vertically into the mirror to be reflected to the origin. (See Figure 7.15.) Our problem is to translate this into a mathematical statement about ϕ, and then determine ϕ.

Suppose that a vertical ray from the sun strikes the mirror at a point P, and is then reflected to the origin. Let $P = (x, y)$, and draw the tangent line at P, a line L of slope dy/dx. Using the fact that the angles of incidence and reflection must be equal, and some elementary geometry, we arrive at two facts (see Figure 7.16):

$$\tan(2\theta) = \frac{x}{y} \tag{7.62}$$

$$\tan\theta = \frac{1}{dy/dx} \tag{7.63}$$

If we use the standard identity

$$\tan(2\theta) = \frac{2\tan\theta}{1-(\tan\theta)^2}$$

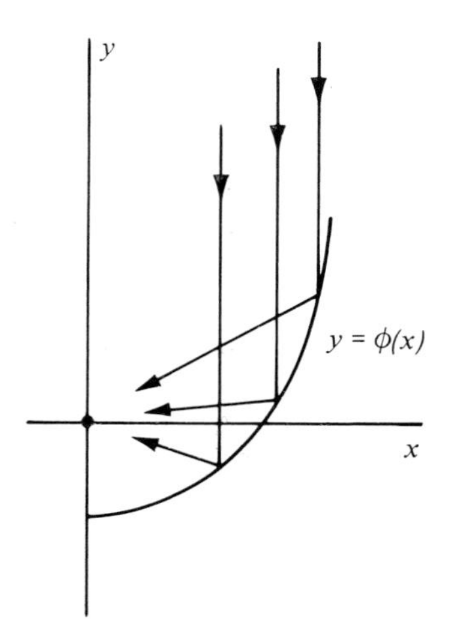

Figure 7.15

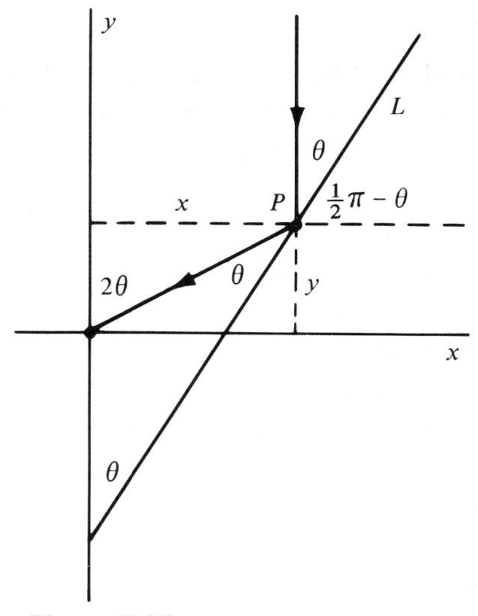

Figure 7.16

then (7.62) and (7.63) lead to the equation

$$\left(\frac{dy}{dx}\right)^2 - \frac{2y}{x}\frac{dy}{dx} - 1 = 0$$

We solve this for dy/dx, electing to use the solution

$$\frac{dy}{dx} = \frac{y}{x} + \sqrt{\frac{y^2}{x^2} + 1} \tag{7.64}$$

This is a "homogeneous" first-order equation, of the sort discussed in Section 2.1. Hence, we make the substitution $y = ux$ in (7.64) and obtain

$$\frac{du}{dx} = \frac{\sqrt{1+u^2}}{x}$$

This is a separable equation, whose solution is

$$\log\left(u+\sqrt{1+u^2}\right) = \log x + \text{const}$$

leading to

$$u + \sqrt{1+u^2} = Ax$$

In this, we replace u by y/x, and obtain the final solution

$$y = \frac{A}{2}x^2 - \frac{1}{2A}$$

This is a parabola; we have therefore shown that the parabola is the only curve having the desired reflection properties.

The Optimal Ski Slope

Suppose we ask the question: What shape should a hill have to make the fastest ski slope? Specifically, can we determine the curve $y = \phi(x)$ which connects the points $(0, 1)$ and $(1, 0)$, and which gives the quickest ski slide from top to bottom?

This is a classical problem in disguise. The search for the brachistochrone (shortest-time) path began with Galileo, who believed, incorrectly, that the correct path was an arc of a circle. In 1696, John Bernoulli, who had found the correct solution, proposed the problem as a challenge to his contemporaries, and in the course of the following year, solutions were also obtained by Newton, Leibniz, and John's brother James. The problem is now recognized as one of the starting points in the creation of the **calculus of variations**; we give a brief treatment of this problem and its solution.

We think of the curve as fixed, initially, and our first objective is to calculate the total time T it will take for a unit mass to slide down the curve under the force of gravity, assuming that the hill is frictionless. It is convenient to think of the sliding mass as a point P with coordinates (x, y) and lying on the curve $y = \phi(x)$, and to think of x and y as unknown functions of t. The position of P is also determined by giving the length of the arc of the curve from A down to P, and this length s is also a function of t. We shall have occasion to use two familiar relations from elementary calculus:

$$\frac{ds}{dt} = \sqrt{\left(\frac{dx}{dt}\right)^2 + \left(\frac{dy}{dt}\right)^2} \tag{7.65}$$

$$\frac{ds}{dx} = \sqrt{1 + \left(\frac{dy}{dx}\right)^2} \tag{7.66}$$

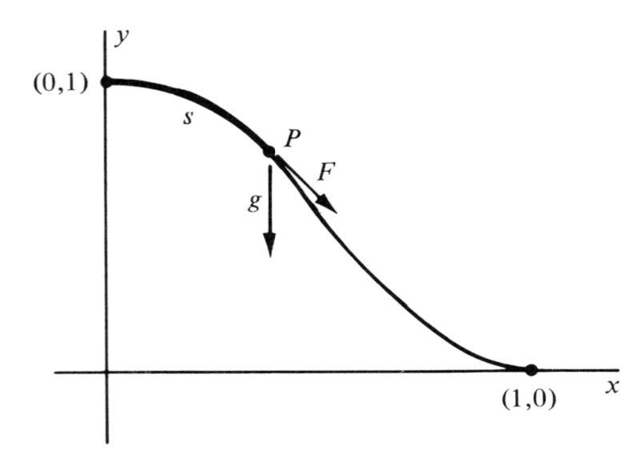

Figure 7.17

We note that $v = ds/dt$ is the speed of the mass along the curve at time t, and d^2s/dt^2 is its acceleration. This acceleration is caused by the tangential component of the gravitational force at P, as shown in Figure 7.17, and the magnitude of this component is $g \sin\theta$, as can be seen from the diagram in Figure 7.18. This leads to the following relation.

$$\frac{d^2s}{dt^2} = g \sin\theta \tag{7.67}$$

Since θ is the angle from the horizontal direction down to the tangent to the curve $y = \phi(x)$ at P, we have $\tan\theta = -dy/dx$. We then have

$$\begin{aligned}\sin\theta &= \frac{\tan\theta}{\sqrt{1+(\tan\theta)^2}} = \frac{-dy/dx}{\sqrt{1+(dy/dx)^2}} \\ &= \frac{-dy/dt}{\sqrt{(dx/dt)^2+(dy/dt)^2}} = -\frac{dy/dt}{ds/dt}\end{aligned}$$

(It is easy to keep the signs straight by noting that dx/dt and ds/dt are positive, and dy/dt is negative; see Figure 7.18.) Using this calculation in (7.67), we obtain

$$\frac{d^2s}{dt^2} = -g\,\frac{dy/dt}{ds/dt}$$

Setting $v = ds/dt$, this becomes

$$v\,\frac{dv}{dt} = -g\,\frac{dy}{dt}$$

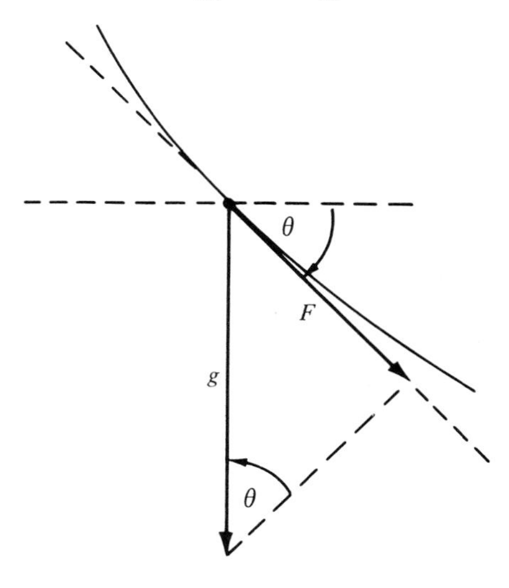

Figure 7.18

which can be immediately integrated to obtain

$$\tfrac{1}{2}v^2 = -gy + \text{const}$$

In our problem, the skier starts at the top of the hill with zero velocity, so that $v = 0$ when $y = 1$. Using this, we arrive at the equation

$$v = \sqrt{2g}\sqrt{1-y} \tag{7.68}$$

We note that this tells us that the terminal speed of the skier is $\sqrt{2g}$, no matter what the shape of the hill!

Suppose that the total time taken in the descent is T. Since the bottom of the hill is to be the point $(1,0)$, we see that $t = T$ must correspond to $x = 1$ and $y = 0$, while $t = 0$ corresponds to $x = 0$ and $y = 1$. Accordingly,

$$T = \int_0^T dt = \int_0^1 \frac{dt}{dx}\,dx = \int_0^1 \frac{dt}{ds}\frac{ds}{dx}\,dx$$
$$= \int_0^1 \frac{ds}{dx}\frac{1}{v}\,dx$$

and using (7.66) and (7.68), we have

$$T = \frac{1}{\sqrt{2g}}\int_0^1 \frac{\sqrt{1+(dy/dx)^2}}{\sqrt{1-y}}\,dx$$

Since we have assumed that the equation of the curve is $y = \phi(x)$, we can write this formula as

$$T = \frac{1}{\sqrt{2g}}\int_0^1 \sqrt{\frac{1+(\phi')^2}{1-\phi}}\,dx \tag{7.69}$$

This displays in an explicit way how the time of descent T is determined by the function ϕ which describes the curve.

We are now ready to examine the second part of the problem, namely, to determine what choice of the function ϕ will minimize the number T. Problems of this general sort are usually called *extremal problems* or *optimization problems*; what makes this one difficult is that we are looking for the minimum of a function $T = \mathcal{F}(\phi)$ where the variable is the function ϕ and not an ordinary numerical variable.

This is a case where it seems easier to present a sketch of a general method for treating such problems than to concentrate only on the immediate problem before us. Suppose, therefore, that we have a formula such as that in (7.69) which enables us to associate a value $\mathcal{F}(\phi)$ with any choice of a function ϕ

that obeys $\phi(0) = 1$, $\phi(1) = 0$. To be explicit, we assume that we have

$$\mathscr{F}(\phi) = \int_0^1 F(\phi, \phi') \tag{7.70}$$

where F is some specific function of two variables. In the special case we are most interested in, $F(u, v) = \sqrt{(1+v^2)/(1-u)}$. We want to find a way to identify the choice of ϕ which minimizes $\mathscr{F}(\phi)$. We use a device that was first discovered by Euler. Suppose that ψ is any function such that $\psi(0) = \psi(1) = 0$, and construct the function $\phi_0 + \lambda\psi$, where ϕ_0 is the function that minimizes $\mathscr{F}(\phi)$. Since $\phi_0 + \lambda\psi$ is another function in the class of those that obey the two restrictions $\phi(0) = 1$, $\phi(1) = 0$, we must have

$$\mathscr{F}(\phi_0) \leq \mathscr{F}(\phi_0 + \lambda\psi)$$

for every choice of the real number λ.

To restate in a more useful way, we have observed that $\mathscr{F}(\phi_0 + \lambda\psi)$, treated as a numerical-valued function of the real number λ, must have its minimum value for $\lambda = 0$, no matter how we have chosen the special function ψ.

But elementary calculus tells us that at a minimum point, the derivative vanishes. Hence, we have learned that

$$\text{when } \lambda = 0 \qquad \frac{d}{d\lambda}\mathscr{F}(\phi_0 + \lambda\psi) = 0 \tag{7.71}$$

for every choice of ψ.

Let us try this out on the formula for $\mathscr{F}$ given in (7.70). We have

$$\mathscr{F}(\phi_0 + \lambda\psi) = \int_0^1 F(\phi_0 + \lambda\psi,\ \phi_0' + \lambda\psi')$$

If we differentiate this with respect to the parameter λ, we obtain

$$\frac{d}{d\lambda}\mathscr{F}(\phi_0 + \lambda\psi) = \int_0^1 F_1(\phi_0 + \lambda\psi,\ \phi_0' + \lambda\psi')\,\psi + \int_0^1 F_2(\phi_0 + \lambda\psi,\ \phi_0' + \lambda\psi')\,\psi'$$

Setting $\lambda = 0$ and using (7.71), we obtain an integral identity which must hold for the minimizing function ϕ_0 and for every choice of the special function ψ.

$$\int_0^1 F_1(\phi_0, \phi_0')\,\psi + \int_0^1 F_2(\phi_0, \phi_0')\,\psi' = 0 \tag{7.72}$$

We now invoke the following theorem, whose proof is easy to follow but hard to discover.

THEOREM 7.2 *If (7.72) holds for every smooth function obeying* $\psi(0) = \psi(1) = 0$, *then the expression*

$$F(\phi_0, \phi_0') - \phi_0' F_2(\phi_0, \phi_0') \tag{7.73}$$

must be constant on the interval $[0, 1]$.

Proof Because of the special assumptions on the function ψ, it is true that

$$\int_0^1 (\psi F_2)' = \psi F_2 \Big|_{x=0}^{x=1} = (0) F_2 - (0) F_2 = 0$$

Differentiating ψF_2, we have

$$\frac{d}{dx}\{\psi F_2\} = \psi' F_2 + \psi \frac{d}{dx} F_2$$

so that we have, by integration,

$$0 = \int_0^1 \psi' F_2 + \int_0^1 \psi \frac{d}{dx} F_2$$

Combining this with (7.72), we see that

$$\int_0^1 \left[F_1(\phi_0, \phi_0') - \frac{d}{dx} F_2(\phi_0, \phi_0') \right] \psi(x)\, dx = 0 \tag{7.74}$$

This relationship holds for every admissible choice of the function ψ. This can only happen if the function in brackets inside the integral is identically zero on $[0, 1]$:

$$F_1(\phi_0, \phi_0') - \frac{d}{dx} F_2(\phi_0, \phi_0') = 0 \tag{7.75}$$

For if there were a point x_0 where the expression in (7.75) was different from zero, an admissible choice of ψ could be made which would be zero everywhere except near x_0, and the integral in (7.74) could not be zero.

To deduce (7.73) from (7.75), we use the chain rule to calculate the derivative of the expression in (7.73):

$$\frac{d}{dx}\{F(\phi_0, \phi_0') - \phi_0' F_2\} = F_1 \phi_0' + F_2 \phi_0'' - \phi_0'' F_2 - \phi_0' \frac{d}{dx} F_2$$

$$= \left(F_1 - \frac{d}{dx} F_2 \right) \phi_0'$$

But this is zero on $[0, 1]$ by (7.75), so that (7.73) must have been constant.

Let us apply Theorem 7.2 to the particular problem we are interested in. We have $F(u,v) = \sqrt{(1+v^2)/(1-u)}$, so that

$$F_2(u,v) = \frac{v}{\sqrt{1+v^2}}\frac{1}{\sqrt{1-u}}$$

Returning to (7.73), we have

$$\begin{aligned} F - \phi_0' F_2 &= \sqrt{\frac{1+(\phi_0')^2}{1-\phi_0}} - \phi_0' \frac{\phi_0'}{\sqrt{1+(\phi_0')^2}\sqrt{1-\phi_0}} \\ &= \frac{1}{\sqrt{1+(\phi_0')^2}\sqrt{1-\phi_0}} \end{aligned}$$

By Theorem 7.2, this must be constant on the interval $[0,1]$. Hence, the minimizing function ϕ_0 can only be a solution of the differential equation

$$(1+(\phi')^2)(1-\phi) = C \tag{7.76}$$

which is easily converted into the equation

$$\phi' = \sqrt{\frac{A+\phi}{1-\phi}}$$

or, since $y = \phi(x)$, the equation

$$\frac{dy}{dx} = \sqrt{\frac{A+y}{1-y}} \tag{7.77}$$

The constant A can be determined, along with the arbitrary constant arising during the solution of (7.77), by using the fact that the curve must go through the points $(0,1)$ and $(1,0)$.

Since the equation (7.77) is separable, there is no insurmountable difficulty in solving it, although the integrations needed are messy. For our purposes, it is sufficient merely to exhibit the family of solutions, and leave the verification of this as an exercise. The solutions of (7.77) turn out to be, in parametric form, the curves

$$\begin{cases} x = a(\beta - \sin\beta) \\ y = 1 - a(1-\cos\beta) \end{cases} \tag{7.78}$$

These are recognizable as **cycloids**, the curves traced out by a point on the edge of a rolling circle (see Figure 7.19). The fact that the solution to the brachistochrone problem turned out to be a curve with such a celebrated history both amazed and delighted the Bernoulli brothers, who saw in this another demonstration of the artistry of the world in which we live.

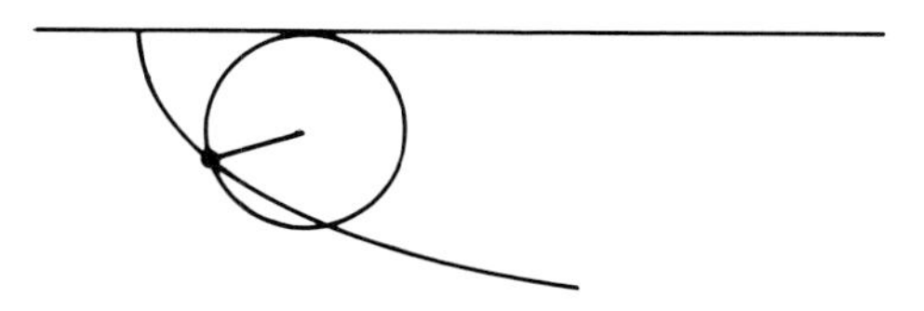

Figure 7.19 The cycloid

Exercises

1. Sketch the water clock obtained as a final answer to the uniform marking problem.
2. How should the markings be placed on a conical water clock?
3. Verify the details in the solution of the differential equation (7.64)—the mirror problem.
4. Find the times of descent for the following two ski slopes.

   (a) $y = 1 - x$

   (b) $y = \begin{cases} 1 - 3x & 0 \le x \le \frac{1}{4} \\ \frac{1}{3}(1-x) & \frac{1}{4} \le x \le 1 \end{cases}$

5. Verify that the cycloid (7.78) satisfies the equation (7.76).

7.4 Analysis Is Still Useful

In previous chapters, we have repeatedly examined the simple differential equation

$$\frac{dy}{dt} = t^2 + y^2, \qquad y(0) = 1 \tag{7.79}$$

For example, in Section 3.4, we applied the Picard iteration method, and derived the approximating polynomial

$$P(t) = 1 + t + t^2 + \tfrac{4}{3}t^3 + \tfrac{7}{6}t^4 + \tfrac{6}{5}t^5 + \tfrac{37}{30}t^6 + \tfrac{404}{315}t^7$$

without finding out much about the accuracy of this approximation except for values of t quite near to $t = 0$. In Section 3.5, we tried for a power-series solution and found the same terms as in the Picard polynomial, but we were able neither to find a general formula for the coefficients that would assure us that the power series converged, nor to determine the accuracy of the approximation.

In Chapter 4, we applied several numerical methods to this equation. For example, in Section 4.1, we applied the Euler method with several different step sizes, and discovered that y became extremely large for values of t near $t = 1$. This led us to conjecture that the exact graph of $y(t)$ might have a vertical asymptote somewhere near $t = 1$.

In the present section, we will show how this can be verified by analytic methods. The methods are not general, but similar techniques can often be used to discover properties of the solutions of an intractable differential equation, and thus to understand better the numerical results obtained by applying methods such as the classical Runge-Kutta.

THEOREM 7.3 *If $y(t)$ is the solution of* (7.79), *then $y(t) \to \infty$ as t increases on the interval* $[0,1]$.

Proof We turn the equation (7.79) upside down, obtaining

$$\frac{dt}{dy} = \frac{1}{t^2+y^2}, \qquad y = 1, \quad t = 0 \tag{7.80}$$

and view it as an equation to be solved for the function $t(y)$, with y as the independent variable, and obeying the initial condition $t(1) = 0$. Since it is evident that dt/dy is positive, $t(y)$ is an increasing function, and since $t(1) = 0$, $t(y) \geq 0$ for all $y \geq 1$. Thus, for $y \geq 1$, $t^2+y^2 \geq y^2$. Hence,

$$\frac{dt}{dy} \leq \frac{1}{y^2}, \qquad y \geq 1$$

If we integrate this inequality between $y = 1$ and y, we obtain

$$t(y) - t(1) \leq 1 - \frac{1}{y}$$

Using the fact that $t(1) = 0$ and $y > 0$, we discover that we have shown that

$$\frac{1}{1-t} \leq y$$

If we now return to the viewpoint that $y = y(t)$, and let t approach 1, we see at once that y must be unbounded.

This argument shows that the function $y(t)$ does become infinite at some point t_0 in the interval $[0, 1]$, but it does not give much information about its location. The next result not only gives a method for finding t_0 but also gives a useful formula for the function $y(t)$.

THEOREM **7.4** *The solution of* (7.79) *is a quotient of two power series that converge everywhere:*

$$y = \frac{1 + \frac{1}{3}t^3 - \frac{1}{4}t^4 - \frac{1}{3\cdot 4\cdot 7}t^7 + \frac{1}{4\cdot 5\cdot 8}t^8 + \cdots}{1 - t - \frac{1}{3\cdot 4}t^4 + \frac{1}{4\cdot 5}t^5 + \frac{1}{3\cdot 4\cdot 7\cdot 8}t^8 - \frac{1}{4\cdot 5\cdot 8\cdot 9}t^9 - \cdots}$$

The vertical asymptote of $y(t)$ occurs at t_0, the first zero of the power series in the denominator.

Proof In the equation (7.79), put $y = -u'/u$, where $u = u(t)$ is an unknown function of t. Since we then have

$$y' = -\frac{u''}{u} + \left(\frac{u'}{u}\right)^2 = -\frac{u''}{u} + y^2$$

equation (7.79) becomes merely the second-order linear equation

$$u'' = -t^2 u \tag{7.81}$$

Impose the initial condition $u(0) = 1$. Then, since $y = 1$ when $t = 0$, we must also have $u'(0) = -1$. We now solve equation (7.81) for $u(t)$ by power series, and easily find

$$u = 1 - t - \frac{1}{3\cdot 4}t^4 + \frac{1}{4\cdot 5}t^5 + \frac{1}{3\cdot 4\cdot 7\cdot 8}t^8 - \frac{1}{4\cdot 5\cdot 8\cdot 9}t^9 - \cdots$$

where the law of formation of the coefficients is one which permits us to find as many as we want (or in fact to find a general formula). Since $y = -u'/u$, we obtain the desired formula for y, as given in the theorem.

It is also possible to obtain rather accurate estimates for the first zero of the function u, which is of course the number t_0 that determines the location of the vertical asymptote of $y(t)$. The series for $u(t)$ converges very rapidly, since the coefficients decrease so quickly. If we estimate $u(t)$ by the first few terms of the series and then ask that $u(t_0) = 0$, we find that $t_0 = 0.9703$ is fairly close.

Methods of this sort can often be used to estimate the behavior of solutions of a differential equation which is hard to solve in other ways.

Exercises

1. Show that the solution of (7.79) obeying the initial condition $y(0) = 0$ is given by the formula

$$y = \frac{t^3}{3}\left(\frac{1 - \dfrac{1}{4\cdot 7}t^4 + \dfrac{1}{4\cdot 7\cdot 8\cdot 11}t^8 - \cdots}{1 - \dfrac{1}{3\cdot 4}t^4 + \dfrac{1}{3\cdot 4\cdot 7\cdot 8}t^8 - \cdots}\right)$$

2. Solve the equation $dy/dt = t^2 - y^2$, with $y(0) = 0$, and show that the solution is well behaved for all values of t.

3. (a) In the equation $dy/dt = \sqrt{t} + \sqrt{y}$, with $y(0) = 0$, make the substitutions $y = u^2$ and $t = s^4$ to obtain the equation

$$u\frac{du}{ds} = 2s^5 + 2s^3u$$

(b) Then, put $u = s^3v$, and show that the resulting equation for v has a power-series solution of the form $v = \sqrt{\frac{2}{3}} + \frac{2}{7}s + \cdots$ and thus obtain the solution

$$y = t^{3/2}\left(\tfrac{2}{3} + \tfrac{4}{7}\sqrt{\tfrac{2}{3}}\,t^{1/4} + \cdots\right)$$

4. Show that there is no real value of t that is a zero for the power series

$$1 - t^2 + 2t^4 - 4t^6 + 8t^8 - 16t^{10} + \cdots$$

(in spite of the alternating signs!).

Miscellaneous Word Problems

In the exercises that follow, we have included a variety of word problems which lead to differential equations and systems of equations. All can be solved by the methods given in this text. In some cases, the central problem will be to devise an appropriate model. We have also included some of the classical problems. It may be better to use these exercises as a basis for group research projects than as individual assignments. If nothing else, they illustrate the variety of areas to which one can apply elementary methods of analysis.

1. The motion of the end of a certain spring is governed by the equation

$$\frac{d^2s}{dt^2} + 3\frac{ds}{dt} + 2s = 0$$

   (a) Describe the motion that results from the initial conditions $t = 0$, $s = 1$, $ds/dt = -3$. [*Hint:* A sketch of the graph of s against t may help.]

   (b) How is the motion altered if a driving force of magnitude $10 \cos t$ is added to the right-hand side of the equation?

2. A series RLC circuit is governed by the differential equation

$$L\frac{d^2q}{dt^2} + R\frac{dq}{dt} + \frac{q}{C} = 0$$

   Find a general solution when $L = 1$, $R = 3$, and $C = \frac{1}{2}$.

3. A constant EMF, $E = 5$, is inserted into the circuit described in Exercise 2, so that the differential equation becomes

$$\frac{d^2q}{dt^2} + 3\frac{dq}{dt} + 2q = 5$$

Find a general solution.

4. It began to snow sometime in the evening. At midnight the plow started out and by 1 a.m. it had plowed a distance of 2 miles down the main road. During the next hour, however, it could cover only another mile. When did it start to snow? Assume it continues to snow at a constant rate and that the plow removes snow at a constant rate.*

5. The setup for a complicated mixing system is shown in Figure M.1. We have three tanks interconnected as indicated by pipes whose capacity of flow is 5 gal/min. Initially, tank I contains 20 gallons of red paint, tank II contains 30 gallons of yellow paint, and tank III contains 40 gallons of blue paint. What will be the mixture of paints in each tank at the end of 5 minutes?

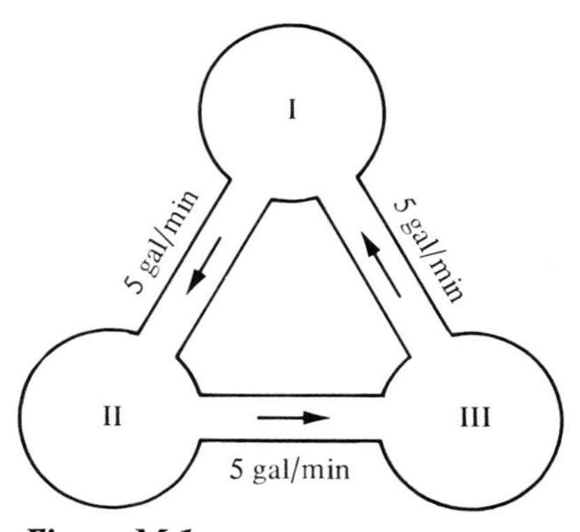

Figure M.1

6. The speed with which water will issue through a small hole in the bottom of a vessel is proportional to $\sqrt{2gh}$, where $g = 32$ ft/sec^2 and h is the depth of water above the hole. Compare the length of time it takes a conical tank to empty through a small hole in the apex of the cone, and through a hole of the same size in the flat top. (See Figure M.2.)

* This exercise comes from Professor R. P. Agnew, Cornell University.

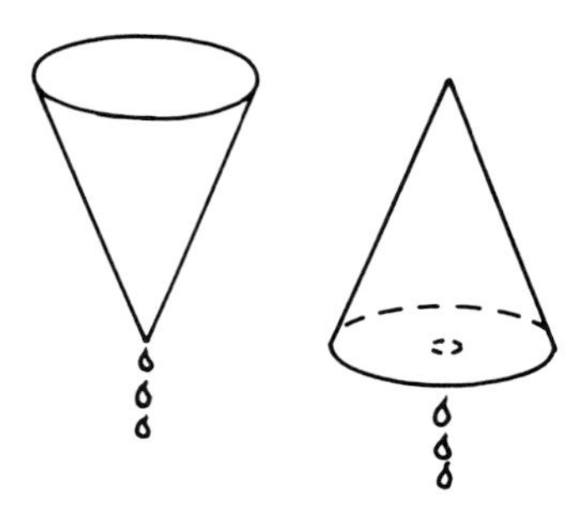

Figure M.2

7. A certain parasite is hatched only from eggs deposited in a host, and the host always dies. If x and y are the number of hosts and parasites alive at time t, then we assume that the number of eggs deposited is proportional to the product xy. Take account of the birth and (normal) death rates of the host and the death rates of the parasites, and show that a model for this situation is the system

$$\frac{dx}{dt} = ax - kxy$$

$$\frac{dy}{dt} = kxy - by$$

Assume that $a > 0$, $b > 0$, and solve the system. Then, plot the point (x, y) in the plane and examine the path of this point as t increases. What does this mean about the biological behavior of the system?

8. An electron of mass m and charge e moves in an electromagnetic field with electric intensity E and magnetic intensity H, starting from rest at the origin. If the position of the electron at time t is (x, y), then

$$m\frac{d^2x}{dt^2} + He\frac{dy}{dt} = Ee$$

$$m\frac{d^2y}{dt^2} - He\frac{dx}{dt} = 0$$

Show that the path of the electron is a cycloid of the form $x = A[1 - \cos(\beta t)]$, $y = A[\beta t - \sin(\beta t)]$.

9. Two tanks are connected as shown in Figure M.3. Each contains 50 liters of water. The first (I) contains 20 grams of dissolved salt, and the second (II) contains 4 grams. Pure water at 5 liters/sec flows into the system and a mixture flows out at the same rate. Assuming uniform mixing, what is the greatest amount of salt that is ever present in tank II?

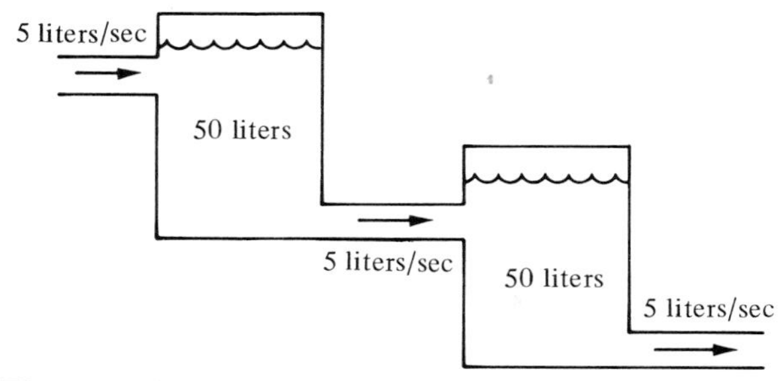

Figure M.3

10. When a large object sinks slowly into a liquid, the resistance is approximately proportional to the velocity of the object. If the object starts from rest, describe the motion.

11. A block of metal of unknown specific heat and weighing 150 pounds is heated to a uniform temperature of 200°. It is then dropped into an insulated container holding 300 pounds of water at 40° (specific heat 1). After 1 minute the water temperature is 45°, and after 2 minutes it is 48°. How long will it take until the metal is 64°? Show that the ultimate temperature of both is 52.5°.

12. Let P be the population of the United States at time t. A frequently used model for population growth assumes that there is a maximum size M that can be tolerated, and that the rate of population growth dP/dt is proportional to the product of P and $M-P$. Some census data are given in the accompanying table. Use the data for 1900, 1920, and 1940 to estimate the value of M, and compare this with the value of M which can be obtained from the data for 1940, 1950, 1960. What does this suggest about the suitability of the proposed model? What value of M would you obtain from the data for 1930, 1940, 1950?

Year of census	1900	1920	1930	1940	1950	1960
Population (10^6)	76	106	123	132	151	180

13. A tank which can hold 100 gallons is half full of a brine solution with a concentration of $\frac{1}{3}$ pound of salt per gallon. Two pipes lead into it at the top, one supplying a brine solution at $\frac{1}{2}$ lb/gal and the other pure water. Each pipe has a flow of 4 gal/min. One pipe leads out at the bottom and removes the mixture at 3 gal/min. What is the concentration of the mixture which first flows out the overflow pipe at the top? Assume uniform mixing.

14. It takes Jupiter about 12 years to go around the sun. What can you deduce about its distance from the sun? (The distance from the earth to the sun is about 93,000,000 miles.)

15. A cylindrical tank with *horizontal* axis is 8 feet long and 4 feet in diameter. If initially it is full of water, find the time required to empty it if there is a 1-inch circular hole in the bottom. Use $(0.6)\pi r^2\sqrt{2gh}$ as the rate of the volume of water in unit time flowing through the hole.

16. Consider the equation $dy/dt = y^2 - 4t$ with $y(0) = 1$. Is $y(1)$ positive or negative?

17. A series RLC circuit satisfies the equation

$$L\frac{d^2q}{dt^2} + R\frac{dq}{dt} + \frac{q}{C} = 0$$

(a) What is the general solution when $L = 1$, $R = 2$, and $C = \frac{1}{2}$?
(b) Find $q(t)$ if the initial conditions are $q = 1$, $dq/dt = 0$, when $t = 0$.

18. An EMF is inserted into the circuit of Exercise 17 which varies with time according to $E(t) = 1 + t$ so that the differential equation has $E(t)$ on the right-hand side. What is the general solution?

19. The shape of a slightly flexible rotating shaft, held by bearings at both ends, is governed by the equation

$$\frac{d^4y}{dx^4} = k\omega^2 y$$

where k depends on the material and ω is the angular velocity of the shaft. The appropriate boundary conditions are

$$x = 0,\ y = 0,\ \frac{dy}{dx} = 0$$

$$x = L,\ y = 0,\ \frac{dy}{dx} = 0$$

Show that the shaft remains straight except at certain critical choices of the velocity ω. Can you check this experimentally?

*20. Find all continuous (not necessarily differentiable) functions $f(t)$ such that

$$[f(t)]^2 = \int_0^t f(s)\,ds, \qquad t \geq 0$$

21. Consider a pair of 50 gallon tanks. (See Figure M.4). Initially tank I is full of compound B and tank II of compound C. We start to introduce compound A into each tank at the rate shown in the figure.
    (a) Draw curves showing what you would predict for the behavior of the amount of each of the three substances in each of the two tanks.
    (b) Solve the system of differential equations that govern this, and compare the results with your prediction.

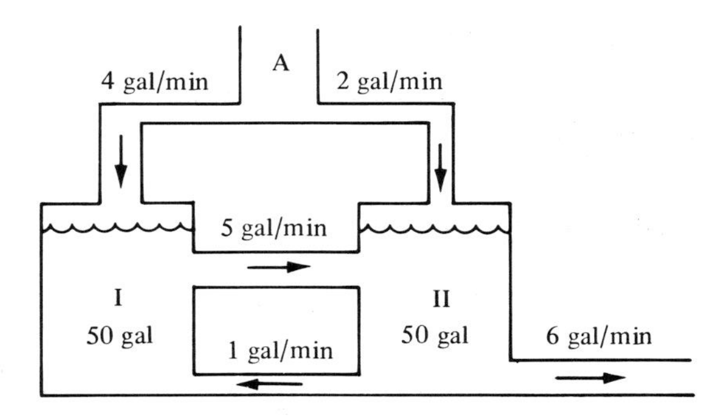

Figure M.4

22. The inside temperature of a cabin is 20 °C and the temperature outside is 0 °C. Inside the cabin is a large bathtub in which there is water at 50 °C. The fire in the cabin has just gone out and the temperature of the house and the water begin to fall. Assume Newton's law of temperature and set up a system of differential equations governing the behavior of the temperature of the air in the house and the water in the tub. In a $\frac{1}{2}$ hour, the air falls to 10 °C and the water to 28 °C. What can you expect in another half hour?

23. A large tank with a square bottom 10 feet on a side has a hole in the bottom through which water can drain out at a rate proportional to the square root of the height of the liquid above it. When the tank is empty, a valve is opened admitting water at 10 cubic feet per second. In one minute the water level is 4 feet. What will it be 4 minutes later? What will happen as time goes on, that is, as t becomes very large?

24. Consider the circuit shown in Figure M.5, having two capacitances and an inductance and resistor. Let V_1 and V_2 be the voltages across C_1 and C_2, respectively, and let I be the current through the inductance L, and let i_s be the current supplied by the source. Circuit theory leads to the

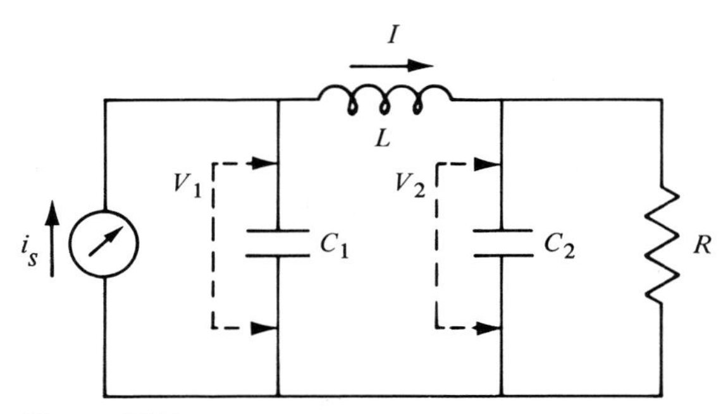

Figure M.5

following system of equations:

$$C_1 \frac{dV_1}{dt} = -I + i_s$$

$$C_2 \frac{dV_2}{dt} = I - RV_2$$

$$L \frac{dI}{dt} = V_1 - V_2$$

Solve this with $C_1 = \frac{5}{3}$, $C_2 = \frac{1}{6}$, $L = \frac{3}{5}$, $R = 1$, and $i_s = 2\cos(\beta t)$, and estimate the behavior for large t.*

* This exercise was suggested by Charles DeSoer.

Appendix 1

Laplace Transform

In Chapter 5, we showed that the solution of a homogeneous linear differential equation with constant coefficients

$$\frac{d^3y}{dt^3} + 3\frac{dy}{dt} - 4y = 0$$

depended upon our being able to find the algebraic roots of the associated polynomial equation $s^3+3s-4=0$, since these determine the null space $\mathcal{N}_T$ of the operator $T=D^3+3D-4$. Knowledge of these roots, in turn, leads (at least in theory) to the solution of the nonhomogeneous equation

$$\text{(A.1)} \qquad \frac{d^3y}{dt^3} + 3\frac{dy}{dt} - 4y = g(t)$$

for *any* choice of a continuous function on the right-hand side, since the methods discussed in Theorem 5.14, reduce everything to an exercise in indefinite integration.

However, as explained in the earlier parts of Section 5.4, there is a collection of choices for $g(t)$ which permit a particular solution of (A.1) to be found much more easily; samples of such choices are e^{-3t}, $(t^2-t)e^t$, and $e^{-2t}\sin(3t)$. They can be characterized as functions which themselves arise as members of the null space of some linear operator with constant coefficients. For obvious reasons, engineers prefer (if they have the choice) to work with differential equations that are easy to solve. It is fortunate that so many important physical problems can be treated in terms of such equations for which the right-hand function $g(t)$ is a linear combination of functions in this convenient class. The methods of Chapter 5 are entirely adequate to solve such problems.

The Laplace Transform provides another way to solve such problems. Although it has its origins in pure mathematics, the techniques have been discovered and exploited largely by electrical engineers during the period 1920–1950, possibly under the initial prodding of Charles Steinmetz, who introduced the use of complex numbers and other more abstract mathematical tools into the profession. The key idea behind the utility of the Laplace Transform in the solution of such differential equations is that it enables the user to follow a standard algorithm which converts the original differential equation into an algebraic problem, from which the solution of the original equation can then be found. [We remark that software programs are being written which largely automate this process, so that most of the solution can be done by a computer, and a person with access to such a program need only input the associated polynomial $p(s)$ determined by the linear differential operator, and the function $g(t)$—assuming that it is one of the special class mentioned above.]

Unfortunately, use of the Laplace Transform is confined to the solution of *linear* differential equations and *linear* systems, and many of the problems now being treated by scientists are strongly nonlinear. However, enough linear ones remain that the techniques are useful.

The Laplace Transform is also useful for advanced electrical engineering, since it provides physical insight into, and analogs for, certain mathematical procedures involved in the design of specialized equipment.

The Laplace Transform is an operator, denoted here by $\mathfrak{L}$, which can be applied to any function $f(t)$ in a class $\mathcal{E}$ of functions that do not grow too rapidly, and which sends $f(t)$ into some other function $F(x)$ in a different class. The formal definitions are as follows.

DEFINITION A.1 $\mathcal{E}$ is the class of continuous or piecewise continuous functions $f(t)$, defined at least for $t > 0$, which are of **exponential type**, meaning that they each obey a growth restriction of the form $|f(t)| \leq Ae^{Bt}$, $t > 0$, for some choice of A and B.

DEFINITION A.2 If f belongs to the class $\mathcal{E}$, then its Laplace Transform is a function $F(x)$, defined at least for all sufficiently large values of x by the integral formula

$$F(x) = \int_0^\infty e^{-xt} f(t)\, dt \tag{A.2}$$

For simplicity, we write $F = \mathfrak{L}\{f\}$ or $F(x) = \mathfrak{L}\{f(t)\}$.

To illustrate this, suppose $f(t) = e^{\beta t}$. This clearly belongs to the class $\mathcal{E}$. We compute $\mathfrak{L}\{f\}$ as follows:

$$\begin{aligned} F(x) &= \int_0^\infty e^{-xt} e^{\beta t}\, dt = \int_0^\infty e^{-(x-\beta)t}\, dt \\ &= (-1)\frac{e^{-(x-\beta)t}}{x-\beta}\Bigg|_{t=0}^{t=\infty} \\ &= (-1)\frac{0-1}{x-\beta} = \frac{1}{x-\beta} \end{aligned}$$

where, at the third step, we assumed that x was larger than β so that $\lim_{t\to\infty} e^{-(x-\beta)t} = 0$.

Summarizing this computation, we would say that the Laplace Transform of $e^{\beta t}$ is $1/(x-\beta)$, and write $\mathfrak{L}\{e^{\beta t}\} = 1/(x-\beta)$. By carrying out such integrations for a variety of choices of the function $f(t)$, one can build up an extensive table of pairs of functions $f(t)$ and $F(x)$ which are related by $F = \mathfrak{L}\{f\}$. The beginning of such a table is shown here as Table A.1; a more extensive table is given at the end of this section, and a reference to a very extensive table is given in the Bibliography, reference 20.

Table A.1

$f(t)$	$F(x) = \mathfrak{L}\{f(t)\}$
t	$\frac{1}{x^2}$
$e^{\beta t}$	$\frac{1}{x-\beta}$
$te^{\beta t}$	$\frac{1}{(x-\beta)^2}$
$\sin(\beta t)$	$\frac{\beta}{x^2+\beta^2}$
$t\cos(\beta t)$	$\frac{x^2-\beta^2}{(x^2+\beta^2)^2}$

The Laplace Transform is an integral operator, rather than a differential operator such as those discussed in Chapter 5. However, it shares with them the property of being a *linear* operator; if $\mathcal{L}\{f_1\} = F_1$ and $\mathcal{L}\{f_2\} = F_2$, then for any constants c_1 and c_2, $\mathcal{L}\{c_1 f_1 + c_2 f_2\} = c_1 F_1 + c_2 F_2$. For example, using Table A.1, we see that

$$\mathcal{L}\{3e^{-t} + 5te^{2t}\} = \frac{3}{x+1} + \frac{5}{(x-2)^2}$$

The key to the usefulness of the Laplace Transform in ordinary differential equations is the effect it has on the derivative of a function. This is given in the following theorem, whose proof we postpone to the end of this section.

THEOREM A.1 *Let f be a function such that f and f' are in $\mathcal{E}$. Then,*

$$\mathcal{L}\left\{\frac{d}{dt} f(t)\right\} = \mathcal{L}\{Df\} = -f(0) + x\mathcal{L}\{f\} \tag{A.3}$$

If we write $F(x) = \mathcal{L}\{f(t)\}$, then this relation can be written

$$\mathcal{L}\{Df\} = xF(x) - f(0) \tag{A.4}$$

If f is such that f'' also lies in $\mathcal{E}$, then the process can be repeated, leading to another relation:

$$\mathcal{L}\{D^2 f\} = \mathcal{L}\left\{\frac{d^2}{dt^2} f(t)\right\} = x^2 F(x) - xf(0) - f'(0) \tag{A.5}$$

To see how this follows from (A.4), we write $f'' = Df'$, and then apply Theorem A.1, obtaining

$$\mathcal{L}\{f''\} = \mathcal{L}\{Df'\} = -f'(0) + x\mathcal{L}\{f'\}$$

If we write $f' = Df$ and use (A.4) to calculate the right-hand side, we have

$$\begin{aligned} \mathcal{L}\{f''\} &= -f'(0) + x(xF(x) - f(0)) \\ &= x^2 F(x) - xf(0) - f'(0) \end{aligned}$$

In the same way, one finds that

$$\mathcal{L}\left\{\frac{d^3}{dt^3} f(t)\right\} = x^3 F(x) - x^2 f(0) - xf'(0) - f''(0)$$

The general case is given by the corollary that follows.

COROLLARY 1 *If f and its first n derivatives belong to $\mathcal{E}$, then*

$$\mathcal{L}\{D^n f\} = x^n \mathcal{L}\{f\} - [f(0)x^{n-1} + f'(0)x^{n-2} + \cdots + f^{(n-1)}(0)] \tag{A.6}$$

The importance of this is that it enables one to compute the Laplace Transform of any derivative of a function in a simple and routine way, if one knows the transform of the function itself. (The limitation that all the derivatives of f must belong to $\mathcal{E}$ is a severe one, and much of the more advanced work with Laplace Transform methods deals with ways to overcome this; this more recent approach is called the *theory of distributions*.)

If the function $f(t)$ were to be such that $f(0) = f'(0) = \cdots = f^{(n-1)}(0) = 0$, formula (A.6) would become simply

$$\mathcal{L}\{D^n f\} = x^n \mathcal{L}\{f\}$$

In this case, we can obtain a very useful generalization of the result, which we state as follows.

COROLLARY 2 *Let T be a differential operator with constant coefficients:*

$$T = D^n + a_1 D^{n-1} + a_2 D^{n-2} + \cdots + a_n$$

Let f and all its derivatives belong to $\mathcal{E}$, and obey the special requirement that $f(0) = f'(0) = \cdots = f^{(n-1)}(0) = 0$. Then, $\mathcal{L}\{Tf\} = p(x)F(x)$, where $F = \mathcal{L}\{f\}$ and where $p(x)$ is the associated polynomial for T given by

$$p(x) = x^n + a_1 x^{n-1} + \cdots + a_n$$

Stated in words rather than symbols, this says that for a special class of functions $f(t)$, the Laplace Transform converts the effect of a differential operator T into multiplication by a polynomial $p(x)$, which in fact is the polynomial that is associated with T. (Unfortunately, this useful property is restricted to operators with *constant* coefficients; thus, no such simple relation holds for the case of linear operators with variable coefficients, which again limits the usefulness of Laplace Transform methods considerably.)

How is all this used in solving a differential equation? Suppose we want to solve the second-order equation

$$y'' + y = 3e^{-2t} \tag{A.7}$$

Since $T = D^2 + 1$ and the associated polynomial has roots i and $-i$, we know that the null space can be generated by $\{\sin t, \cos t\}$. To complete the solution, we need one particular solution of the nonhomogeneous equation (A.7). Let us assume that $y(t)$ is such a solution, and that it and its derivatives belong to $\mathcal{E}$, and that it obeys the initial conditions $y(0) = y'(0) = 0$. Although we do not know the function $y(t)$, we denote its Laplace Transform $Y(x) = \mathcal{L}\{y(t)\}$. We take the Laplace Transform of both sides of (A.7), using Table A.1 to calculate the right side and Corollary 2 to calculate the left. Accordingly, we have

$$(x^2+1)\,Y(x) = \frac{3}{x+2}$$

We solve this simple algebraic equation for the unknown function $Y(x)$, and arrive at

(A.8) $$Y(x) = \frac{3}{(x^2+1)(x+2)}$$

Since $Y(x) = \mathcal{L}\{y(t)\}$, we can read (A.8) as saying that our desired solution $y(t)$ is a function whose Laplace Transform is the rational function given on the right-hand side of (A.8).

If we had an infinitely long complete list of *all* Laplace Transforms, we could search through it and eventually find exactly one match, namely,

(A.9) $$y(t) = \tfrac{3}{5}e^{-2t} + \tfrac{6}{5}\sin t - \tfrac{3}{5}\cos t$$

[It is easy to check that this is the desired solution of (A.7), either by substituting it into (A.7), or by solving (A.7) using the methods of Chapter 5.] Of course, it is not possible to list *all* possible Laplace Transforms. Instead, as in the familiar table of integrals, one finds only certain standard forms, and then one learns methods by which these can be combined to yield others. One of the procedures that is most often used is **partial fraction decomposition**. This is usually learned in connection with the indefinite integration of rational functions or, in a class in modern algebra, in connection with the study of the field of rational functions. In the example before us, we would rewrite the right-hand side of (A.8) as

$$\frac{3}{(x^2+1)(x+2)} = \frac{A}{x^2+1} + \frac{Bx}{x^2+1} + \frac{C}{x+2}$$

and find the values of A, B, and C that make this an identity. If we multiply both sides by $(x^2+1)(x+2)$ and collect like powers of x, and then equate coefficients, we arrive at a set of algebraic linear equations for A, B, and C, namely,

$$2A + C = 3$$

$$A + 2B = 0$$

$$B + C = 0$$

Solving these, we find $A = \frac{6}{5}$, $B = -\frac{3}{5}$, $C = \frac{3}{5}$. Hence, the desired partial fraction decomposition is

$$\frac{3}{(x^2+1)(x+2)} = \left(\frac{6}{5}\right)\frac{1}{x^2+1} - \left(\frac{3}{5}\right)\frac{x}{x^2+1} + \left(\frac{3}{5}\right)\frac{1}{x+2}$$

If we now refer to the abbreviated table of Laplace Transforms at the end of this section, we see that these terms appear as the transforms of the function given in (A.9), giving us $y(t)$ as asserted.

At this point, it is mathematically important to observe that we have made use of a basic theorem about Laplace Transforms: *Two functions that have the same transform are identical.* It is this fact which allows us to conclude that the unknown function $y(t)$ whose transform was given in (A.8) must be exactly the function given in (A.9). Unfortunately for teachers and elementary textbook writers, this basic result does not have an elementary proof and so must be taken on faith. (See reference 20 in the Bibliography.)

We can now complete the solution of the differential equation (A.7) by combining the particular solution (A.9) with the null space found earlier. Every solution of (A.7) is given in the form

$$y = \tfrac{3}{5}e^{-t} + \tfrac{6}{5}\sin t - \tfrac{3}{5}\cos t + C_1 \sin t + C_2 \cos t$$

The second and third expressions in this sum can be combined with the last two since C_1 and C_2 are arbitrary constants, thus obtaining the simpler form

$$y = \tfrac{3}{5}e^{-2t} + c_1 \sin t + c_2 \cos t$$

However, we have not explained one of the advantages of the transform method. Suppose we had been interested only in finding a solution of the differential equation obeying a specific set of initial conditions—for example, a solution $y(t)$ with $y(0) = 1$, $y'(0) = -1$. If we go through the transform method again, using Corollary 1 instead of Corollary 2, we can go directly to the desired y without first finding the general solution of the differential equation.

As before, let $Y(x) = y(t)$. Then, by (A.5),

$$\begin{aligned} \mathfrak{L}\{y''(t)\} &= x^2 Y(x) - (y(0)x + y'(0)) \\ &= x^2 Y(x) - x + 1 \end{aligned}$$

Hence,

$$\begin{aligned} \mathfrak{L}\{y'' + y\} &= x^2 Y(x) - x + 1 + Y(x) \\ &= (x^2 + 1) Y(x) - x + 1 \end{aligned}$$

and so the equation (A.7) leads first to

$$\mathfrak{L}\{y'' + y\} = \mathfrak{L}\{3e^{-2t}\}$$

and then to

$$(x^2 + 1) Y(x) - x + 1 = \frac{3}{x+2}$$

from which we obtain

$$Y(x) = \frac{3}{(x^2+1)(x+2)} + \frac{x-1}{x^2+1}$$

The method of partial fractions, applied to this and combined with the table of Laplace Transforms, gives the desired solution:

$$y(t) = \tfrac{3}{5}e^{-2t} + \tfrac{1}{5}\sin t + \tfrac{2}{5}\cos t$$

As this example may suggest, the use of the Laplace Transform does not often lead to a dramatic saving in arithmetical calculations. Indeed, in many cases the arithmetic processes are just about the same, whether one uses the methods of Chapter 5 or the Laplace Transform, except that the order in which they are done may be different. For example, to solve the equation

$$\text{(A.10)} \qquad \frac{d^3y}{dt^3} + 3\frac{dy}{dt} - 4y = e^{-t}$$

by the methods of Chapter 5, we need the roots of the associated polynomial $p(s) = s^3+3s-4$, and must therefore solve a cubic equation. (The roots turn out to be 1, and $-\frac{1}{2} \pm \sqrt{15}\,i$.)

If, instead, we use Laplace Transform, we convert (A.10) into the algebraic equation

$$(x^3+3x-4)\,Y(x) = \frac{1}{x+1}$$

and easily find

$$Y(x) = \frac{1}{(x^3+3x-4)(x+1)}$$

However, to find $y(t)$ by partial fraction decomposition of the right-hand side of this, we have to factor the denominator, and this again requires that we solve the same cubic equation, $x^3+3x-4 = 0$.

Transform methods are more useful when it comes to solving *systems* of equations with constant coefficients. In Chapter 6, we discussed several methods for doing this, some involving the determination of eigenvalues and eigenvectors for a matrix. For the simplest cases, the Laplace Transform does have some utility. Let us consider the following problem.

Find $u(t)$ and $v(t)$, if

$$\text{(A.11)} \qquad \begin{aligned} \frac{du}{dt} &= 2u - 3v + \sin(2t) \qquad & u(0) &= 0 \\ \frac{dv}{dt} &= 3u - 4v - 2\cos t \qquad & v(0) &= 3 \end{aligned}$$

We apply the transform $\mathcal{L}$ to both sides of each equation, writing $U(x)$ for $\mathcal{L}\{u(t)\}$ and $V(x)$ for $\mathcal{L}\{v(t)\}$. Using formula (A.4) to calculate the transforms of du/dt and dv/dt, and the table at the end of this section for the transforms of $\sin(2t)$ and $\cos t$, we obtain a pair of algebraic equations:

$$xU(x) - 0 = 2U(x) - 3V(x) + \frac{2}{x^2+4}$$

$$xV(x) - 3 = 3U(x) - 4V(x) - 2\frac{x}{x^2+1}$$

We rewrite these in the form

$$(x-2)\,U(x) + 3V(x) = \frac{2}{x^2+4}$$

$$-3U(x) + (x+4)\,V(x) = 3 - \frac{2x}{x^2+1}$$

and then solve them for $U(x)$ and $V(x)$, obtaining

(A.12)
$$U(x) = \frac{1}{(x+1)^2}\left(\frac{2x+8}{x^2+4} + \frac{6x}{x^2+1} - 9\right)$$
$$V(x) = \frac{1}{(x+1)^2}\left(\frac{6}{x^2+4} + 3(x-2) - \frac{2x^2-4x}{x^2+1}\right)$$

The method of partial fractions can now be applied to each of these, in order to make it possible to use the transform table to find $u(t)$. Thus, we find that the first equation can be rewritten as

$$U(x) = \frac{0.88}{x+1} - \frac{10.8}{(x+1)^2} - \frac{0.32}{x^2+4} - \frac{0.88x}{x^2+4} + \frac{3}{x^2+1}$$

Using the transform table, we obtain

(A.13)
$$u(t) = 0.88e^{-t} - 10.8te^{-t} - 0.16\sin(2t) - 0.88\cos(2t) + 3\sin t$$

In a similar fashion, we can obtain

(A.14)
$$v(t) = 4.48e^{-t} - 10.8te^{-t} - 0.36\sin(2t) - 0.48\cos(2t) + 2\sin t - \cos t$$

and these are the unique solutions of the system (A.11), obeying the required initial conditions.

The exponential terms in these solutions are called "transients," since their effect is extremely small for large values of t because of the negative exponent [which of course is the eigenvalue of the coefficient matrix of the system (A.11)]. Thus, the most important terms in the solutions for $u(t)$ and $v(t)$ are the "steady state" portions toward which they converge as t increases. These are

$$u_0(t) = -0.16 \sin(2t) - 0.88 \cos(2t) + 3 \sin t$$

$$v_0(t) = -0.36 \sin(2t) - 0.48 \cos(2t) + 2 \sin t - \cos t$$

With experience the transform technique enables one to find these directly from the transformed equations by determining the correct coefficients for only the terms $1/(x^2+4)$, $x/(x^2+4)$, $1/(x^2+1)$, and $x/(x^2+1)$ in the partial fraction decomposition of $U(x)$ and $V(x)$, as given by (A.12).

We conclude this brief outline of the use of the Laplace Transform by giving a proof for Theorem A.1 and mentioning several other useful properties of $\mathcal{L}$.

The proof of Theorem A.1 follows immediately from (A.2) by an integration by parts. Alternatively, one may proceed as follows. First, observe that

$$\frac{d}{dt}(e^{-xt}f(t)) = -xe^{-xt}f(t) + e^{-xt}f'(t)$$

Then, integrate this between zero and ∞, obtaining

$$\int_0^\infty \frac{d}{dt}(e^{-xt}f(t))\,dt = \int_0^\infty -xe^{-xt}f(t)\,dt + \int_0^\infty e^{-xt}f'(t)\,dt$$

Since the left-hand side is merely the integral of a derivative, while each of the integrals on the right resembles (A.2), this identity becomes

$$e^{-xt}f(t)\Big|_{t=0}^{t=\infty} = -x\mathcal{L}\{f(t)\} + \mathcal{L}\{f'(t)\} \tag{A.15}$$

To evaluate the left-hand side of (A.15), we use the fact that f belongs to the class $\mathcal{E}$. In particular, we have the estimate

$$|e^{-xt}f(t)| \leq e^{-xt}Ae^{Bt} = Ae^{(B-x)t}$$

so that if $x > B$, $\lim_{t\to\infty} e^{-xt}f(t) = 0$. Returning to (A.15), we have

$$0 - e^0 f(0) = -x\mathcal{L}\{f(t)\} + \mathcal{L}\{f'(t)\}$$

or

$$\mathcal{L}\{f'\} = x\mathcal{L}\{f\} - f(0)$$

verifying (A.3).

It is natural to ask if there are other identities involving the operator $\mathcal{L}$ which permit one to find Laplace Transforms or inverse Laplace Transforms more easily.

THEOREM **A.2** *If f belongs to the class $\mathcal{E}$, and $\mathcal{L}\{f\} = F$, then*

$$\mathcal{L}\{t^m f(t)\} = (-1)^m \frac{d^m}{dx^m} F(x) \tag{A.16}$$

THEOREM **A.3** *Let f and g belong to $\mathcal{E}$, and let h be their* **convolution**, *$h = f * g$, defined by the formula*

$$h(t) = \int_0^t f(t-s)g(s)\,ds \tag{A.17}$$

Then,

$$\mathcal{L}\{h\} = \mathcal{L}\{f\}\mathcal{L}\{g\} \tag{A.18}$$

Both of these results are used frequently, both to calculate the transform of some function and to recognize the function having a known transform. For example, you may have reached the stage in which you have found a formula for the transform of an unknown function $y(t)$, and have

$$\mathcal{L}\{y(t)\} = Y(x)$$

You notice that $Y(x)$ can be expressed as the product of two standard transforms, so that you have

$$\mathcal{L}\{y(t)\} = F(x)\,G(x)$$

where $F = \mathcal{L}\{f\}$ and $G = \mathcal{L}\{g\}$. Then, Theorem A.3 tells you at once that $y(t)$ is the convolution of f and g, and can therefore be calculated by means of (A.17). As an illustration, suppose we have

$$\mathcal{L}\{y(t)\} = \frac{x}{(x^2+1)(x^2+4)} \tag{A.19}$$

Write the right-hand side as

$$\frac{1}{2}\left(\frac{x}{x^2+1}\right)\left(\frac{2}{x^2+4}\right)$$

which we note is the product of the transforms of two standard functions, $f(t) = \frac{1}{2}\cos t$ and $g(t) = \sin(2t)$. Hence, we have $y(t) = (f*g)(t)$, and using (A.17), we write

$$y(t) = \tfrac{1}{2}\int_0^t \cos(t-s)\sin(2s)\,ds$$

This can be integrated, using for example some of the familiar trigonometric identities such as

$$\sin A \cos B = \tfrac{1}{2}(\sin(A+B) + \sin(A-B))$$

The final result is

$$y(t) = \tfrac{1}{3}\cos t - \tfrac{1}{3}\cos(2t)$$

which can be seen to confirm (A.19).

Table A.1 *An Abbreviated Table of Laplace Transforms:* $F(x) = \mathfrak{L}\{f(t)\}$

$f(t)$	$F(x)$	$f(t)$	$F(x)$
0	0	$t\sin(\beta t)$	$\dfrac{2\beta x}{(x^2+\beta^2)^2}$
1	$\dfrac{1}{x}$		
t	$\dfrac{1}{x^2}$	$t\cos(\beta t)$	$\dfrac{x^2-\beta^2}{(x^2+\beta^2)^2}$
t^n	$\dfrac{n!}{x^{n+1}}$		
$\dfrac{1}{\sqrt{t}}$	$\dfrac{\sqrt{\pi}}{\sqrt{x}}$	$\sin(\beta t) - \beta t\cos(\beta t)$	$\dfrac{2\beta^3}{(x^2+\beta^2)^2}$
t^β	$\dfrac{\Gamma(\beta+1)}{x^{\beta+1}}$		
$e^{\beta t}$	$\dfrac{1}{x-\beta}$	$(3-\beta^2t^2)\sin(\beta t) - 3\beta t\cos(\beta t)$	$\dfrac{8\beta^5}{(x^2+\beta^2)^3}$
$te^{\beta t}$	$\dfrac{1}{(x-\beta)^2}$		
$t^n e^{\beta t}$	$\dfrac{n!}{(x-\beta)^{n+1}}$	$\dfrac{\sin(\beta t)}{t}$	$\arctan\dfrac{\beta}{x}$
$\sin(\beta t)$	$\dfrac{\beta}{x^2+\beta^2}$		
$\cos(\beta t)$	$\dfrac{x}{x^2+\beta^2}$	$J_0(\beta t)$ (Bessel function)	$\dfrac{1}{\sqrt{x^2+\beta^2}}$
$e^{\alpha t}\sin(\beta t)$	$\dfrac{\beta}{(x-\alpha)^2+\beta^2}$		
$e^{\alpha t}\cos(\beta t)$	$\dfrac{x-\alpha}{(x-\alpha)^2+\beta^2}$	$\delta(t)$ (Dirac delta function)	1

Exercises Using Laplace Transform

1. Use partial fractions to find the function $f(t)$ whose transform is

   (a) $F(x) = \dfrac{2x+3}{(x+1)(x+4)}$ (b) $F(x) = \dfrac{x^2+3x+1}{(x^2+9)(x+2)}$

2. Solve the equation $y''+5y'+6y = 5 \sin t$, with $y(0) = 1$, $y'(0) = -2$.

3. Solve $y''+2y'+5y = 3e^{-2t}+4$, with $y(0) = 0$, $y'(0) = 1$.

4. Solve the system $du/dt = 2u-4v+e^{-t}$, $dv/dt = 3v-e^{-2t}$, with $u(0) = 1$, $v(0) = 2$.

5. Show that if you know the transform of $f(t)$, it is easy to find the transform of $e^{\beta t}f(t)$.

6. Verify that the functions given in (A.13) and (A.14) in fact satisfy the system (A.11).

7. Solve the system (A.11) by the methods developed in Chapter 6.

8. Obtain formula (A.16) in Theorem A.2 by repeated differentiation of formula (A.2).

9. Use Theorem A.2, and knowledge of the transform of $\cos(\beta t)$, to calculate the transform of $t \cos(\beta t)$ and $t^2 \cos(\beta t)$.

10. Use Theorem A.3 to find the function whose transform is

    (a) $x^3/[(x^2+4)(x^2+9)]$ (b) $x(x+1)^{-2}(x^2+4)^{-1}$

11. Show that if f'' belongs to the class $\mathcal{E}$, so does f.

*12. Prove Theorem A.3 by substituting (A.17) in the formula for $\mathcal{L}\{h(t)\}$.

*13. Do you think that the function $F(x) = x^3$ could be the transform of some function $f(t)$ in the class $\mathcal{E}$?

Appendix 2

Numerical Solutions

There are many situations in which one must find the approximate roots of a polynomial or more complicated function. For example, this is an essential step in solving a linear differential equation. In such cases, several simple numerical algorithms are often useful in finding real roots. The first step in solving $f(x) = 0$ is to find two numbers a and b such that $f(a)f(b) < 0$ and such that f is continuous on $[a, b]$. You are then sure that there is at least one root $\bar{x}$ of the equation $f(x) = 0$ in the interval $[a, b]$. (These numbers a and b sometimes must be found by trial and error.)

The Bisection Method

This consists merely in calculating the value of f at a sequence of points x_n and observing the sign of $f(x_n)$, and using this information to choose a sequence of subintervals of $[a, b]$ in such a way that you can be sure that they form a nested decreasing sequence containing $\bar{x}$. Each succeeding x_n is often chosen as the midpoint of the previously chosen subinterval. Clearly, the sequence $\{x_n\}$ will converge to $\bar{x}$.

The Secant Method

Start with a pair of estimates for $\bar{x}$, labeled x_1 and x_2. Define a sequence $\{x_n\}$ by the recursion

$$x_{n+1} = \frac{x_n f(x_{n-1}) - x_{n-1} f(x_n)}{f(x_{n-1}) - f(x_n)}$$

Depending upon the nature of f and the location of x_1 and x_2, the sequence $\{x_n\}$ will often converge to a root $\bar{x}$. The rationale for this algorithm comes from Figure A.1.

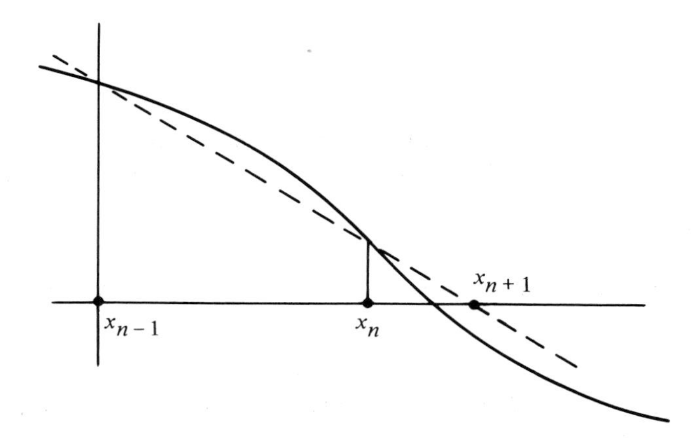

Figure A.1

Newton's Method

Start with one estimate x_1 for the root $\bar{x}$, and define a sequence $\{x_n\}$ by the recursion

$$x_{n+1} = x_n - \frac{f(x_n)}{f'(x_n)}$$

Depending on the nature of f and the location of x_1, the sequence $\{x_n\}$ will often converge to $\bar{x}$.

This is a limiting case of the secant method, obtained by allowing x_n and x_{n-1} to coincide. It can also be obtained from Figure A.2; it is also clear that problems will arise if any term x_n is near a zero of f'.

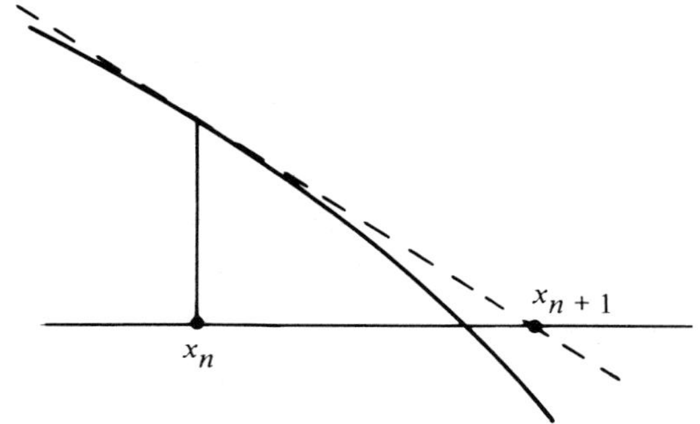

Figure A.2

Iteration

The given equation $f(x) = 0$ can be rewritten, with a different function F, in the form $x = F(x)$. Start with an estimate x_1 for the desired root $\bar{x}$, and define a sequence $\{x_n\}$ by

$$x_{n+1} = F(x_n)$$

If $|F'(x)| < 1$ and x_1 is sufficiently close to $\bar{x}$, the sequence $\{x_n\}$ will converge to $\bar{x}$.

As illustration, we apply each of these methods to find the roots of the equation $x^3 - 5x + 3 = 0$. (This has three real roots; $\alpha = -2.490863615$, $\beta = 0.656620431$, $\gamma = 1.834243184$.) Our first step is to evaluate the function $f(x) = x^3 - 5x + 3$ at a number of values of x.

x	-3	-2	-1	0	1	2	3
$f(x)$	-9	5	7	3	-1	1	15

From this table, we conclude that there is a root in each of the intervals $[-3, -2]$, $[0, 1]$, and $[1, 2]$.

If we apply the Secant Method on $[1, 2]$ with initial terms $x_1 = 1$ and $x_2 = 2$, we obtain the following data:

x_n	$f(x_n)$
1	-1
2	1
1.5	-1.125
1.7647	-0.32791
1.8735995	0.2090397

cont.

x_n	$f(x_n)$
1.8312058	−0.0154195
1.8341181	−0.0006369
1.8342436	0.0000021
1.8342432	0.000000...

Within the limits of the calculations, we have found a root.

If we next apply Newton's Method, using $f(x) = x^3 - 5x + 3$ and $f'(x) = 3x^2 - 5$, with $x_1 = 1$ as starting point, the result is shown in the following table.

x_n	$f(x_n)$
1	−1
0.5	0.6250
0.6470588	0.0356198
0.6565728	0.0001766
0.65662043	0.00000...

Thus, we have the root in [0, 1]. If we instead choose $x_1 = 2$ as the start, we obtain the following results:

x_n	$f(x_n)$
2	1
1.8571429	0.1195335
1.8347874	0.0027733
1.8342435	0.0000016
1.8342432	0.00000...

Thus different starts lead to different roots. If we had chosen $x_1 = -2$, the sequence would have converged to the third root.

To apply Iteration, we must rewrite the original equation to match the form $x = F(x)$. It is easily seen that one choice for F is $F(x) = \frac{1}{5}(x^3 + 3)$. With this selection, and with any choice of x_1 not too far away from 0.6, the resulting sequence converges to the root in [0, 1]. For example, if $x_1 = 1$, we obtain the sequence $x_2 = 0.8$, $x_3 = 0.7024$, $x_4 = 0.669308$, $x_5 = 0.6599664$, $x_6 = 0.6574904$, and eventually $x_{12} = 0.6566207$. Similarly, if $x_1 = -1$, the sequence is $x_2 = 0.4$, $x_3 = 0.6128$, $x_4 = 0.646024$, and $x_{14} = 0.6566204$. However, starting points that are near to the other roots do not produce convergent sequences but instead sequences that oscillate wildly. For example, if $x_1 = 2$, then $x_2 = 2.20$, $x_3 = 2.7296$, $x_4 = 4.6675$, $x_5 = 20.94$, $x_6 = 1836.1$,

$x_7 = 1.24 \times 10^9$. A similar behavior is found for the sequence starting with $x_1 = -2.5$.

The equation $x^3 - 5x + 3 = 0$ can also be rewritten in the form $x^3 = 5x - 3$, and thus put into the form $x = F(x)$ by choosing $F(x) = (5x-3)/x^2$. Iteration of this acts differently. For example, if $x_1 = 2$, then $x_2 = 1.75$, $x_3 = 1.87755$, $x_{10} = 1.833834$, $x_{15} = 1.8342578$, and $x_{20} = 1.8342427$.

Of these methods, Newton's usually converges faster, but calculation of both $f(x_n)$ and $f'(x_n)$ can be time consuming. In addition, special difficulties can arise due to multiple roots or zeros of f'. The simple bisection method has great advantages, although it is often slower to converge. Iteration is sometimes the easiest to use, but may not converge, as shown above.

In the introduction to Chapter 4, we mentioned that students who have access to a programmable computer, and who wish to test their ability to use the algorithms discussed in Chapter 4 and elsewhere to solve differential equations, were encouraged to do so. In the pages to follow, we give numerical results which we obtained in solving a number of equations, using the following methods: Midpoint, Classical RK, AB_1 with Midpoint starter, AB_2 with Classical RK starter, PC_1 with RK starter, and the algorithm for solving second-order equations given in Chapter 4.

These will permit a student who has written his own program for one of these algorithms to compare his numerical results with ours; small differences are to be expected. Our answers were obtained on an HP65 pocket calculator.

Teaching computer programming is not an objective of this book and we are reluctant to give lengthy programs. If a student is prepared to write his own FORTRAN program for the RK methods, he does not need instruction from us. We have chosen to include the RK program we used for the HP65 for the benefit of those who are as much a novice as we, and who would like to see a simple embodiment of this algorithm. Since we believe that, in the future, many more students will have access to this class of minimal computers, we encourage others to experiment with them as an aid to understanding mathematics.

Equation: $\dfrac{dy}{dt} = y^2, \qquad 0 \le t \le 1.4$

$y(0) = 1$

Method: Midpoint, with $h = 0.1$

t	y
0	1.00000
0.1	1.11025
0.2	1.24758
0.3	1.42325
0.4	1.65567
0.5	1.97706
0.6	2.44903
0.7	3.20469
0.8	4.58719
0.9	7.76736
1.0	19.39673
1.1	165.38488
1.2	2.35×10^{5}
1.3	7.65×10^{17}
1.4	8.55×10^{67}

Equation: $\dfrac{dy}{dt} = \dfrac{3y}{t}, \qquad 1 \le t \le 3$

$y(1) = 1$

Method: Midpoint, with $h = 0.1$

t	y	t	y
1.0	1.0000	2.1	9.1834
1.1	1.3286	2.2	10.5563
1.2	1.7224	2.3	12.0598
1.3	2.1875	2.4	13.6997
1.4	2.7297	2.5	15.4821
1.5	3.3549	2.6	17.4128
1.6	4.0692	2.7	19.4778
1.7	4.8784	2.8	21.7430
1.8	5.7885	2.9	24.1542
1.9	6.8054	3.0	26.7378
2.0	7.9351		

Equation: $\dfrac{dy}{dt} = t^2 + y^2, \qquad 0 \le t \le 1$

$y(0) = 1$

Method: Classical RK, with $h = 0.1$

t	y
0	1.00000
0.1	1.11146
0.2	1.25302
0.3	1.43967
0.4	1.69610
0.5	2.06696
0.6	2.64386
0.7	3.65220
0.8	5.84201
0.9	14.02182
1.0	735.09915

Equation: $\dfrac{dy}{dt} = 2ty - t^2, \qquad 0 \le t \le 3$

$$y(0) = \begin{cases} 0.44 \\ 0.45 \end{cases}$$

Method: Classical RK, with $h = 0.5$

t	y	
0	0.44000	0.45000
0.5	0.51900	0.53184
1.0	0.68110	0.70823
1.5	0.86289	0.95662
2.0	0.94769	1.46854
2.5	−0.10307	4.44622
3.0	−17.74787	43.03379

Equation: $\dfrac{dy}{dt} = \dfrac{2ty}{t^2 - y^2}, \qquad 0 \le t \le 2$

$y(0) = 2$

Method: Classical RK, with $h = 0.2$

t	y
0	2.00000
0.2	1.97980
0.4	1.91651
0.6	1.79999
0.8	1.59991
1.0	1.06013
1.2	0.59382
1.4	0.94512
1.6	2.06173
1.8	5.48234
2.0	5.32169

Equation: $\dfrac{dy}{dt} = 3t^2 y \cos(t^3), \qquad 0 \le t \le 3$

$y(0) = 1$

Method: Classical RK, with $h = 0.5$, 0.25, 0.1

t	y		
	$h = 0.5$	$h = 0.25$	$h = 0.1$
0	1.00000	1.00000	1.00000
0.5	1.13205	1.13274	1.13278
1.0	2.29420	2.31801	2.31974
1.5	0.57442	0.75520	0.79413
2.0	−0.13045	1.69253	2.66668
2.5	0.87074	−3.45251	0.83565
3.0	13.57290	1.84194	0.45631
3.5	−2938.65		

Equation: $\dfrac{dy}{dt} = 3t^2 y \cos(t^3), \qquad 0 \le t \le 3$

$y(0) = 1$

Method: Classical RK, with $h = 0.2$

t	y
0	1.000000
0.2	1.008030
0.4	1.066039
0.6	1.238990
0.8	1.632012
1.0	2.319075
1.2	2.683718
1.4	1.466591
1.6	0.440873
1.8	0.636752
2.0	2.368537
2.2	0.219606
2.4	0.900971
2.6	2.852760
2.8	−6.984984
3.0	14.750214

Equation: $\dfrac{dy}{dt} = y^2, \qquad 0 \le t \le 1$

$y(0) = 1$

Method: AB_1 with Midpoint starter, $h = 0.1$

t	y
0	1.00000
0.1	1.11025
0.2	1.24758
0.3	1.43420
0.4	1.68651
0.5	2.04329
0.6	2.59085
0.7	3.50315
0.8	5.24662
0.9	9.45887
1.0	24.77337

Equation: $\dfrac{dy}{dt} = t^2 + y^2, \qquad 0 \leq t \leq 1$

$y(0) = 1$

Method: AB_1 with classical RK starter, $h = 0.1$

t	y
0	1.00000
0.1	1.11146
0.2	1.25302
0.3	1.43726
0.4	1.68990
0.5	2.05232
0.6	2.60290
0.7	3.52147
0.8	5.28739
0.9	9.57225
1.0	25.28681

Equation: $\dfrac{dy}{dt} = y^2, \qquad 0 \le t \le 1$

$y(0) = 1$

Method: AB_2 with Midpoint starter, $h = 0.1$

t	y
0	1.0
0.1	1.11025
0.2	1.24758
0.3	1.42321
0.4	1.65527
0.5	1.97521
0.6	2.44206
0.7	3.17906
0.8	4.48354
0.9	7.23741
1.0	15.01774

Equation: $\dfrac{dy}{dt} = 2ty - t^2, \qquad 0 \leq t \leq 3.5$

$y(0) = 0.45$

Method: AB_2 with classical RK starter, $h = 0.25$

t	y
0	0.450000
0.25	0.473689
0.50	0.531676
0.75	0.608531
1.00	0.700650
1.25	0.805516
1.50	0.924481
1.75	1.066669
2.00	1.260643
2.25	1.591121
2.50	2.318549
2.75	4.288035
3.00	10.402676
3.25	31.213841
3.50	107.234327

Equation: $\dfrac{dy}{dt} = t^2 + y^2, \qquad 0 \le t \le 1$

$y(0) = 1$

Method: PC_1 with classical RK starter, $h = 0.1$

t	y
0	1.0
0.1	1.11146
0.2	1.25302
0.3	1.43954
0.4	1.69562
0.5	2.06548
0.6	2.63902
0.7	3.63298
0.8	5.72817
0.9	12.38146
1.0	78.21338

Equation: $\dfrac{dy}{dt} = 2ty - t^2, \qquad 0 \le t \le 3.5$

$y(0) = 0.45$

Method: PC_1 with classical RK starter, $h = 0.25$

t	y
0	0.450000
0.25	0.473689
0.50	0.531675
0.75	0.611183
1.00	0.706922
1.25	0.818187
1.50	0.951756
1.75	1.132375
2.00	1.438880
2.25	2.132930
2.50	4.150883
2.75	11.133247
3.00	38.478039
3.25	157.005227
3.50	720.508969

Equation: $\dfrac{d^2y}{dt^2} = 2yy', \qquad 0 \le t \le 2$

$y(0) = 0,\ y'(0) = 1$

(Exact solution $y = \tan t$)

Method: Second-order equation, with $h = 0.1$

t	y	y'
0	0.0	1.0
0.2	0.202697	1.040774
0.4	0.422636	1.177051
0.6	0.683374	1.462018
0.8	1.026738	2.039196
1.0	1.546292	3.338972
1.2	2.518254	7.079976
1.4	5.256407	25.482281
1.6	34.045596	494.708206
1.8	4.0×10^7	1.2×10^9
2.0	1.7×10^{63}	5.1×10^{64}

Classical RK Method as HP 65 Program

LBL	B	×
A	STO 5	+
STO 1	RCL 3	RCL 5
g R ∨	×	2
STO 2	2	×
g R ∨	÷	+
STO 3	RCL 2	RCL 4
g R ∨	+	+
g R ∨	RCL 1	6
LBL	B	÷
1	STO 6	RCL 3
B	RCL 3	×
STO 4	×	RCL 2
RCL 3	RCL 2	+
×	+	STO 2
2	RCL 1	RCL 1
÷	RCL 3	R/S
RCL 2	2	g R ∨
+	÷	R/S
RCL 3	+	g R ∧
2	STO 1	go to
÷	B	1
RCL 1	RCL 6	LBL
+	2	B
STO 1		

This program leaves about 25 places for insertion of the program for $f(t, y)$, at B. The procedure for using it is the same as that for most of the packaged HP programs. After loading the program (by card or directly), move to B, switch to the write/prgm mode, and insert the program for the particular $f(t, y)$ being used, assuming that t is in the x-register and y in the y-register, and ending with RTN. Return to RUN mode, enter h, y_0, and then t_0 followed by A. You will see t_1. Next, R/S produces y_1 and then t_2, and then y_2, etc.

Bibliography

1. Morris Kine, *Mathematical Thought from Ancient to Modern Time*. Oxford University Press, New York, 1972.
2. S. D. Conte and Carl W. de Boor, *Elementary Numerical Analysis: An Algorithmic Approach*, rev. 2nd ed. McGraw-Hill, New York, 1972.
3. C. W. Gear, *Numerical Initial Value Problems in Ordinary Differential Equations*. Prentice-Hall, Englewood Cliffs, N.J., 1971.
4. R. W. Hamming, *Numerical Methods for Scientists and Engineers*, 2nd ed. McGraw-Hill, New York, 1973.
5. A. Ralston, *A First Course in Numerical Analysis*. McGraw-Hill, New York, 1965.
6. F. J. Murray and K. S. Miller, *Existence Theorems for Ordinary Differential Equations*. New York University Press, New York, 1954.
7. E. A. Coddington and N. Levinson, *Theory of Ordinary Differential Equations*. McGraw-Hill, New York, 1955.
8. R. C. Buck, *Advanced Calculus*, 2nd ed. McGraw-Hill, New York, 1965.
9. W. Kaplan, *Advanced Calculus*, 2nd ed. Addison-Wesley, Reading, Mass., 1973.
10. Tom Apostol, *Mathematical Analysis*, 2nd ed. Addison-Wesley, Reading, Mass., 1974.
11. Ben Noble, *Applied Linear Algebra*. Prentice-Hall, Englewood Cliffs, N.J., 1969.
12. H. G. Jacob and D. W. Bailey, *Linear Algebra*. Houghton Mifflin, Boston, Mass., 1971.
13. C. C. Lin and L. A. Segel, *Mathematics Applied to Deterministic Problems in the Natural Sciences*. Macmillan, New York, 1974.
14. Ben Noble, *Applications of Undergraduate Mathematics in Engineering*. Macmillan, New York, 1967.

15. Rufus Oldenburger, *Mathematical Engineering Analysis.* Dover, New York, 1950.
16. I. S. Sokolnikoff and R. M. Redheffer, *Mathematics of Physics and Modern Engineering*, 2nd ed. McGraw-Hill, New York, 1966.
17. A. H. Taub, ed., *Studies in Applied Mathematics*, No. 7, MAA Studies. Prentice-Hall, Englewood Cliffs, N.J., 1971.
18. A. J. Lotka, *Elements of Mathematical Biology.* Dover, New York, 1957.
19. E. Kamke, *Differentialgleichungen, Lösungsmethoden und Lösungen*, Leipzig, 1959; Chelsea, New York, 1971.
20. *Handbuch der Laplace-Transformation*, Vols. I, II, III. Birkhauser, Basel, 1950–1956.

Solutions to Selected Exercises

Section 1.1

1. (a) A particle, at first stationary, moves forward, stops, returns to its initial position and stops.
   (b) Two moving particles approaching each other collide and cease to exist.
   (c) A stolen car, trapped between two police cars, each initially at rest, attempts to escape and is hemmed in and captured.

3. At $(-3,-2,5)$, $v=(3,5,-6)$, $|v|=\sqrt{70}$, $a=(0,-8,2)$.
   At $(6,4,-4)$, $v=(3,8,0)$, speed $=|v|=\sqrt{73}$, $a=(0,10,2)$.

7. Equation (1.10) yields $V_0=\sqrt{2}\sqrt{gL}$ and (1.14) yields $V_0=\frac{1}{2}\pi\sqrt{gL}$.

9. $d^2\theta/dt^2=-(MgL_0/I)\sin\theta$, where $M=M_1+M_2$, L_0 is the distance from the pivot to the center of gravity of the weights, the pendulum is displaced to an angle θ, and I is the moment of inertia of the system about the pivot.

11. $\phi'(t)=s-(r/V)\phi(t)$, $\phi(0)=A$, where r is the rate water enters the lake, s is the rate the pollutant enters, V is volume of lake, and $\phi(t)$ amount of pollutant at time t.

13. $A'(t)=-kA(t)$,
    $B'(t)=kA(t)$.

15. (a)

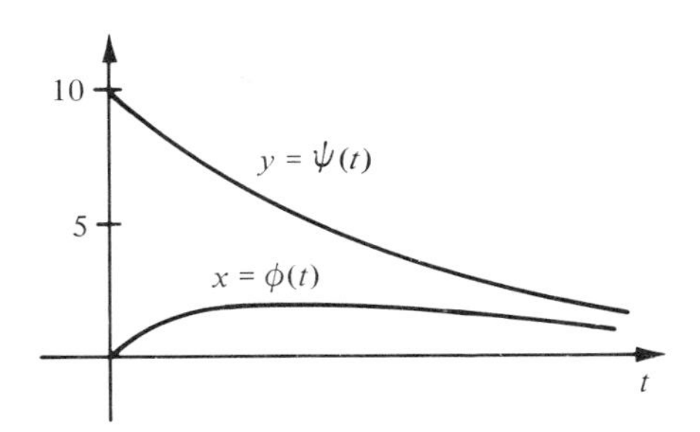

(b)

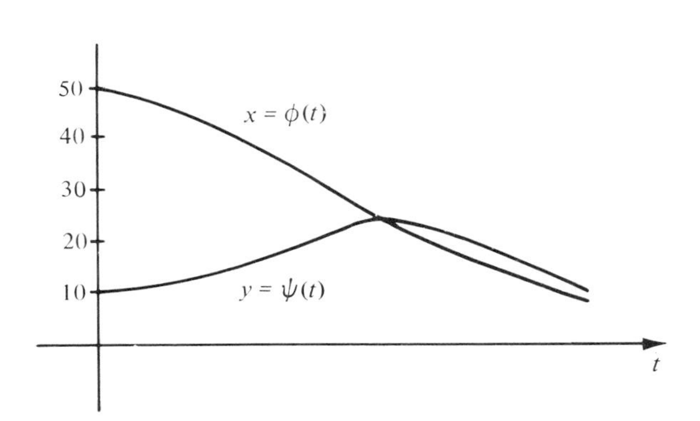

(c) The mixing is much faster.

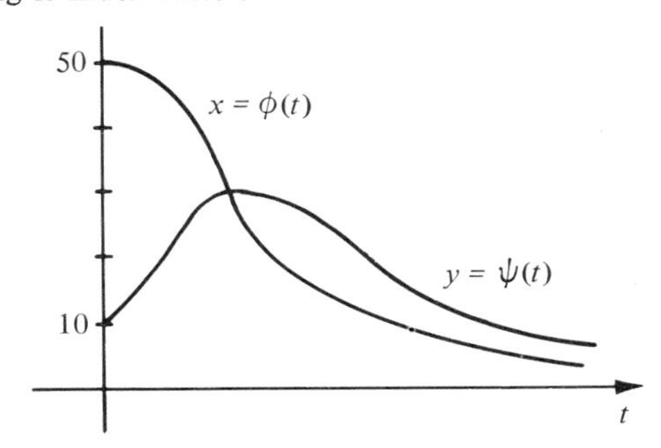

Section 1.2

1. Both $y = \sqrt{C - t^2}$ and $y = -\sqrt{C - t^2}$ are solutions.

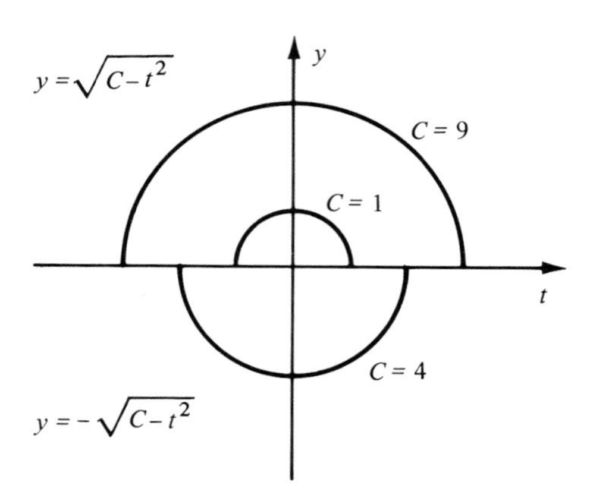

3. Note that requirement $1 - Ce^t \neq 0$ excludes values $t = -\log C$ when $C > 0$.

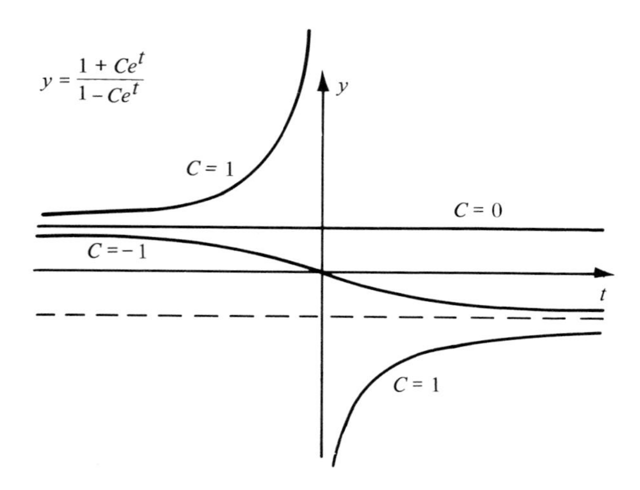

5.

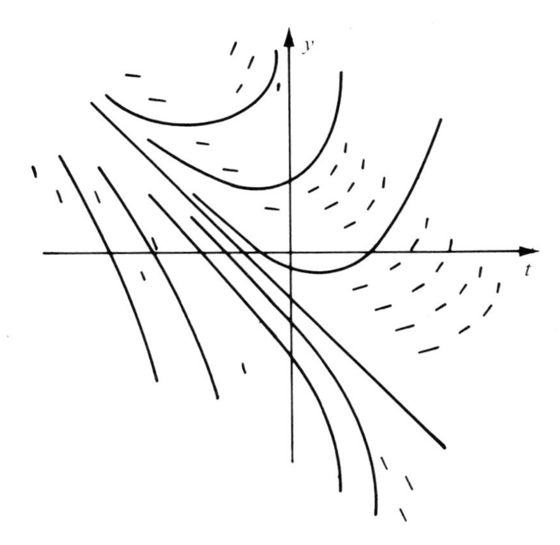

7. Careful sketches will reveal that solution curves seem to have vertical asymptotes to the right of the origin.

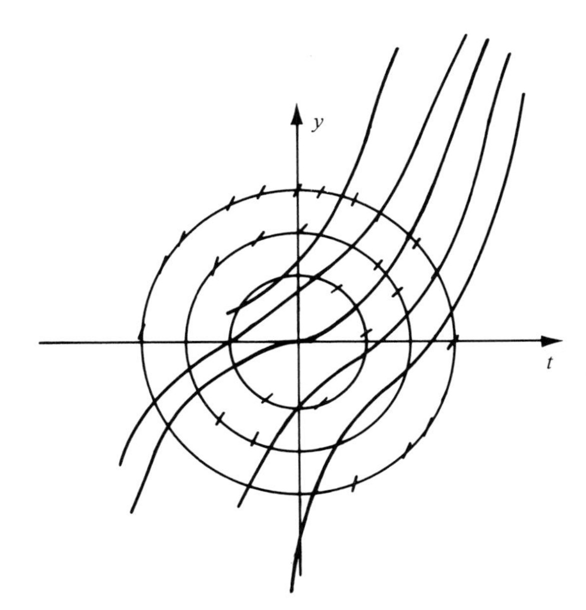

9. $C = 2/(3e)$, $C = \frac{1}{3}$.

13. The given solutions cannot be the only solutions of the system. Others can be found by observing that $dy/dx = 3/(x^2y^2)$.

Section 1.3

1. (a) $$24y - 2\frac{dy}{dt} = 24Ae^{4t} + 24Be^{-6t} - 8Ae^{4t} + 12Be^{-6t}$$
$$= 16Ae^{4t} + 36Be^{-6t}$$
$$= \frac{d}{dt}(4Ae^{4t} - 6Be^{-6t}) = \frac{d^2y}{dt^2}$$

   (b) $A = \frac{15}{10} = \frac{3}{2}$ and $b = \frac{1}{2}$.

3. (a) If $y(t) = t^\alpha$, then $y'(t) = \alpha t^{\alpha-1}$ and $y''(t) = \alpha(\alpha-1)t^{\alpha-2}$.

   If $$y'' = \frac{5}{2t}y' - \frac{3}{2t^2}y$$

   then $\alpha(\alpha-1)t^{\alpha-2} = (\frac{5}{2}\alpha - \frac{3}{2})t^{\alpha-2}$, $2\alpha^2 - 7\alpha + 3 = 0$, or $\alpha = 3$, $\alpha = \frac{1}{2}$. Hence let $\phi(t) = t^3$, $\psi(t) = t^{1/2}$ $(t \geq 0)$.

   (b) Thus $A = \frac{1}{5}$ and $B = \frac{14}{5}$.

5. $A = -\frac{3}{2} = B$.

9. $(D+5)(D-3)x = (D+5)(2y) = 2(D+5)y = 2(7x) = 14x$.
   Hence $(D^2+2D-15)x = 14x$, or $(D^2+2D-29)x = 0$.

11. Set $u_1 = y$, $$u_2 = \frac{dy}{dt}, \ldots, u_n = \frac{d^{n-1}y}{dt^{n-1}}$$

   Then
$$\frac{du_1}{dt} = u_2$$
$$\frac{du_2}{dt} = u_3$$
$$\vdots$$
$$\frac{du_{n-1}}{dt} = u_n$$
$$\frac{du_n}{dt} = F(t, u_1, u_2, \ldots, u_n)$$

Section 1.4

1. $d^4y/dx^4 = 0$
3. $\dfrac{dy}{dx} = \dfrac{y+1}{x+\frac{1}{2}}$
5. $ww'' = (w')^2$
7. $\dfrac{dy}{dx} = \dfrac{y^2+2xy-x^2}{x^2+2xy-y^2}$

9. $4y - 4xy' + (y')^2 = 0$. Yes.

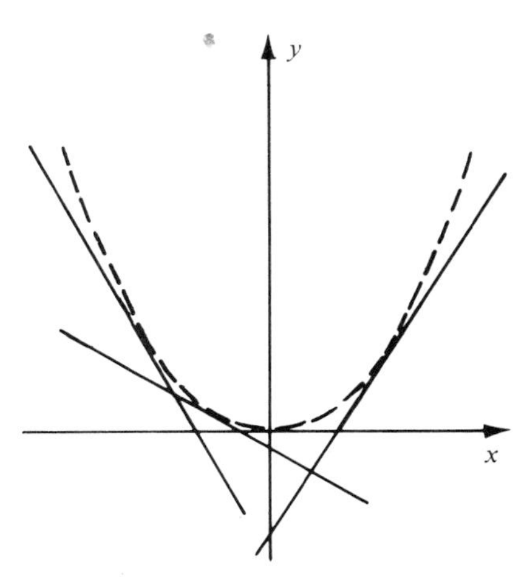

15. (a) $\dfrac{dy}{dx} = \dfrac{-x}{y-2}$ (b) $\dfrac{dy}{dx} = \dfrac{y-2}{x}$

(c) Orthogonal trajectories must be the radial lines through the center of the circles, that is all lines through $(0, 2)$, namely $y - 2 = Ax$.

Section 1.5

1. We can take the region D to be the TY-plane. Any point may be used as an initial point.
3. If $-\infty < t < b$, then $dy/dt = 0 = 3y^{2/3}$; if $b \leq t < \infty$, then $dy/dt = 3(t-b)^2 = 3[(t-b)^3]^{2/3} = 3y^{2/3}$.
5. With $f(t, y, y') = \dfrac{y}{[4-(y')^2]}$,

$$f_2 = \frac{1}{[4-(y')^2]} \quad \text{and} \quad f_3 = \frac{2yy'}{[4-(y')^2]^2}$$

Each is continuous at all points (t, y, y') with $y' \neq \pm 2$. The only excluded initial conditions are those with $y' = \pm 2$.

7. If the equation does not have a solution, this might indicate the reaction cannot take place. Our confidence in this conclusion should be only as great as our belief that the model is an accurate description of reality. If the reaction does indeed take place, we should conclude that the model is not a useful one.

If the equation has a solution, this does not guarantee that the reaction does take place, for the model may not be an accurate description of physical reality. For example, in the equations for the motion of a pendulum, we saw that the model predicted the pendulum would never stop swinging, which certainly is not an accurate description of reality.

9. No. Theorem 1.1 demands that f_y be continuous in D for a unique solution to exist; $f_y = (x^2 - y)^{-1/2}$ which does not exist on the curve $y = x^2$. Thus Theorem 1.1 does not apply to points on the parabola $y = x^2$. (See exercise 9, Section 1.4.)

Section 2.1

1. $y = 2e^{(t^3-1)/3}$

3. $\phi(t) = \sqrt{t^2+8}$, $\phi(t) = \sqrt{t^2+1}$, $\phi(t) = -\sqrt{t^2+8}$

5. $y = 1/(1-\frac{1}{2}t)$, $-\infty < t < 2$, and $y = 1, 2 \le t < \infty$
7. $\phi(t) = 1 + 2e^{(t^3-1)/3}$, $\phi(t) = 1 + e^{(t^3-1)/3}$, $\phi(t) = 1 - e^{(t^3-1)/3}$, $\phi(t) = 1 + e^{t^3/3}$
11. $2t^3 - 3ty^2 = A$
13. $x^2 + Ax + y^2 - Ay = 0$. This is the family of circles whose centers are on the line $y = -x$ and which pass through $(0,0)$. Note that on the line $y = x$, $dy/dx = 1$ so that another solution is $y = x$, the line to which the family of solution curves is tangent.
15. $y^4 - x^4 = C$
17. $y^4 x = A$
19. At the end of 2 months, $45,000 will have been contributed.
21. 78.1 grams after 8 days.
23. At 9 a.m. the vat will be $\frac{4}{5}$ full; the vat will be full after 2.1614 hours, that is at 9:09:41 a.m.
25. It drops to a depth of 3 inches in 6 days; in 12 days, it will be empty.
27. If α is the density of the moisture in the air, and r the radius of the raindrop at time t, $dr/dt = \frac{1}{4}\alpha v$, where v is the speed of the falling raindrop.
29. Newton's law of cooling becomes $du/dt = -k(u - T)$; 42.5°.
31. If M is the maximum size, k constant of proportionality, then

$$x = M\frac{Ae^{kMt}}{1 + Ae^{kMt}}$$

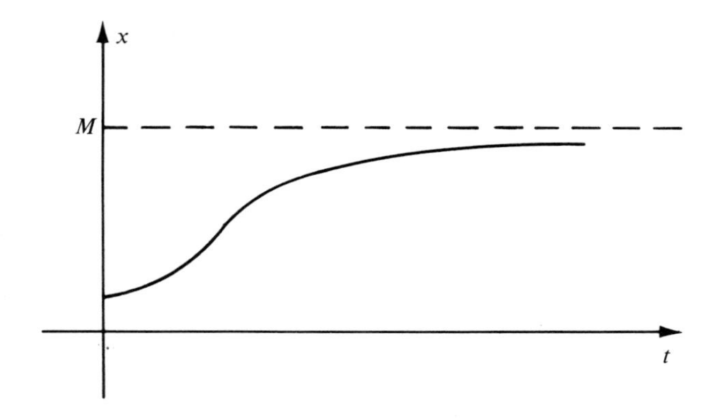

33. $14\frac{2}{7}$ grams of C after a total of 10 minutes.

Section 2.2

1. $\phi(t) = \frac{1}{3}e^t + \frac{2}{3}e^{-2t}$ is the solution through $(0,1)$, defined for all t.
3. $\phi(t) = (t+1)e^t$.
5. $\phi(t) = 1/t$ is a solution through $(1,1)$ for $0 < t < \infty$, $\phi(t) = 1/t - 1/t^2$ is a solution through $(1,0)$ for $0 < t < \infty$.
7. $\phi(t) = (t^3 + Ct)^2$ and $\phi(t) \equiv 0$ are the solutions.
9. $y = \dfrac{1}{Ce^x + x + 1}$ and $y \equiv 0$ are the solutions.
11. $y = (2e^x + Ce^{-2x})^{-1/2}$ and $y \equiv 0$.
13. $y = (t - \frac{1}{2} + Ce^{-2t})^{1/2}$
15. $y = \dfrac{1 - 2At^3}{t(1 + At^3)}$
17. $y = 8/x$ crosses $x = 1$ at $y = 8$.
19. (a) Tank is half full 100 seconds after the valves are opened and the chlorine is at a 0.75% solution.
    (b) The final concentration is 0.875%.

Section 2.3

1. $y = \frac{1}{3}e^t - \frac{1}{2}C_1 e^{-2t} + C_2$
3. $y = \frac{1}{2}C_1 x^2 + C_2$
5. $y = \frac{1}{4}x^2 - \frac{1}{4}x - \frac{1}{2}C_1 e^{-2x} + C_2$
7. $y = C_1(x - e^{-x}) + C_2$
9. $y = a\tan(at + b)$, $y = C$, and $y = a\dfrac{1 + be^{2at}}{1 - be^{2at}}$
11. $x = (At + B)^2$ and $x = k \neq 0$
13. $y = \sqrt{16 - (1 - \frac{1}{2}t)^2}$ and $y(4) = \sqrt{15}$
15. The equation becomes $(d/dy)(e^{-2By}v) = 2Ay\,e^{-2By}$ which is solved by integration. Then $v = u^2$ and $u = dy/dt$ which reduces to a problem in integration.
17. In (2.32) let $x = 64R$, $R \approx 4000$.
19. Solve $x'' = -gRx^{-1}$
21. Depth reached is 1000 feet. (Solve $v' = -k\sqrt{v}$.)
23. The bullet reaches 1522 feet. (Solve $x'' = -g - 0.001v^2$.)

Section 2.5

1. $y = 2[2Cx - x(\log x)^2]^{-1}$
3. $y^2 - x^2 = Ax^3$
5. $y = Ce^t - 2t + D$
7. $y = (Cx^4 - 4x^4 \log x)^{-1/2}$
9. $y = -\frac{3}{2}t^3 + Ct^5$
11. $y = \sqrt{2t + 1 + Ce^{2t}}$
13. $y = Ce^{-x}$ and $y = -\frac{1}{2}x^2 + C$

15. Through $(1, e^2)$, $y = e^{2t}$; through $(-1, e^2)$, $y = e^{-2t}$. There is no solution through $(1, -2)$ because $\log y$ is not defined if $y = -2$.
17. (a) After 50 minutes, there are 100 gallons in the tank.
    (b) There are 175 pounds of dissolved salt in the tank then.
19. 838 feet.
21. The tank overflows after 32.19 minutes (1931.47 sec).
23. The bullet was traveling 300 ft/sec.
25. If the pool is first drained to half volume and then flushed (that is, pure water enters while the mixture is draining), until only 1 pound of impurity remains, and the pool is then refilled with pure water, it will take 215.12 minutes. Without draining to half volume but flushing with constant volume, it would take 299.57 minutes.
27. The leak is approximately $30\frac{1}{4}$ feet below the top.

Section 3.1

1. 2, 7, 17, 37, 77, 157, 317, 637, 1277, 2557, 5117
3. 1, 1, 2, 6, 24, 120, 720, 5040, 40320, 362880, ($x_n = n!$)
5. 2, 2.667, 4.148, 8.501, 29.757, 315.003, 33285.549, ...
   0.5, 0.4167, 0.3356, 0.2613, 0.1970, 0.1442, 0.1031, 0.0723, 0.0499, ...
9. $y_n = 3$ when n is even; $y_n = -1$ when n is odd
11. $x_n = (\frac{12}{5})6n - \frac{7}{5}$
13. (a) Any sequence of the form $\frac{5}{3}n^3 - n^2 - \frac{2}{3}n + D$ where D is any constant.
    (b) Any sequence of the form $\frac{1}{2}n^4 - \frac{11}{6}n^3 + 2n^2 + Dn + E$, where D and E are constants.
15. (a) 1, 2, $\frac{3}{2}$, $\frac{7}{4}$, $\frac{13}{8}$, $\frac{27}{16}$, $\frac{53}{32}$, $\frac{107}{64}$, $\frac{213}{128}$, $\frac{427}{256}$
    (b) 2, 3, 2, 1, 1, 2, 3, 2, 1, 1, 2, 3, 2, $x_{19} = 1$
19. 2, 5, 13, 35, 97, 275, 793; $x_n = 2^n(A-1) + 3^n$
21. If M is the given matrix, then

$$M^n = \begin{pmatrix} 2^n & A_n \\ 0 & 4^n \end{pmatrix}$$

    where $A_n = 3(2^n - 1)2^{n-1}$
23. (a) $x_8 = 21$ (b) $x_8 = 9$
25. $x_5 = 2.89414$

Section 3.2

1. Series diverges; $a_n \to \frac{1}{10}$
3. Series converges; compare $\sum a_n$ with $\sum(n^{-3/2})$
5. Series converges; compare $\sum a_n$ with $\sum(13n^{-2})$
7. Series converges; use Theorem 3.12.
9. Series diverges; $|a_n| \to \frac{1}{2}$
11. Convergent for $|x| \leq 1$
13. Convergent for $-6 < x < -4$
15. Convergent for $-1 < x < \frac{1}{3}$
17. Converges for all x

19. Converges for $-1 \leq x < 1$

21. (a) $\cos x = 1 - \dfrac{x^2}{2} + \dfrac{x^4}{4!} - \dfrac{x^6}{6!} + \cdots + \dfrac{(-1)^n x^{2n}}{(2n)!} + \cdots$

    (b) $\cos x = -(x - \frac{1}{2}\pi) + \dfrac{(x-\frac{1}{2}\pi)^3}{3!} + \cdots + \dfrac{(-1)^{n+1}(x-\frac{1}{2}\pi)^{2n+1}}{(2n+1)!} + \cdots$

23. $-x/2 - 3x^2/4 - 3x^3/8 - \cdots - 3x^n/2^n - \cdots$
27. $x/(1-x^3)$
31. (a) The sum of the series is 1.
    (b) The sum of the series is $\frac{1}{2}$.

Section 3.3

5. $x_1 = 1+2i,\ x_2 = -4-2i$
7. $\frac{5}{3} + \frac{1}{2}i$
9. $1+i$
17. Since the derivative of each side of the equation is $1/(1+x^2)$, and both sides are equal to 0 at $x = 0$, and there is a unique solution to the equation $y' = 1/(1+x)^2$ through $(0,0)$, the two functions must be the same.

Section 3.4

5. $y \approx 1 + 2t + 3t^2 + \frac{13}{3}t^3 + \frac{25}{4}t^4 + \frac{541}{60}t^5 + \frac{1561}{120}t^6$
7. $y = 3t$
9. $y \approx 2 + t + \frac{1}{3}t^3 + \frac{1}{6}t^4 + \frac{1}{90}t^6$
11. $y \approx 1 + 2(t-1) + \frac{5}{2}(t-1)^2 + \frac{11}{6}(t-1)^3 + \frac{13}{12}(t-1)^4 + \frac{7}{12}(t-1)^5 + \frac{5}{18}(t-1)^6$
13. $y_P = 1+t+t^2+t^3$. The exact solution: $y = 1/(1-t)$.

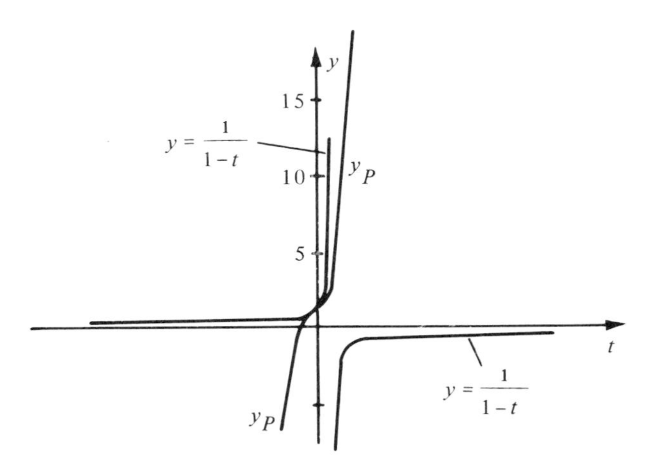

15. Show that $y \leq 2$ on $[0, \frac{1}{6}]$ and thus $R_4 < (2190/4!)(\frac{1}{6})^4$
17. $y = -e^t + 3e^{2t}$
19. $y = 2/(2-t^2)$

Section 3.5

1. $y = -2 - 2t - t^2 + 3e^t = 1 + t + \frac{1}{2}t^2 + 3\left(\frac{t^3}{3!} + \frac{t^4}{4!} + \cdots + \frac{t^n}{n!} + \cdots\right)$

3. $y \approx C\left(1 + t + \frac{t^2}{2!} + \frac{t^3}{3!} + \frac{t^4}{4!} + \frac{t^5}{5!} + \frac{t^6}{6!}\right)$
$$+ t\left(1 + t + \frac{4t^2}{3!} + \frac{10}{4!}t^3 + \frac{34}{5!}t^4 + \frac{154}{6!}t^5\right)$$

5. $y \approx 1 + t + \frac{1}{2}t^2 + \frac{2}{3}t^3 + \frac{1}{2}t^4 + \frac{7}{15}t^5 + \frac{31}{72}t^6$

7. $y = 1 + \frac{1}{3!}t^3 + \frac{4}{6!}t^6 + \frac{4\cdot 7}{9!}t^9 + \cdots + \frac{4\cdot 7\cdot 10\cdot(3n-2)\,t^{3n}}{(3n)!} + \cdots$

9. The next term is
$$-\frac{1}{C_0}[C_4 C_0^3 - C_1^4 - C_0^2(C_2^2 + 2C_1C_3) + 3C_0 C_1^2 C_2]\left(\frac{t}{C_0}\right)^4$$

11. $y \approx 1 + 3t - \frac{3}{2}t^2 + \frac{23}{6}t^3$

Section 3.6

1. $\phi_1 = 2, \qquad \phi_2 = t^3 - \frac{1}{2}t^2 - 2t + 2$
$\phi_3 = -\frac{1}{4}t^4 + \frac{7}{6}t^3 + \frac{1}{2}t^2 - 2t + 2$
$$\phi_4 = \frac{t^5}{5\cdot 4} - \frac{7t^4}{6\cdot 4} + \frac{5t^3}{2\cdot 3} + \frac{t^2}{2} - 2t + 2$$
$$\phi_5 = \frac{-t^6}{6\cdot 5\cdot 4} + \frac{7t^5}{6\cdot 4\cdot 5} - \frac{5t^4}{2\cdot 3\cdot 4} + \frac{5t^3}{6} + \frac{t^2}{2} - 2t + 2$$

3. $\phi_3 = \frac{1}{63}t^7 + \frac{2}{15}t^5 + \frac{1}{6}t^4 + \frac{2}{3}t^3 + t^2 + t + 1$

5. $\phi_1 = 1$, $\phi_2 = 10t + 1$, $\phi_3 = \frac{1000}{3}t^3 + 100t^2 + 10t + 1$. Coefficients increase enormously. The reason is that as $t \to 0.1$, y increases without bound. (Actual solution is $y = (1-10t)^{-1}$.)

7. (a) $\phi_1 = t, \qquad \phi_2 = \frac{1}{6}t^3 + t, \qquad \phi_3 = \frac{t^5}{5!} + \frac{t^3}{3!} + t$
$$\phi_4 = \frac{t^7}{7!} + \frac{t^5}{5!} + \frac{t^3}{3!} + t$$
These seem to indicate that
$$\phi_n = t + \frac{t^3}{3!} + \frac{t^5}{5!} + \cdots + \frac{t^{2n-1}}{(2n-1)!}$$

9. $\phi_1 = t \qquad \phi_2 = 3t - 2 \qquad \phi_3 = 9t - 6(\log t) - 8$
$\phi_4 = 3[9t - 3(\log t)^2 - 8(\log t)] - 26$

Section 4.1

1. (a) $y(1) \approx 0.0778$, $y(2) \approx 0.0060$
(b) exact values: $y(1) = 0.1353*$, $y(2) = 0.0183*$

3. $y(1) \approx y_{20} = 0.1216$

5. (a)

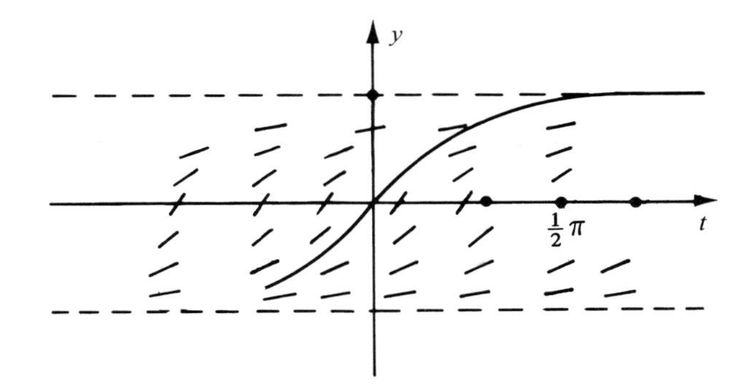

(c) After a few applications of the Euler method, $y_k > 1$, and then y' does not exist; thus, method "blows up."

7. In both cases, $y'' > 0$ and the tangent lines therefore are below the actual curve when $t \geq 0$.

9. (a) $y_1 = 1.2$, $y_2 = 1.488$, $y_3 = 1.9308$, $y_4 = 2.6764$, $y_5 = 4.1091 \approx y(1)$, $y_6 = 7.4861 \approx y(1.2)$

(b) Because y increases without bound as $t \to 1$, strange behavior can be expected from the Euler method in the neighborhood of $t = 1$.

Section 4.2

1. If $y(T)$ is to be obtained, then $T = 1 + nh$. Applying Euler method yields

$$y_n = \frac{t_n t_{n+1}}{t_0 t_1} = \frac{(1+nh)(1+h+nh)}{1+h}$$

and

$$\lim_{n\to\infty} y_n = \lim_{h\downarrow 0} \frac{T(T+h)}{1+h} = T^2 = y(T)$$

3. (b) $S_{1000} = 1.35789$, $S_{5000} = 1.35889$, whereas $\frac{1}{2}e = 1.35914*$.

5. Using (4.10), $N > 272{,}990.7$, or approx 273,000 and so $h = 0.00001465$. If $h = 0.00001$, $N = 400{,}000$.

Section 4.3

1. $y_1 = 1.22$, $y_2 = 1.488$, $y_3 = 1.816$, $y_4 = 2.215$, $y_5 = 2.703$
3. $y(2) \approx 7.0857$
5. $y(1) \approx 0.36803$, $y(1) = e^{-1} = 0.367879*$
7. (b) The actual solution is $y = e^{-2t}$, so that

$$\begin{aligned} y(t_{n+1}) &= e^{-2(n+1)h} = e^{-2h}e^{-2nh} = e^{-2h}y(t_n) \\ &= (1 - 2h + 2h^2 - \tfrac{4}{3}h^3 + \tfrac{2}{3}h^4 - \cdots)\, y(t_n) \end{aligned}$$

9. (a) $\frac{1}{2}(m_P + m_R)$ has about half the error that m_Q has as an estimate of m; both errors are of the order $O(h^2)$

(b) If $\bar{m} = (m_P + 4m_Q + m_R)/6$, then

$$\bar{m} = \phi'(c) + \tfrac{1}{6}h^2\phi'''(c) + \tfrac{1}{72}h^4\phi^{(5)}(c) + \cdots$$
$$m = \phi'(c) + \tfrac{1}{6}h^2\phi'''(c) + \tfrac{1}{120}h^4\phi^{(5)}(c) + \cdots$$

so that error in using $\bar{m}$ is $h^4(\tfrac{1}{72} - \tfrac{1}{120})\phi^{(5)}(c) + \cdots = O(h^4) \ll O(h^2)$

11. $y(0.9) \approx y_3 = 5.627503$
13. $y(0.9) \approx y_3 = 11.614964$

Section 4.4

1. $y_2 = 0.744$, $y_3 = 0.7024$, $y_4 = 0.70304$, $y_5 = 0.74118$, $2e^{-1} = 0.735759*$
3. $y_3 = 1.4266$, $y_4 = 1.6597$, $y_5 = 1.9815$
7. $y_1 = 0.8400$, $y_2 = 0.7384$, $y_3 = 0.6766$, $y_4 = 0.6961$, $y_5 = 0.7505$

Section 4.5

1. $7.38866 \approx y(2) = e^2 = 7.389056*$; $20.08342 \approx y(3) = e^3 = 20.085554*$
3 $26.94110 \approx y(3) = 3^3 = 27$.
5. (a) The function $y(t)$ has a discontinuity somewhere after $t = 0.5$.
   (b) The exact solution: $y = 1/(1-9t^4)$ does not exist at $t = \frac{1}{3}\sqrt{3}$.

Section 5.1

1. (a) and (b) belong to $\mathcal{C}^1$ but (c) does not.
   $f(t) = |\cos t|$ does not have a derivative at $t = (2n+1)(\frac{1}{2}\pi)$.
4. $\alpha t^2 + \beta t + \gamma = A(2t+1) + B(t^2+2) + C(3t+t^2)$ if $A = \frac{1}{7}(-6\alpha + 2\beta + 3\gamma)$, $B = \frac{1}{7}(3\alpha - \beta + 2\gamma)$, and $C = \frac{1}{7}(4\alpha + \beta - 2\gamma)$
7. (a) Space of all polynomials of degree at most 3 which vanish at $t = 0$.
   (b) Space of all polynomials of degree at most 3 which vanish at $t = -1$.
   (c) Space of all polynomials of degree at most 3.
9. $15t^2 - 7t + 16$, $-3t^2 - 4t + 5$, $9t^4 + 24t^3 + 10t^2 + t + 6$, $12t^2 - 2t$.
11. (a) $6t^4$, $6t^3$, t^6
   (b) T_1 is not linear; T_2 and T_3 are.

15. $\sin^2(3t+5) = \dfrac{1}{2} + \dfrac{\sin(10)}{2}\sin(6t) - \dfrac{\cos(10)}{2}\cos(6t)$

Section 5.2

1. Unless the single function is identically zero, it is independent.
3. If a function f has two representations, $\sum_{i=1}^{N} c_i\phi_i = f = \sum_{i=1}^{N} d_i\phi_i$ then $\sum_{i=1}^{N}(c_i - d_i)\phi_i \equiv 0$. But since $\{\phi_i\}$ are independent, $c_i - d_i = 0$, for all i.
5. $W(t) = 0$. Moreover, $2t^2 + 3t + 1 = 2(t^2+2) + 3(t-1)$.
7. $W(t) \neq 0$ if $\alpha \neq \beta$
9. $W(t) = e^{(\alpha+\beta)t}(t\beta - 1 - t\alpha) \not\equiv 0$
11. Look for c_1 and c_2 so that $c_1g_1 + c_2g_2 = 0$, for all t. If $t = 1$, $c_1 + 2c_2 = 0$. If $t = -1$, $-c_1 - 3c_2 = 0$. Thus $c_1 = c_2 = 0$ and functions are independent.

13. No. For example, take $g_1(t) = (t^2-1)^3$ for all t; $g_2(t) = 2(t^2-1)^3$, $|t| \leq 1$; $g_2(t) = 3(t^2-1)^3$, $|t| \geq 1$. On $-1 < t < 1$, $g_2(t) = 2g_1(t)$, and thus they are linearly dependent. But for all t, $c_1 g_1 + c_2 g_2 = 0$ only if $c_1 = c_2 = 0$. In the case of functions dependent on $(-\infty, \infty)$, then there are c_1 and c_2 not equal to zero such that $c_1 g_1 + c_2 g_2 = 0$ for all t and then certainly for every subinterval.

Section 5.3

1. (b) The null space of D^n is the space of all polynomials of degree at most $(n-1)$.
5. Observe that $Ae^{\alpha t} + Bte^{\alpha t} + Ct^2 e^{\alpha t} = e^{\alpha t}(A + Bt + Ct^2)$ and this can equal zero for at most two values of t unless $A = B = C = 0$.
9. (a) all functions of the form $C_1 e^{(1+\sqrt{2})t} + C_2 e^{(1-\sqrt{2})t}$
   (b) all functions of the form $C_1 e^{-t}\cos(\sqrt{2}\,t) + C_2 e^{-t}\sin(\sqrt{2}\,t)$
   (c) $C_1 e^{-3t/2}\cos\left(\frac{3\sqrt{7}}{2}t\right) + C_2 e^{-3t/2}\sin\left(\frac{3\sqrt{7}}{2}t\right)$
11. $x = C_1 e^t + e^{-t/2}(C_2 e^{\sqrt{17}\,t/2} + C_3 e^{-\sqrt{17}\,t/2})$
13. $x = e^{-t}\left(\frac{3\sqrt{2}}{2}\sin(\sqrt{2}\,t) + \cos(\sqrt{2}\,t)\right)$
15. $\{1, t, t^2, \cos(2t), \sin(2t)\}$
17. $\{e^{-t}, te^{-t}, t^2e^{-t}, e^t, te^t\}$
25. Make use of Exercise 24. If T is of order 3, t and e^t could be in its null space. For example, $1, t, e^t$ are in null space of $T = D^2(D-1)$.
27. $x = Ae^{1.2t}\cos(1.47t + \beta)$
29. $y = C_1 e^{-1.8933t} + C_2 e^{0.9466t}\sin(0.8297t + \gamma)$

Section 5.4

1. $y = \frac{2}{3}e^{-2t} + \frac{1}{3}e^t - e^{-t}$
3. $y = -e^{2it} + 2e^{it}$
5. (a) $y = \frac{1}{2}e^{-2t} + C_1 + C_2 e^{-t}$ (b) $y = \frac{1}{2}e^{-2t} + \frac{9}{2} - 3e^{-t}$
11. $y = C_1 e^{-3t} + C_2 e^t + \frac{1}{10}\sin t - \frac{1}{5}\cos t$
13. $x = C_1 + C_2 e^{-4s} + \frac{1}{4}s^2 - \frac{3}{8}s$
15. $x = e^{-t} - \frac{3}{2}t\cos(2t) + C_1\cos(2t) + C_2\sin(2t)$
17. $y = -s - 2 - \frac{1}{2}e^{-s} + C_1 e^{-2s} + C_2 e^s$
19. $x = C_1 e^{2s} + C_2 se^{2s} + \frac{1}{2}s^2e^{2s} + \frac{1}{4}s^2 + \frac{1}{2}s + \frac{3}{8}$
21. $y = C_1 e^t + C_2 e^{-t} - \frac{1}{4}te^{-t} - \frac{1}{4}t^2e^{-t}$
23. $x = C_1\sin t + C_2\cos t + \frac{3}{4}t\sin t - \frac{3}{4}t^2\cos t$
25. $x = e^{4t}[C_1\sin(3t) + C_2\cos(3t)] + \frac{3}{160}\sin t + \frac{1}{160}\cos t$
27. $y = C_1 e^{2t} + C_2 e^t + te^{2t} + \frac{1}{12}e^{-2t}$
29. (b) $y = \frac{2}{3}t^{-1}$
31. $y = C_1\sin t + C_2\cos t - 2\cos t\log(\sec t + \tan t)$

Section 5.5

1. $x = 0$ at all $t = n\pi$, or when $\cos t = -\frac{1}{4}$.

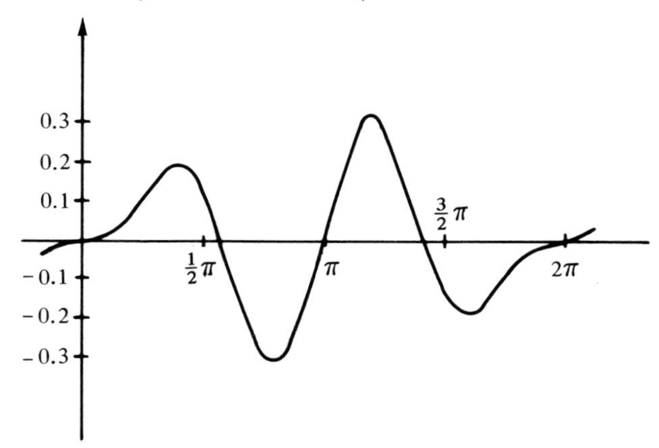

3. $x = 0$ only if $t = [1/(b-a)] \log(-C_2/C_1)$, which is uniquely defined only when $C_1 C_2 < 0$. If $C_1 C_2 > 0$, $x \neq 0$.
5. Motion is asymptotically a periodic one, with same frequency as that of the supporting rod, with amplitude $1/2k$, but with a phase change since sine is replaced by cosine.

Section 6.1

1. $x = 2e^{2t}$, $y = 4e^{2t}$
3. $u = 4e^t - 2e^{-t} + 1$, $v = 4e^t + 2e^{-t} - 2$
5. $x = e^{5t}$, $y = 2e^{5t}$
7. $x = e^{t-1} - e^{1-t} + t^2 + 1$, $y = e^{t-1} + e^{1-t} + t$
9. (a) $x = 5(e^{-0.025t} + e^{-0.095t})$, $y = 8.62(e^{-0.025t} - e^{-0.095t})$
   (b) Maximum concentration in tank II is reached at 19.07 seconds and $y = 3.943$.
11. $x = \frac{4}{9}e^{-2t} + \frac{5}{9}e^t + \frac{4}{3}te^t$
    $y = \frac{2}{9}e^{-2t} + \frac{7}{9}e^t - \frac{4}{3}te^t$
    $z = -\frac{8}{9}e^{-2t} + \frac{17}{9}e^t + \frac{4}{3}te^t$

Section 6.2

5. (a) $\dfrac{dx_1}{dt} = x_1 - tx_2, \qquad \dfrac{dx_2}{dt} = x_2 + tx_1$

   (b) Let $u_1 = x_1$, $u_2 = x_1'$, $u_3 = x_2$, $u_4 = x_2'$ and let

$$U = \begin{bmatrix} u_1 \\ u_2 \\ u_3 \\ u_4 \end{bmatrix} \qquad \text{Then,} \qquad \frac{dU}{dt} = G(t, U) = \begin{bmatrix} u_2 \\ 3u_2 - tu_3 + u_1 \\ u_4 \\ 3u_4 + u_3 + tu_1 \end{bmatrix}$$

7. Let $x_1 = u$, $x_2 = u'$, $x_3 = v$, $x_4 = v'$, and

$$X = \begin{bmatrix} x_1 \\ x_2 \\ x_3 \\ x_4 \end{bmatrix}, \qquad \frac{dX}{dt} = F(t, X) = \begin{bmatrix} x_2 \\ -5/(x_1{}^2 + x_3{}^2) \\ x_4 \\ -3/(x_1{}^2 + x_3{}^2) \end{bmatrix}$$

9. If $x_1 = y$, $x_2 = y'$,

$$X = \begin{bmatrix} x_1 \\ x_2 \end{bmatrix}$$

then

$$\frac{dX}{dt} = \begin{bmatrix} x_2 \\ -4x_1 \end{bmatrix}$$

and $x_1 = A\cos(2t) + B\sin(2t)$, $x_2 = -2A\sin(2t) + 2B\cos(2t)$. The solution curves are a set of ellipses: $4x_1{}^2 + x_2{}^2 = 4(A^2 + B^2)$.

Section 6.3

5. $u(0.4) \approx 0.415$, $v(0.4) \approx -2.641$

Section 6.4

1. $$\begin{bmatrix} 2 & 1 & 1 & 1 \\ 1 & -2 & -3 & -3 \\ 0 & 1 & 2 & 5 \\ 0 & 1 & 1 & -4 \\ 1 & 1 & 2 & 4 \end{bmatrix} \begin{bmatrix} x_1 \\ x_2 \\ x_3 \\ x_4 \end{bmatrix} = \begin{bmatrix} 3 \\ -4 \\ -2 \\ 11 \\ 1 \end{bmatrix},$$

3. (a) $A'(t) = \begin{bmatrix} 6t^2e^{-t} - 2t^3e^{-t} & 3 + 2t \\ 2 & e^t \end{bmatrix}$

(b) $\det\{A(t)\} = -6t^2$

(c) $A(t)^{-1} = \begin{bmatrix} -\dfrac{e^t}{6t^2} & \dfrac{3+t}{6t} \\ \dfrac{1}{3t} & -\dfrac{te^{-t}}{3} \end{bmatrix}$

7. (b) $X = \begin{bmatrix} 2e^t + te^t \\ 3e^t + te^t \end{bmatrix}$

9. (a) $\dfrac{du}{dt} = \begin{bmatrix} 0 & 1 \\ 4 & 0 \end{bmatrix} u$, where $u = \begin{bmatrix} x \\ y \end{bmatrix}$

(c) $x = 2e^{2t}$, $y = 4e^{2t}$

13. $x_1 = \frac{2}{3}t^5$, $x_2 = t^3$

Section 6.5

5. (a) $\lambda_1 = 6, \quad v_1 = \begin{bmatrix} 1 \\ 1 \end{bmatrix}, \quad \lambda_2 = 4, \quad v_2 = \begin{bmatrix} 1 \\ 3 \end{bmatrix}$

   (b) $\lambda_1 = -1, \quad v_1 = \begin{bmatrix} 1 \\ -1 \end{bmatrix}, \quad \lambda_2 = 7, \quad v_2 = \begin{bmatrix} 3 \\ 1 \end{bmatrix}$

7. $\lambda_1 = 2, \quad v_1 = \begin{bmatrix} 1 \\ 0 \\ 2 \end{bmatrix}, \quad \lambda_2 = -1 + i, \quad v_2 = \begin{bmatrix} 5 \\ -7 - i \\ 2 + 6i \end{bmatrix},$

   $\lambda_3 = -1 - i, \quad v_3 = \begin{bmatrix} 5 \\ -7 + i \\ 2 - 6i \end{bmatrix}$

9. $X = C_1 \begin{bmatrix} 1 \\ 0 \\ 0 \end{bmatrix} e^{2t} + C_2 \begin{bmatrix} -1 \\ 1 \\ 0 \end{bmatrix} e^{3t} + C_3 \begin{bmatrix} 17 \\ 3 \\ -12 \end{bmatrix} e^{-t}$

11. $X = C_1 \begin{bmatrix} 1 \\ 0 \\ 0 \end{bmatrix} e^{-2t} + C_2 \begin{bmatrix} 0 \\ 1 \\ 0 \end{bmatrix} e^{-2t} + C_3 \begin{bmatrix} 3 \\ 1 \\ 3 \end{bmatrix} e^{t}$

13. $X = C_1 \begin{bmatrix} 3 \\ 2 \\ -7 \end{bmatrix} e^{3t} + C_2 \begin{bmatrix} t + 1 \\ -2t - 4 \\ -t - 4 \end{bmatrix} e^{-t} + C_3 \begin{bmatrix} 1 \\ -2 \\ -1 \end{bmatrix} e^{-t}$

15. $X = \frac{9}{14} \begin{bmatrix} 1 \\ 1 \end{bmatrix} e^{5t} + \frac{2}{7} \begin{bmatrix} 3 \\ -4 \end{bmatrix} e^{-2t} + \begin{bmatrix} -\frac{1}{2} \\ \frac{3}{2} \end{bmatrix} e^{-t}$

17. $X = C_1 \begin{bmatrix} 1 \\ 0 \\ -1 \end{bmatrix} e^{3t} + C_2 \begin{bmatrix} 1 \\ 2 \\ 0 \end{bmatrix} e^{9t} + C_3 \begin{bmatrix} t \\ 2t + 1 \\ -1 \end{bmatrix} e^{9t}$

21. $A^n = \begin{bmatrix} 2^n & (-3)2^{n-1} \\ 0 & 0 \end{bmatrix} \qquad e^{At} = \begin{bmatrix} e^{2t} & \frac{3}{2} - \frac{3}{2}e^{2t} \\ 0 & 1 \end{bmatrix}$

25. (a) Yes; the probability that the determinant of a random matrix has *any* particular value is zero.
    (b) Yes; $\det\{\lambda I - A\}$ is a polynomial of tenth degree in λ. The chance that a random polynomial curve of degree 10 is tangent to the x-axis is zero.

29. $U = \begin{bmatrix} x \\ y \end{bmatrix} = \begin{bmatrix} -\frac{20}{7}e^{-0.095t} + \frac{20}{7}e^{-0.025t} \\ 5e^{-0.095t} + 5e^{-0.025t} \end{bmatrix}$

31. $\frac{dX}{dt} = AX, \quad \text{where} \quad X = \begin{bmatrix} x_1 \\ x_2 \\ x_3 \\ y_1 \\ y_2 \\ y_3 \end{bmatrix}, \quad X(0) = \begin{bmatrix} 0 \\ 10 \\ 0 \\ 0 \\ 0 \\ 20 \end{bmatrix}, \quad \text{and}$

$$A = \begin{bmatrix} -0.08 & 0.03 & 0.01 & 0 & 0 & 0 \\ 0.08 & -0.08 & 0 & 0 & 0 & 0 \\ 0 & 0.05 & -0.05 & 0 & 0 & 0 \\ 0 & 0 & 0 & -0.08 & 0.03 & 0.01 \\ 0 & 0 & 0 & 0.08 & -0.08 & 0 \\ 0 & 0 & 0 & 0 & 0.05 & -0.05 \end{bmatrix}$$

Section 7.1

1. Kepler's second law said that $dA/dt = C$. Then $ds/dt = 2C/r$, which shows that the speed ds/dt is inversely proportional to the distance from the sun.

Section 7.2

3. The wave is compressed as it moves along; that is, the distance between the maxima decreases as t increases.
9. The general term of the series is $[(-1)^k \omega^{2k} r^{2k}]/[(k!)^2 4^k]$

Section 7.3

1.

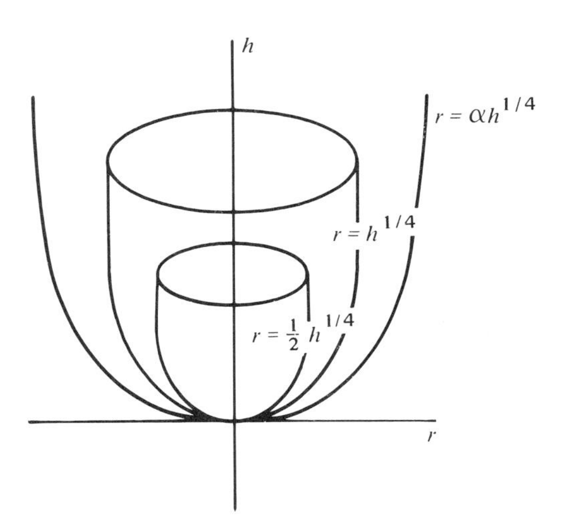

4. (a) $2/\sqrt{g} = 0.3536$ (b) $(\sqrt{20} - \sqrt{\frac{20}{3}})/\sqrt{g} = 1.89/\sqrt{g} = 0.3341$

Appendix 1

1. (a) $f(t) = \frac{1}{3}e^{-t} + \frac{5}{3}e^{-4t}$
   (b) $f(t) = \frac{11}{39}\sin(3t) + \frac{14}{13}\cos(3t) - \frac{1}{13}e^{-2t}$
3. $y(t) = \frac{1}{5}e^{-t}[2\sin(2t) - 7\cos(2t)] + \frac{4}{5} + \frac{3}{5}e^{-2t}$

5. $\mathcal{L}[e^{\beta t}f(t)] = F(x-\beta)$

9. $\mathcal{L}[t^2 \cos(\beta t)] = \dfrac{2x(x^2-3\beta^2)}{(x^2+\beta^2)^3}$

11. First show that if $|f'(t)| \leq Ae^{Bt}$ for $t \geq 0$, then $|f(s)| \leq (A/B)e^{Bs} + |f(0)|$ and thus $f(t)$ is bounded by Me^{Bt} for a suitable M. Repeating this, if f'' belongs to $\mathcal{E}$, so does f' and hence f.
13. No, because if $f \in \mathcal{E}$, then its Laplace transform $F(x)$ must obey the condition $\lim_{x\to\infty} F(x) = 0$, which x^3 clearly fails to do.

Index

Index of Symbols

ABCDEFGHIJ-H-79876